# Cognitive Systems Monographs

Volume 21



For further volumes:
http://www.springer.com/series/8354

*About this Series*

The Cognitive SystemsMonographs (COSMOS) publish new developments and advances in the fields of cognitive systems research, rapidly and informally but with a high quality. The intent is to bridge cognitive brain science and biology with engineering disciplines. It covers all the technical contents, applications, and multidisciplinary aspects of cognitive systems, such as Bionics, System Analysis, System Modelling, System Design, Human Motion, Understanding, Human Activity Understanding, Man-Machine Interaction, Smart and Cognitive Environments, Human and Computer Vision, Neuroinformatics, Humanoids, Biologically motivated systems and artefacts Autonomous Systems, Linguistics, Sports Engineering, Computational Intelligence, Biosignal Processing, or Cognitive Materials as well as the methodologies behind them.Within the scope of the series are monographs, lecture notes, selected contributions from specialized conferences and Workshops.

Paolo Arena · Luca Patanè
Editors

# Spatial Temporal Patterns for Action-Oriented Perception in Roving Robots II

## An Insect Brain Computational Model

Springer

*Editors*
Paolo Arena
Luca Patanè
Dipartimento di Ingegneria Elettrica
Elettronica e Informatica
Università di Catania
Catania
Italy

ISSN 1867-4925 ISSN 1867-4933 (electronic)
ISBN 978-3-319-02361-8 ISBN 978-3-319-02362-5 (eBook)
DOI 10.1007/978-3-319-02362-5
Springer Cham Heidelberg New York Dordrecht London

Library of Congress Control Number: 2013954845



Printed on acid-free paper

Springer is part of Springer Science+Business Media (www.springer.com)

# Preface

In the bio-inspired robotics field, robots can be used to reproduce animal behavior in order to study their interaction with the environment. Robots help to improve the understanding of animal behavior and, on the other side, animal behavior and neurobiological inspection help to create efficient and robust robotic systems. The study of animal brains is interesting from an Engineering perspective since it can allow to introduce new control systems that can endow robots with orienting capabilities in complex environments, new decision making strategies, and dangerous missions solving capabilities, to achieve an ever increasing autonomy level. Robotic implementation of biological systems could also lead to the introduction of new models for basic sciences, in particular when investigating the emergent properties of models. Several attempts are present in the literature related to algorithms or bio-inspired networks able to mimic the functionalities of parts of the brain. A lot of work has been done in several animal species belonging to mammals, mollusks, and insects. Looking into the insect world different research groups around the world are trying to design models that are able to reproduce interesting behaviors shown by insects: cooperation mechanisms in ants, navigation strategies in bees, looming reflex in locusts, homing mechanisms in crickets, central pattern generator and obstacle climbing in cockroaches, reflex-based locomotion control in the stick insect, just to cite some examples. It is evident that the effort is focused on specific peculiarities associated with the different insect species that can be also useful for robotic applications. Nevertheless, a more challenging task consists in trying to model the main functionalities of an insect brain by analyzing the dynamics from a higher level and trying to identify the mechanisms involved in the sensing-perception-action loop.

In the recent years, biological experiments unraveled details of the *Drosophila* brain, with particular emphasis to Mushroom Bodies (MBs) and Central Complex (CX), and a number of functional blocks, explaining the main functionalities of such centers were designed. The proposed work is focused on the development of an insect brain computational model mainly focused on the *Drosophila melanogaster*, the fruit fly. The insect brain architecture, structured in functional blocks, has been developed in a complete software/hardware framework in order to evaluate the capabilities of this bio-inspired control system on both simulated and experimental robotic platforms. In order to develop a useful and suitable architecture, the proposed framework is flexible and robust and presents a structure able to decouple simulations from control algorithms. The functional separation helps to isolate the application itself from graphic interfaces and the underlying hardware. The main aim is to develop an extensible and general-purpose architecture. The insect brain model has been evaluated in scenarios directly linked to the neurobiological experiments to make a direct comparison. Moreover, the available data on wild-type flies and mutant brain-defective flies allow to identify the main role of each neural assembly in performing specific tasks like visual orientation, olfactory learning, adaptive termination of behaviors, and others. Finally, the main guidelines used for the definition of evaluation criteria and the creation of benchmarking scenarios where the system performance can be evaluated, are also investigated.

The developed software/hardware framework, named RealSim for Cognitive Systems (RS4CS), transparently uses, as a final actor, either a robot (wheeled, legged or hybrid) simulated in a detailed dynamical simulator (included within the simulation environment), or experimental robotic platforms in a real environment. From the hardware point of view, mainly two robotic structures were used to test the architecture performance, one classical dual-drive wheeled robot, the Rover, and one hybrid robot, named Tribot: this is composed of two whegs-based modules and a front module with two legs that can be also used to manipulate objects.

Among the proposed strategies for the design and assessment of robotic experiments, a biologically driven and a robot-oriented approach have been considered. This choice was adopted both for the direct links to the neurobiological results and for the possibility to further extend the robot capabilities by scaling-up the insect brain blocks controlling the robot behavior. The considered scenarios include situations of increasing complexity to show the acquisition of cognitive capabilities (starting from the basic behaviors but going beyond these skills) like multisensory integration, contradictory cues resolution, efficient information storage and retrieval. These capabilities can be tested starting from classical exploration and foraging tasks, going up to tasks involving motivation based on reward functions. The neural structures responsible for these behaviors have been enriched with new substructures inspired by very recent discoveries in the insect counterparts. This gave the opportunity to design and simulate new networks that, via processes that can be described through reaction-diffusion dynamics, are able to lead to the emergence of complex behaviors like attention and expectation. These can be considered the basis of a cognitive behavior, i.e., the capability of planning ahead. The experiments can be tailored to the sensory-motor loop available in each robotic structure. The scenarios were scaled to the model capabilities, using the robots as test beds. Learning in this case is fundamental and should be performed autonomously in an unsupervised way, but cognitive skills could also be acquired by introducing a teacher (a signal or another agent) in the arena, in a supervised or semi-supervised way, as this happens in cooperation tasks.

The activities performed in the directions outlined above, were carried out by scientists from four institutions: the Institute of Neurobiology of the University of Mainz (Germany), the Instituto Pluridisciplinar, Universidad Complutense de Madrid (Spain), the Company Innovaciones Microelectronicas s.l., ANAFOCUS, from Seville (Spain). The Consortium was Coordinated by the Dipartimento di Ingegneria Elettrica Elettronica e Informatica of the University of Catania (Italy).

The book is organized into five parts. *Part I Models of the Insect Brain: From Neurobiology to Computational Intelligence* reports the biological background on *Drosophila melanogaster* with particular attention to the two main centers, the mushroom bodies and the central complex used for the implementation of the insect brain computational models.

*Part II Complex Dynamics for Internal Representation and Locomotion Control* introduces the locomotion control principles used to integrate body and brain. Mathematical bases of the Central Pattern Generator used for the gait generation in a six-legged robot are illustrated looking at the complex dynamics emerging from a network of nonlinear units. Similarly, the Reaction-diffusion principles in non-linear lattices are used to develop a compact internal representation of a dynamically changing environment to plan the best behavior to reach a target while avoiding moving obstacles.

*Part III Software/Hardware Cognitive Architectures* develops upon the basis presented in the previous part. In particular a software/hardware framework conceived to integrate the insect brain computational model in a simulated/real robotic platform is illustrated. The different robots used for the experiments are also depicted analyzing their peculiarities in terms of basic behaviors. Moreover, the problems related to the vision system are herewith addressed proposing robust solutions for object identification and feature extraction. A whole chapter is also dedicated to the design and implementation on FPGA of the compact internal representation model discussed in the previous part.

The last part of the book *Part IV Scenarios and Experiments* is related to the identification of the relevant scenarios used in the experiments to test the capabilities of the insect brain-inspired architecture taking as comparison the biological case. A number of experimental results are finally reported, whose related multimedia material can be found in the SPARK II web page: www.spark2.diees.unict.it

Catania, July 2013

Paolo Arena
Luca Patanè

# Acknowledgments

The editors would like to thank all the partners that contributed with their expertise to the success of the project, with a professional and proactive collaboration.

A sincere gratitude is expressed to the project reviewers, Profs. Ralf Der and Ronald Tetzlaff, for their steering comments that largely contributed to the research activity during the 36 months of research.

A special thank is addressed to the EC Project Officer, Dr. Hans-George Stork, for his continuous, constructive comments and encouragement that were essential in our research project. We think that without such a support from the EC, we would have never been able to work together in such a way, enjoying while working hard.

# Contents

# Contributors

**L. Alba-Soto** AnaFocus (Innovaciones Microelectrónicas S.L.), Av. Isaac Newton 4, Pabellon de Italia, Planta 7, PT Isla de la Cartuja, 41092 Sevilla, Spain, e-mail: Luis.Alba@anafocus.com

**I. Aleo** Department of Electrical, Electronic and Computer Science Engineering, University of Catania, 95125 Catania, Italy

**E. Arena** Department of Electrical, Electronic and Computer Science Engineering, University of Catania, 95125 Catania, Italy

**P. Arena** Department of Electrical, Electronic and Computer Science Engineering, University of Catania, 95125 Catania, Italy, e-mail: parena@diees.unict.it

**D. J. Caballero-Garcia** AnaFocus (Innovaciones Microelectrónicas S.L.), Av. Isaac Newton 4, Pabellon de Italia, Planta 7, PT Isla de la Cartuja, 41092 Sevilla, Spain

**S. De Fiore** Department of Electrical, Electronic and Computer Science Engineering, University of Catania, 95125 Catania, Italy

**E. del Rio** Departamento de Física Aplicada, E.T.S.I. Aeronáuticos, Universidad Politecnica de Madrid, Madrid, Spain, e-mail: ezequiel.delrio@upm.es

**A. Jimenez-Marrufo** AnaFocus (Innovaciones Microelectrónicas S.L.), Av. Isaac Newton 4, Pabellon de Italia, Planta 7, PT Isla de la Cartuja, 41092 Sevilla, Spain

**V. A. Makarov** Instituto Pluridisciplinar, Universidad Complutense de Madrid, Paseo Juan XXII 1, 28040 Madrid, Spain, e-mail: vmakarov@mat.ucm.es

**L. Patanè** Department of Electrical, Electronic and Computer Science Engineering, University of Catania, 95125 Catania, Italy, e-mail: lpatane@diees.unict.it

**M. Pollino** Department of Electrical, Electronic and Computer Science Engineering, University of Catania, 95125 Catania, Italy

**L. Salas-Paracuellos** AnaFocus (Innovaciones Microelectrónicas S.L.), Av. Isaac Newton 4, Pabellon de Italia, Planta 7, PT Isla de la Cartuja, 41092 Sevilla, Spain

**R. Strauss** Institute of Zoology III Neurobiology, Johannes Gutenberg-University, Mainz, Germany, e-mail: rstrauss@uni-mainz.de

**P. S. Termini** Department of Electrical, Electronic and Computer Science Engineering, University of Catania, 95125 Catania, Italy

**M. G. Velarde** Instituto Pluridisciplinar, Universidad Complutense de Madrid, Paseo Juan XXII 1, 28040 Madrid, Spain, e-mail: mgvelarde@pluri.ucm.es

**C. Ventura** Department of Electrical, Electronic and Computer Science Engineering, University of Catania, 95125 Catania, Italy

**J. A. Villacorta-Atienza** Instituto Pluridisciplinar, Universidad Complutense de Madrid, Paseo Juan XXII 1, 28040 Madrid, Spain, e-mail: joseavillacorta@pluri.ucm.es

**A. Vitanza** Department of Electrical, Electronic and Computer Science Engineering, University of Catania, 95125 Catania, Italy

# Part I
# Models of the Insect Brain: From Neurobiology to Computational Intelligence

# Chapter 1
# Neurobiological Models of the Central Complex and the Mushroom Bodies

**R. Strauss**

**Abstract** This study reviews the actual knowledge on functions of the central complex (CX) and the mushroom bodies (MBs) in a genetic model insect, the fly *Drosophila melanogaster*. Ongoing research of UNIMAINZ and respective data are included. Reference is made to other insects, where respective functions are not yet studied in *Drosophila*. Neuroanatomical information is reported with regard to the general flow of information in these central brain neuropils. Particular projection systems and circuits are taken into account where this can be linked to functions. Models are developed.

## 1.1 Introduction

This chapter reviews studies on the functional significance of the central complex (CX) and the mushroom bodies (MBs) in a genetic model insect, the fly *Drosophila melanogaster*. Reference is made to other insects, where respective functions are not yet studied in *Drosophila*. Neuroanatomical information is reported with regard to the general flow of information in these central brain neuropils. Particular projection systems and circuits are taken into account where this can be linked to functions. The four neuropilar regions of the CX have prominent and quite specialized functions. For instance, the protocerebral bridge controls step length to facilitate visual object approach, the fan-shaped body stores information on objects for recognition, and the ellipsoid body performs path integration to bridge missing actual sensory information. Together with their neuroanatomical information, those functions have been integrated into a functional model of the CX. Additional known or hypothesized functional roles were investigated. Among the established ones are celestial $E$-vector orientation, right-left-bargaining between brain sides, front-leg targeting in climbing

R. Strauss (✉)
Institute of Zoology III Neurobiology, Johannes Gutenberg-University, Mainz, Germany
e-mail: rstrauss@uni-mainz.de

P. Arena and L. Patanè (eds.), *Spatial Temporal Patterns for Action-Oriented Perception in Roving Robots II*, Cognitive Systems Monographs 21,
DOI: 10.1007/978-3-319-02362-5_1, 

over chasms, and the maintenance of locomotor activity. Stabilization of orientation behavior, long-term motor skill learning, or humidity-gradient orientation are under study (Berg et al., Unpublished; Krause and Strauss, Unpublished; Zäpf and Strauss, Unpublished).

In the same way the analysis of MB functions is carried out. The following functional properties of MBs are being reviewed: olfactory conditioning and odor learning, context generalization, decision making, adaptive termination of behavior, temperature choice behavior, short-term motor skill learning, sleep regulation and memory consolidation, place memory, sensory discrimination and the integration of sensory perception with motor actions, enhancement of initial motor activity and suppression of long-term locomotor activity, regulation of aggression behavior, and centrophobism and thigmotaxis. The MB model by Gerber et al. [1, 2] explaining the classical conditioning of odors with reward or punishment is reviewed. A global model is then proposed that includes the classical conditioning model in a substructure. The global model can control behavioral activity in interplay between MBs and CX. Drives are ranked in terms of urgency. Always one behavioral option out of an array of options is chosen in the light of the current needs and the actual sensory input. Stimuli of all modalities provide input to the model in pre-processed form. Constellations of stimuli are compared to "look-up tables" holding previous good or bad experience with that constellation. The stored information is partially represented in the MBs (chiefly olfactory information), and partially in the CX (chiefly visual information). The model is able to perform operant learning, that is learning from the results of own actions. In accord with results from our behavioral experiments, the model can also stabilize behavior over extended periods of time and prevent it from frequent flipping via a hysteresis function that increases the probability of the currently executed behavior to be chosen again. The model contains a first approach to integrate motor-skill learning.

## 1.2 A Neurobiological Model of the Central Complex

### *1.2.1 General Introduction to the Central Complex*

#### Shape, Compartments and Location

The central complex (CX) resides in all insect brains at the midline of the brain between the protocerebral hemispheres. It is unpaired but mirror-symmetrical with respect to the mid-sagittal plane and interconnects the hemispheres. The CX is composed of four major regions ([3], Fig. 1.1):

1. The *fan-shaped body* is the largest neuropil comprising eight subunits in the latero-lateral extent, six layers in the dorso-ventral direction and four shells in the anterior-posterior direction.

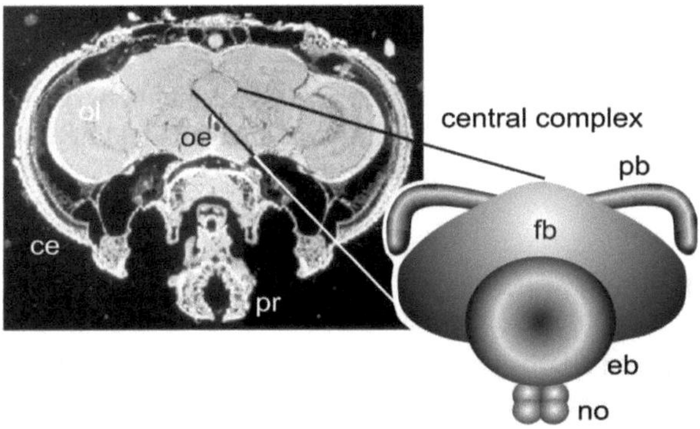


**Fig. 1.1** Position of the central complex (CX) in a frontal brain section of *Drosophila melanogaster*. The head is 0.8 mm wide and the CX is about 0.1 mm. **ce** compound eye; **ol** optic lobes; **pr** proboscis; **oe** oesophagus. The CX is composed of four neuropilar regions; **fb** fan-shaped body; **eb** ellipsoid body; **no** noduli; **pb** protocerebral bridge

2. In a frontal concavity of the fan-shaped body resides the *ellipsoid body* a neuropil of perfect toroid shape with an ellipsoid-body channel forming the main axis. The ellipsoid body possesses 16 segments and it has two radial zones.
3. The paired *noduli* reside ventral to the fan-shaped body just dorsal to the oesophagus.
4. The *protocerebral bridge* is a rod-like neuropil composed of a chain of 16 glomeruli just posterior to the fan-shaped body. The eight glomeruli per brain hemisphere are uninterruptedly connected across the midline of the brain.

### General Internal Neuroarchitecture

The CX can be described as a matrix of *small-field neurons*, which interconnect the four neuropilar regions in various different orderly projections systems usually coming in homologous sets of eight or 16 neurons. The major direction of these connections is dorso-ventral. *Large-field neurons* run perpendicular to the small-field neurons preferentially in the latero-lateral direction. Large-field neurons seem to contact all eight or 16 neurons of a homologous small-field set. A third type of neurons is called *pontine neurons*. They are CX intrinsic and come again in many different homologous sets of eight or 16. They connect for instance adjacent fan-compartments of the fan-shaped body, or different layers within the same fan-compartment, or different glomeruli in the bridge in an ordered fashion [3].

## General Flow of Information

Deduced from the shape of the synaptic specializations the large-field neurons are thought to deliver input into the CX, whereas the small-field neurons are likely the major carriers for the output from the CX [3].

## Input/Output Relation of the CX

The large-field neurons deliver information from many different brain areas. There are no obvious tracts that would allow guessing for instance a major modality represented in the CX (This is different in the case of the mushroom bodies: a thick fiber bundle from the antennal lobes tells us that olfactory processing is a major function of the mushroom bodies). There are, however, two accessory areas of the CX. The lateral triangles are the starting point of input into the ellipsoid body, and the ventral lobes are a major input and output region for the fan-shaped body. The input into the CX seems to be highly preprocessed because there are no known projections from primary sensory areas whatsoever. The ventral lobes are the starting point of descending pathways towards the ventral ganglion in the thorax. The ventral ganglion serves comparable functions as the spinal cord in vertebrates, e.g. leg coordination and many reflexes.

## Sensor Modalities

Excellent experimental evidence from *Drosophila* and many different insect species tell us that visual information [4–6], particularly also polarization-vision information in the locust [7–9] is processed in the CX. For all other modalities there is no clear evidence, but for mechanosensory information. Wind from the front indicates the flying state in the locust CX [10, 11]. Behavioral evidence from *Drosophila* structural CX mutant lines points to mechanosensory information from the legs to be processed in the CX [Strauss, unpublished].

## Cell Numbers

In the fruit fly *Drosophila melanogaster* the CX is composed of 2000–5000 neurons. The entire brain is estimated to contain about 150,000 neurons and the ventral ganglion might add another 20,000 neurons. Most of the CX cell types of *Drosophila* are known and described by an excellent Golgi study [3] but it is unclear how many isomorphic sets of a given cell type exist.

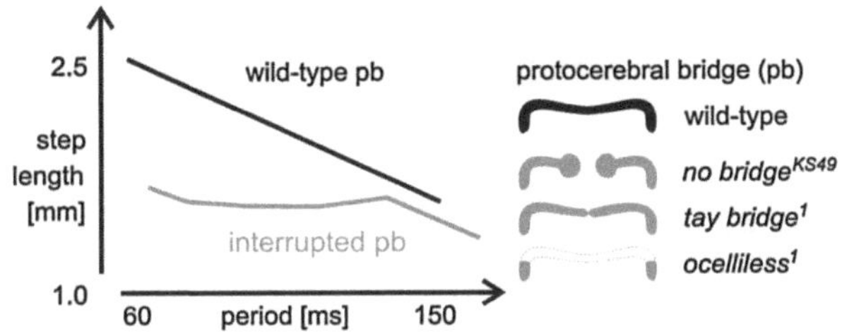


**Fig. 1.2** Structural defects in the protocerebral bridge of mutants alter step length. A period of 60 ms corresponds to 16 steps/s and 150 ms to 6.8 steps/s

## *1.2.2 Functional CX Analyses*

### The Protocerebral Bridge and its Role in Step Length Control

The mutant line *no bridge*$^{KS49}$ was the first one found with a disrupted protocerebral bridge [12]. Steps of these flies are invariably small, regardless of the frequency of stepping. In wild-type flies the strides become larger with the frequency of stepping (Fig. 1.2). Because the swing phases of *no bridge*$^{KS49}$ flies are of normal duration, it was concluded that the protocerebral bridge controls swing speed in order to influence step size. A mosaic analysis provided evidence for a causal relation between the protocerebral-bridge defect and this very distinct behavioral defect. Four more independent mutant lines with bridge disruptions were collected (e.g. [13, 14]) and all of them exhibited the same walking problem. It was hypothesized that the protocerebral bridge might not only be needed for the optimization of walking speed but also for the differential control of step length in order to influence the direction of heading. A molecular analysis of the *tay bridge*$^{1}$ mutation [14] revealed that the bridge is needed indeed for the control of step length when it comes to approaching a visual object (see also [6]).

However, an intact protocerebral bridge was not a prerequisite for optomotor compensation while walking. Thus, Poeck et al. [14] narrowed down the role of the protocerebral bridge to differential step-length control in object approaches. A functional model of object approach is presented in Sect. 1.2.3. It takes the known neuroanatomical details of the interrelation between the bridge and the rest of the CX and the bridge and the non-CX protocerebrum into account.

### The Fan-Shaped Body and Ellipsoid Body and Their Role in Right-Left Bargaining

Three independent CX defective mutant lines were found, which can only walk in circles. A mosaic analysis on one of those lines revealed that the exchange of information between the brain hemispheres through the fan-shaped body and the

**Fig. 1.3** CX-split fly circling for 15 min between attractive objects

ellipsoid body are essential for normal straight walking and flying [15]. As a common feature of all three independent mutant lines, this exchange was found to be disturbed (partial split-brain phenotype; [15, 16]). In such a case one brain hemisphere has usually a stronger influence on the behavior than the other one so that individual flies always circle in the same sense of rotation. Regardless of the behavioral task the steps on one body side become consistently larger than on the other side. Optomotor stimuli of one sense of rotation are being overcompensated and those of the alternative sense are being undercompensated. Flies can reach attractive objects but only in a circling fashion (Fig. 1.3). The exchange of information through the CX turns out to be essential for balancing the brain and body sides.

## The Fan-Shaped Body and its Role in Object Recognition

Liu et al. [4] trained stationary flying flies to avoid heading towards certain visual objects on a surrounding white cylinder. If flies turned towards such an object, a heat beam was automatically activated, strong enough to heat-up the fly to life-threatening temperatures. Wild-type flies effectively learn to avoid those objects but not so the learning mutant *rutabaga*$^{2080}$, defective in the cyclic AMP dependent pathway for plasticity. The dangerous and the harmless objects shown on the cylinder were distinct either with regard to their height over ground, their angular orientation or their size. In a partial-rescue strategy for localizing the memory in the fly brain, small areas of the brain were "repaired" by expressing the wild-type form of the *rutabaga* gene in otherwise *rutabaga*$^{2080}$-defective flies. When plasticity was exclusively restored in the first layer of the fan-shaped body the flies were able to recognize objects distinct

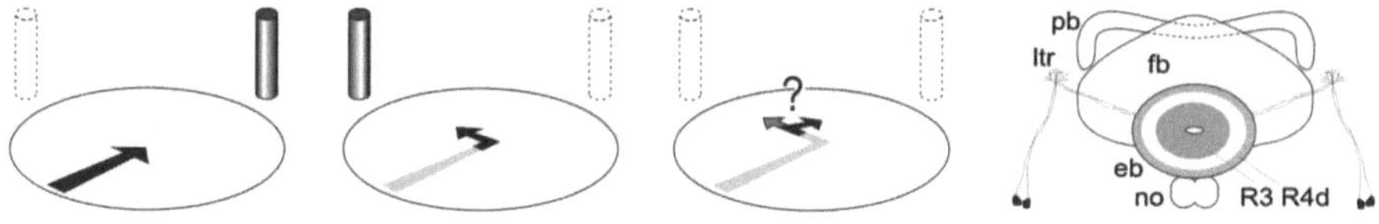


**Fig. 1.4** The detour paradigm after [5]. Flies are lured out of the direct way to a previously chosen, now invisible object. When the distractor disappears, the flies remember the path to their former target. Frontal drawing of the fly CX (**pb** protocerebral bridge; **fb** fan-shaped body; **eb** ellipsoid body; **no** noduli). The location of ring-neuron type R3 and R4d is shown. Their axons project from the lateral triangles (**ltr**), where they form spiny arborization, to the eb

in the height over ground of their centers of gravity. Differences in the angular orientation of the objects didn't help the flies to memorize the dangerous objects. Likewise, if layer five of the fan-shaped body was "repaired" the flies were able to remember the dangerous objects by their orientation, but not by their height over ground. The presumed object features stored in layers 2, 3, 4, and 6 are not yet identified, but wild-type flies are known to be able to recognize visual objects also by their color, size or vertical distributedness [17].

In another partial rescue study Wang et al. [18] showed that the same layers of the fan-shaped body that require functional *rutabaga* need also an intact *foraging* (*for*) gene; *for* encodes for a cyclic GMP kinase. Again, the memory for the center of gravity requires intact *for* in layer one and the memory for the object orientation requires intact *for* in layer five.

The functional model proposed below (Sect. 1.2.3) therefore assumes a "what"-dimension in the layers of the fan-shaped body and a "where"-dimension in the latero-lateral extent of the fan-shaped body.

### The Ellipsoid Body and its Role in Visual Short-Term Orientation Memory

Neuser et al. [5] showed that *Drosophila melanogaster* possesses a short-term spatial memory during locomotion. Flies can remember the position of an object for several seconds after it has been removed from their environment. The memory enables them to continue the approach [19]. In a novel detour paradigm the test flies were temporarily lured away from the direction towards their hidden target, yet they were able to aim for their former target thereafter (Fig. 1.4). This flexible goal-driven orientation requires that the position of a target be stored, especially in case the target moves out of sight. The capability to retain, recall and integrate such positional information into guiding behavior has been summarized under the term spatial working memory. Neurophysiological studies in primates, which possess a comparable memory, indicate that sustained activity of neurons encodes the sensory information even though the object is no longer present [20].

In the study of Neuser et al. [5] we also show that the GABAergic ring neurons of the ellipsoid body in the central brain [3] are necessary and their plasticity is sufficient for a functional spatial orientation memory in flies. The protein kinase S6KII (*ignorant*; [21]) is required in a distinct subset of ring neurons to display this memory. Conditional expression of S6KII in these neurons only in adults did restore the loss of the orientation memory of the *ignorant* mutant. The S6KII signaling pathway therefore seems to be acutely required in a subset of ring neurons of the ellipsoid body for the spatial orientation memory in flies [5].

## The Role of the Central Complex in Stabilizing Approach Behavior

In addition to the short-term memory for directions explained above, there exists also an integrator for the time or the distance during which a fly had approached an attractive visual object. The longer the approach towards an object had already lasted, the longer will be the continued approach towards the former object position after this target became invisible [Berg and Strauss in preparation]. Neurogenetic tools allowed us to localize the underlying visual short-term memory. Likely candidates were the CX and particularly the ellipsoid body, because of the outcome of our previous analysis of the spatial orientation memory. Respective structural brain mutants and flies holding genetic constructs, which silence groups of ellipsoid-body neurons, show no dependency of their run-outs on the in-runs during which the object is present. Existence of short-term memories for other sensory modalities is being tested in an ongoing study.

## The Ellipsoid Body and its Probable Role in Humidity Gradient Orientation

When released in the middle of a humidity gradient from dry to 100 % relative humidity, agitated flies will run towards the extremely dry side (Fig. 1.5, [22]). In a relaxed state their choice depends on their state of dehydration. The sensors have been identified by Liu et al. [23]. In an ongoing study we are testing structural mutants of CX and mushroom bodies (MBs). The incentive is to answer the question, which other sensory modalities might be processed but vision in the CX and olfaction in the MBs. The detour study of Neuser et al. [5] had shown the pivotal role of the ellipsoid body in the short-term storage and retrieval of visual orientation information. When orienting in a humidity gradient the flies might be storing the sensed humidity of one spot to be able to compare it to the sensed humidity of a later spot. Direct sensing of the gradient by comparing the humidity-sensor inputs of the two body sides without a memory seems error prone because the sensors on the two antennae of the flies are very close together (Indeed, Sayeed and Benzer [22] have shown that the ablation of one antenna does not impair the ability of a fly to orient in a humidity gradient).

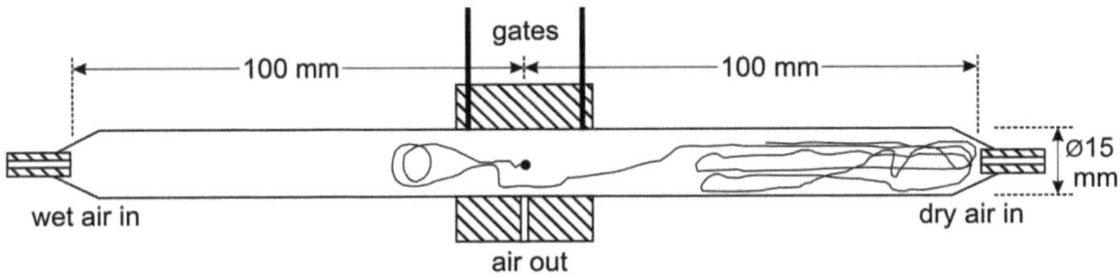


**Fig. 1.5** Humidity choice paradigm adapted from [22]. A fly started in the middle of the setup and went towards the side with wet air first. After a short excursion it turned and went to the dry-air side

## Map-Like Representation of Celestial *E-vector* Orientations in the Central Complex of the Locust

For many insects including locusts, honey bees and ants the polarization pattern of the blue sky serves as a compass cue for spatial navigation. The polarization pattern depends on the position of the sun modulo 180°. The ambiguity is resolved by the animals by distinguishing sun-ward from anti-sun-ward. *E-vector* orientations are detected by photoreceptors in a dorsal rim area of the eye. Insects possess a circadian clock and can take the diurnal changes of the sun position into account. Polarized-light signals from both eyes are finally integrated in the CX. Heinze and Homberg [7] show in an electrophysiological study that a topographic representation of zenithal *E-vector* orientations underlies the columnar organization of the protocerebral bridge in a locust. Map-like arrangements are found also in other parts of the locust CX. They are highly suited to signal head orientation under the open sky. Descending pathways to the motor centers in the thorax are described ([8]; review: [9]).

## The Protocerebral Bridge and its Probable Role in Front-Leg Targeting

Flies can climb over chasms in their walkway using a specific climbing behavior [24]. The size of the chasm is visually estimated and wild-type flies engage in gap-crossing only at gaps of surmountable width. The behavior requires the concerted action of front legs which stretch out to reach the other side, middle legs which lift up the body to achieve a more favorable position for the front leg action, and hind legs which go closer and closer to the proximal edge of the gap.

Several different ways of failing were found in climbing-defective mutant lines; they reveal a modular and distributed structure of the control of gap climbing. The position at which climbing is initiated, the middle leg action that lifts up the body, or the gap-width dependent initiation of climbing can be specifically disturbed. In particular, flies with a defective protocerebral bridge take a correct decision to initiate climbing but then are losing the direction towards the opposite side. Instead, they will start climbing actions in all directions including those with no opposite landing site (Fig. 1.6). When the integrity of the protocerebral bridge was restored in otherwise *tay*

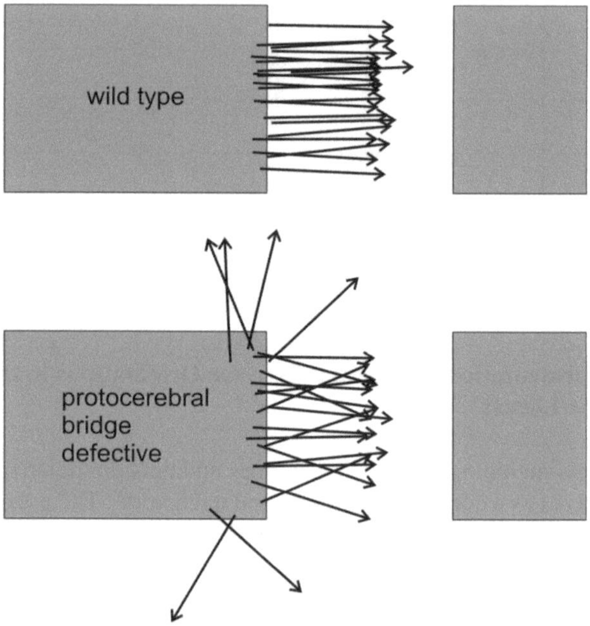


**Fig. 1.6** In contrast to wild-type flies the protocerebral bridge-defective flies are losing the direction towards the opposite side of the cat walk beyond the chasm. Arrows denote the longitudinal body axis at the start of climbing

*bridge* defective flies (partial rescue approach), those flies were able to reach the other side. However, they had a considerably larger variation in their climbing direction than wild-type flies. The conclusion of the study is that the protocerebral bridge is not the seat of the short-term memory for the intended climbing direction, but an indispensable constituent of a targeting system which uses acute visual information. This targeting system steps in as a fall-back solution and replaces the not yet localized memory for the intended climbing direction [6].

## The Central Complex and its Probable Role in Long-Term Motor Skill Learning

An ongoing study is dealing with a motor-learning aspect of the gap-climbing behavior explained above [24]. It turned out that normal flies can get significantly more successful and significantly faster per transition. The observed motor skill memory has two phases [25]. The short-term component lasts for up to an hour. If the next training unit follows before that time and if at least 50 successful crossings were achieved in five consecutive training units on the first day, the flies carry the improvements to the next day, and without noticeable decay also to the longest time span tested up to date, which is 72 h.

This motor skill learning is cAMP-dependent because mutants defective in the adenylyl cyclase (*rutabaga*$^{2080}$) do not improve in their climbing skills. A neurogenetic partial rescue strategy allowed us to restore adenylyl cyclase function in various small areas of the brain whereby the rest of the nervous system remains (*rutabaga*$^{2080}$) defective. The short-term component of motor skill learning has been restored when the adenylyl cyclase function was restored just in the mushroom bodies. The long-term component can be restored presumably in the CX. This latter result is tentative, because the only driver line found to date, that restores the long-term memory, expresses in the MBs (which restores the short-term component), in the CX and in some other parts of the brain ([25], [Kienitz and Strauss, in preparation]).

## Role of the Central Complex Protocerebral Bridge and Fan-Shaped Body for the Maintenance of Locomotion in Flies

Earlier studies in *Drosophila melanogaster* based on structural brain mutants suggested that the CX is a higher control center of locomotor behavior [26]. Continuing this investigation Martin et al. [27] studied the effect of the CX on the temporal structure of spontaneous locomotor activity in the time domain of a few hours. The peculiarities of the wild-type behavior are published by Martin et al. [28]. In an attempt to dissect the internal circuitry of the CX, those authors perturbed a putative local neuronal network connecting the four neuropil regions of the CX, the protocerebral bridge, the fan-shape body, the noduli and the ellipsoid body. Two independent and non-invasive methods were applied: mutations affecting the neuroarchitecture of the protocerebral bridge, and the targeted expression of tetanus toxin in small subsets of CX neurons so that their synaptic output was interrupted. All groups of flies with a disturbed component of this network exhibited a common phenotype—a drastic decrease in locomotor activity. While locomotor activity was still clustered in bouts and these were initiated at the normal rate, their duration was reduced. This finding suggested that the bridge and some of its neural connections to the other neuropilar regions of the CX are required for the maintenance but not the initiation of walking.

## The Distribution of Walking-Time Intervals Requires the Ellipsoid Body in *Drosophila*

The temporal properties of a variety of behavioral traits obey power law distributions, a property often referred to as fractal. Martin et al. [28] showed that the temporal pattern of locomotor activity of the fruit fly *Drosophila melanogaster* follows such a distribution. Although an increasing number of such fractal patterns are being discovered, the responsible brain areas and neuronal networks remain unknown. Martin et al. [29] showed that blocking specifically the output of neurons of the *Drosophila* ellipsoid body, the toroid-shaped substructure of the CX, leads to a loss of the fractal properties in the temporal pattern of locomotor activity. The authors conclude that the temporal fractal pattern of locomotor activity is regulated in the ellipsoid body.

### *1.2.3 A Functional Central Complex Model*

**Representation of Acute Visual Information**

Let us start the proposed model with the proven role of the protocerebral bridge in controlling step length when it comes to approaching objects (Sect. 1.2.2). Let us assume that the current visual target of the fly is represented on the same side of the protocerebral bridge on which its image falls on the retina. The bridge is a linear array of 16 glomeruli (Fig. 1.1). On each side the glomerulus no. 1 is closest to the midline and no. 8 is in the most lateral position [3]. In the proposed model, the bridge holds information on the azimuth position of objects (Fig. 1.7a). Frontal objects are represented in the middle glomeruli, rear ones in the lateral glomeruli. There might be more than one object represented simultaneously on the bridge, but the current target should cause the highest neuronal activity. According to the model, the bridge requires visual input, which contains azimuth information. Indeed, a presumably homologous set of input neurons is described by Hanesch [30], which projects from the posterior optic foci to the glomeruli of the bridge. At the posterior optic foci the visual information can be assumed to be already highly preprocessed.

**Step-Length Control for Course Control and Speed Enhancement**

The output from the bridge relevant for step length control is assumed to occur via the horizontal fiber system [3]. The horizontal fiber system projects from the eight glomeruli of one bridge side to the eight fan-compartments of the fan-shaped body and exits toward the ventral lobe contralateral to the bridge input. The ventral lobes are accessory areas of the CX where also descending neurons to the ventral ganglion have input sites. Because both sides of the bridge, i.e. two times eight glomeruli project to the one time eight fan-compartments of the fan-shaped body, a cross-over pattern is created by the horizontal fiber system (Fig. 1.7a). The model assumes that a representation of the target object on the left retina and consequently the left side of the bridge produces larger steps on the contralateral side of the body. Without the influence of the horizontal fiber system the strides are of a basic step length, which is probably seen in all the protocerebral bridge defective mutant flies [14, 15]. The enhanced step size contralateral to the target object and the basic step size on the body side with the object are causing turning toward the object until it will be in front of the fly (Fig. 1.7b). There, it will be seen by the binocular ranges of both eyes ($\pm 15°$) and be represented by the two innermost glomeruli of the bridge. Consequently the fly—now on target—will enhance step size on both sides of the body and will approach the object with enhanced walking speed.

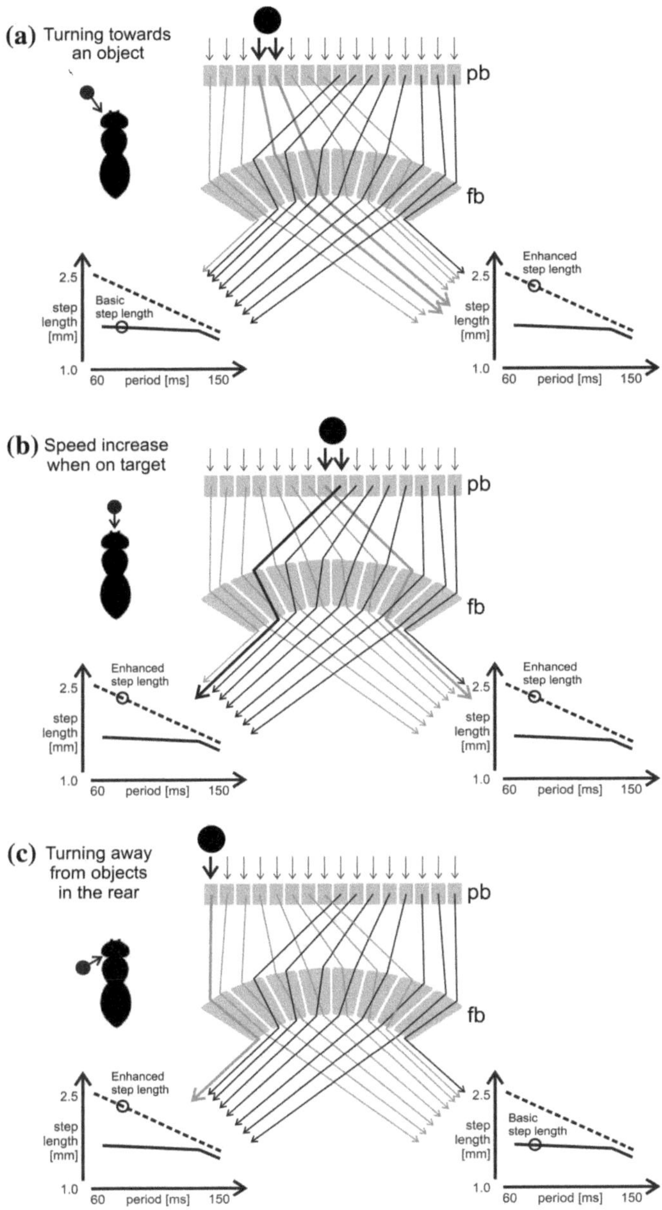


**Fig. 1.7** Functional protocerebral bridge model. **a** An object is represented at the left side of the bridge, transferred via the horizontal fiber system to the right ventral body, where the step lengths of the contralateral legs are increased. The fly turns towards the object. **b** Representation of an object at the middle of the bridge leads to an increase in step lengths of the legs on both sides of the body. The fly increases its walking speed towards the object. **c** The outermost bridge glomeruli 8 project ipsilaterally rather than contralaterally. Objects beyond 100° in the rear of the fly are represented there. The step lengths of ipsilateral legs are increased and the fly turns away from such an object

## Step-Length Control for the Retreat from Objects

Anatomical studies in different insect species have shown that exclusively the fibers emerging from the outermost glomeruli of the bridge exit the fan-shaped body toward the ipsilateral ventral lobe rather than the contralateral one. For the model it is assumed that these outermost glomeruli represent the rearmost visual field of the fly. An object represented on a glomerulus no. 8 will lead to ipsilaterally enhanced steps with respect to the object, so that the fly will turn away from it (Fig. 1.7c). Behavioral analyses by Mronz and Strauss [31] have proven that flies indeed turn away from objects presented in the rear. Attractive objects were presented to flies under various viewing angles and distances. Objects attractive in the frontal and lateral visual field became repulsive, if presented in the rear visual field. This behavior stabilizes retreat from unrewarding objects and prevents immediate re-choice. Flies can walk away from a just-visited object simply by triggering turning after the inspection. This will bring the object into the rear visual field. Consequently, there is no necessity to memorize the abandoned object in order to prevent immediate re-choice.

## A Look-up Table for Previous Experience with Objects

The neurons of the horizontal fiber system pass through the fan-shaped body and make synaptic contact there before they exit to the ventral lobes. For the model it is proposed that previous experience with a given object is stored in the six layers of the fan-shaped body (Fig. 1.1). Liu et al. [32] and Wang et al. [18] had shown that plasticity in the ventral-most layer one of the fan-shaped body is sufficient for flying flies to recall previous heat-punishment, given that the objects can be distinguished by the height over ground of their centers of gravity. Likewise, if plasticity was restored in layer five of the fan-shaped body, the flies were able to distinguish objects by a differing angular orientation. Up to date there is no information on the specificity of the other layers, but size, distance, color and vertical distributedness are known object features that wild-type flies can use to distinguish between objects. Flies can have two principally different experiences with objects—good or bad—and they should therefore be able to approach or avoid a recognized object. Each fan-compartment of the fan-shaped body has connections to the ventral lobes of both brain sides. Conceivably, step lengths can be enhanced either on the contralateral side for approach, or on the ipsilateral side for avoidance of a given object, depending on previous experience (Fig. 1.8). In this model the latero-lateral dimension of the fan-shaped body stands for an azimuth zone under which the object appears. The six layers hold the object features of previously encountered, particularly good or bad objects and facilitate comparison with the current object.

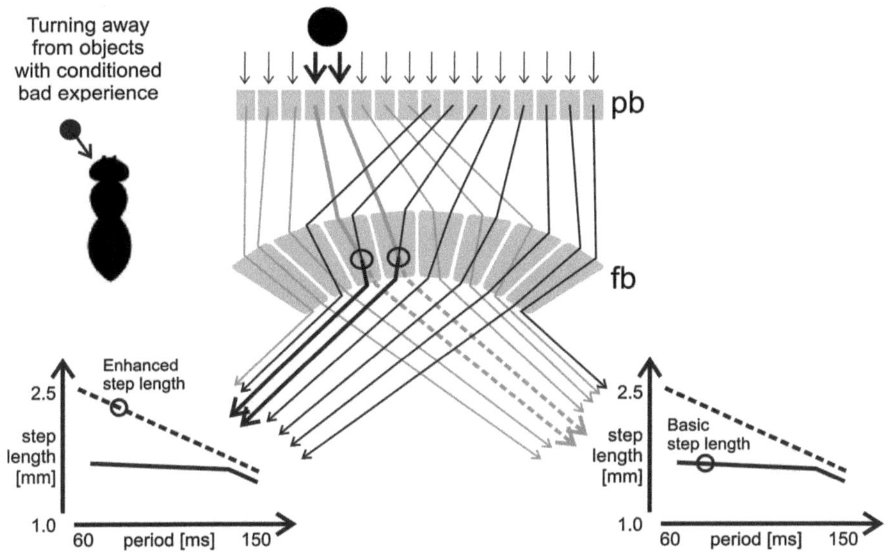


**Fig. 1.8** The fan-shaped body serves object-recognition functions. A memory for good or bad experience might directly switch the motor output. When a similarly looking object was previously "good", the step lengths of the contralateral legs will be increased and the fly turns towards object (*stippled* pathway). When the object was previously "bad", a switch in the fan-shaped body will increase the step lengths of the ipsilateral legs and the fly will turn away from the object (*thick black* pathway)

## Replacement for Acute Visual Information from a Working Memory

The protocerebral bridge is connected not only to the fan-shaped body but also to the ellipsoid body with several one-by-one glomerulus-to-segment projection systems. The idea of an information flow in both directions through several projection systems is supported by anatomical data [3]. For the model it is assumed that acute visual information on the azimuth position of a target object is conveyed from the protocerebral bridge to the ellipsoid body where a short-term working memory forms. Such a memory has been proven to exist in the ellipsoid body of *Drosophila* [5]. Whenever the visual contact to the target object gets lost, the stored positional information in the ellipsoid body can be updated by path-integration and uploaded to the protocerebral bridge. From there it is conveyed to the fan-shaped body along the same pathway as acute visual information (Fig. 1.9).

Anatomically, the ellipsoid body of *Drosophila* consists of an outer and an inner ring. In the model we will assume that one ring will represent body-centered coordinates, the other one world-centered coordinates. With such a system the fly is able to replace the acute visual information by idiothetic information from path integration. Hartmann and Wehner [33] proposed a cybernetical model for the path integration mechanism underlying homing in desert ants. Hartmann and Wehner propose two

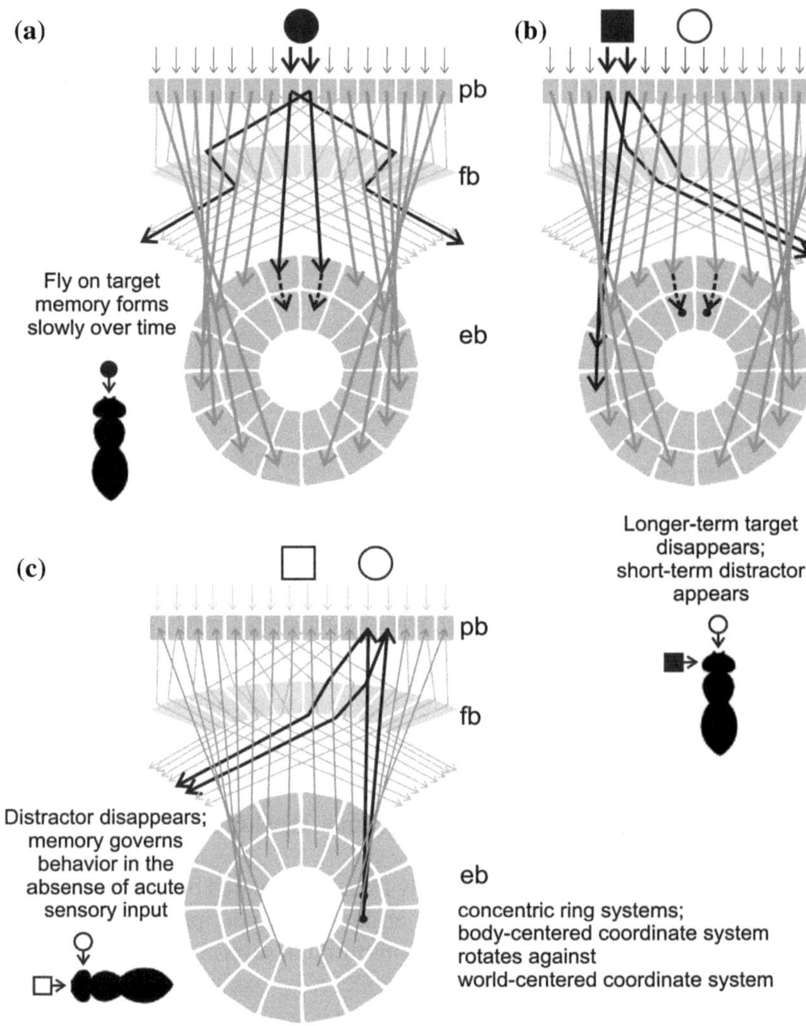


**Fig. 1.9** Short-term storage of visual object positions and replacement of temporarily missing acute information by path integration. **a** An object is represented at the *middle* of the bridge. A memory forms over time in the ellipsoid body. **b** After the first target object disappeared, the fly turns toward an alternative distractor object (appears on the *left* side, steps enhanced on the *right* side). Turning information is integrated in the ellipsoid body. The distractor is presented only for a short time, so that the memory is weaker than that for the first target. **c** An approach of the first, still invisible object is possible with the stored and ideothetically updated information in the ellipsoid body. The step lengths of the left legs are increased. The fly turns back to the first object

concentric circular shift registers which can be taken as a model for the ellipsoid-body function proposed here. In their model, one register holds the integrated information on the path (idiothetic information), the other one the sun-compass information (external world coordinates).

## Refinement of the Model with a Neuronal Fovea

Quite some insect species have an improved resolution in the frontal part of their visual field. Others have task-related foveae like the visual streak of water striders or the eye regions focusing on the tip of the extended proboscis in hummingbird hawk-moths. The compound eyes of *Drosophila* are of uniform resolution over the entire visual field. But there is a specialized eye region, the binocular range of three vertical rows of ommatidia per eye, which are directed towards the contralateral half sphere. Larger flies possess an eye region called "love spot", which allows males to track females in chasing flights, which are part of their courtship ritual. The respective eye region is not any different from its surround, but has additional computation behind it; one might call it a neuronal fovea.

Any blind visual field in the rear or specialized visual field with improved resolution, or data evaluation can justify a memory for the entire 360°. Such a memory will enable animals to analyze the visual world section by section and build up a better-resolved panoramic representation of the surround. Likewise, blind areas can be filled up by scanning section by section. The neuroanatomy of the ellipsoid body is well suited to house such a memory. In addition to replacing currently missing visual information (the function proven by Neuser et al. [5], and modeled in Fig. 1.9) it might additionally serve such an hypothetical extended function. We like to remind the reader that he or she feels to know what is behind him/her. One has previously seen that part of the surround and stored it in a memory. Our actions are based on the knowledge of the entire surround, not just the actual frontal range of acute vision.

## Orientation Using Sun-Compass Information

*Drosophila* is not known for using sun-compass orientation for migration. For a long time, there was only one account on their ability to use a dorsally presented, linear polarization pattern for stabilizing their flight direction in a flight simulator [34]. Much later, it was shown that *Drosophila* can spontaneously aligning with the *E*-vector of linearly polarized light, when stimuli are presented ventrally to walking flies. Distinct photoreceptors where found to mediate the two behaviors: photoreceptors R7/R8 of the dorsal rim are necessary and sufficient for dorsal polarotaxis, whereas ventral responses are mediated by combinations of R1-6 and R7/R8 photoreceptors [35]. The locust studies show without doubt that polarized light information is processed in the CX of these insects. Orientation based on normal light (objects, motion) is not much studied with regard to the role of the locust CX. On the other hand, there is firm evidence in *Drosophila* for orientation functions of the fly CX, when dealing with normal light. Long-range orientation, for which the polarization pattern of the sky is particularly helpful, seems not to be important for the life style of a fruit fly. We propose that both modalities are present in the model CX and that the model organism can switch between visual modalities depending on the task and the availability of visual information. This seems important for the locust as well, as the polarization information can get lost on cloudy days. A model for the locust CX is proposed in the supplement to Heinze and Homberg [7].

### Central Complex and Locomotor Activity Enhancement

A rich body of investigations point to the fact that the locomotor activity of insects decreases, if the CX is compromised. A review of the earlier work on different species is found in Homberg [10]. *Drosophila* loses activity as well, if the CX is genetically compromised [26]; just interference with the ellipsoid body seems to be activity-neutral. But is the CX of insects itself the source of behavioral activity? That might be the case, but the experimental data are consistent also with the idea that behavioral activity comes from other sources and is just canalized by the CX. Very likely, the CX integrates the inputs of two largely independent brain hemispheres and creates consistent behavior [15]. Very likely, the CX also integrates different visual qualities, such as recognized objects, motion vision, and polarization-pattern vision. And it might integrate other sensory modalities but vision, again to achieve consistent behavior at any given time. Mechanosensation is the next best proven modality to be represented in the CX. On the contrary, there is close to no evidence for olfaction to be represented in the insect CX. Humidity and temperature orientation is currently studied with regard to an involvement of the CX. The obvious inactivity phenotype in CX-defective experimental animals might well result from the un-coordinated activities of the two relatively independent brain hemispheres. Based on its individual sensory input each brain side might want to do something, just each of the two sides something else. The generation of behavioral activity depends on drives (search for water, food, mates, oviposition places, the right temperature, shelter, saving of energy, or curiosity) which are up- or down-regulated by internal states. With the exception of very basic reflexes, activity is generated by the animal and is not a consequence of external stimuli. Sensory information shapes and channels rather than creates behavioral activity. An activity controller is a necessity in any brain model but it is proposed to leave it open whether it is within or outside the insect CX. In fact, evidence is accumulating that such functions are served by the mushroom bodies rather than by the CX, particularly when it comes to switching adaptively between behavioral options in the light of incoming sensory information.

### Right-Left Bargaining Through the Central Complex

The evidence reviewed in Sect. 1.2.2 on partial split-brain mutant lines and the neuroanatomy of the CX proposes its function in right-left bargaining. The goal of a brain must be to produce consistent behavior at any time. Experiments at the flight simulator, in which contradictory visual cues were given to the two eyes of stationary

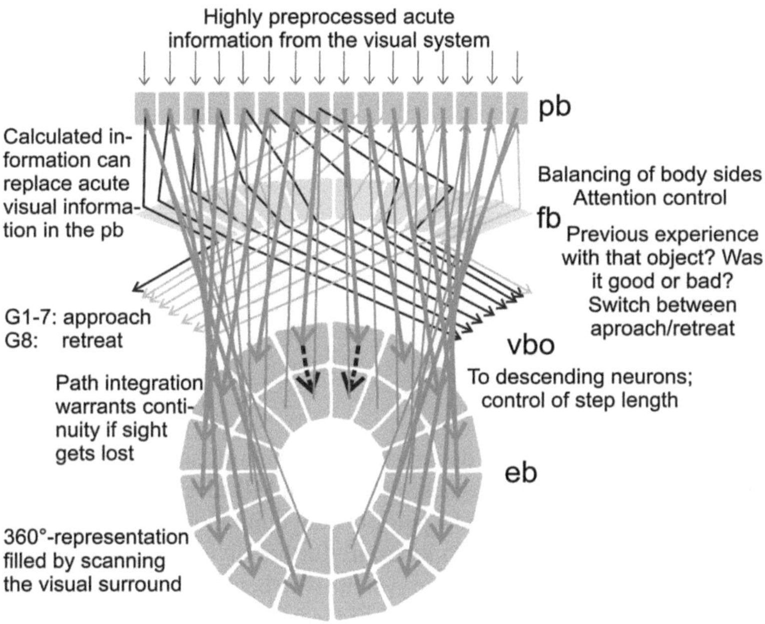


**Fig. 1.10** A summary of the proposed CX model. **pb** protocerebral bridge; **fb** fan-shaped body; **eb** ellipsoid body; **vbo** ventral lobes; horizontal fiber system; two pb-eb and eb-pb projection systems

flying *Drosophila* resulted in switching between the two sides. At any given time there was consistent behavior and the flies seemed to switch their attention between sides (Heisenberg M., Wolf R., personal communication). The latency for reactions to experimentally introduced changes in the visual stimulation of one side was significantly higher, when the attention was on the other side. The fan-shaped body should be modeled accordingly; it is the likely stage at which the inputs of the two sides might be balanced (Fig. 1.10).

## 1.3 A Neurobiological Model of the Mushroom Bodies

### *1.3.1 General Introduction to the Mushroom Bodies*

#### Position and Shape

The mushroom bodies (MBs) are a paired structure of the insect protocerebral hemispheres extending in three dimensions. The MBs "frame" the CX without known direct neuronal connections (Fig. 1.11). A bulgy calyx resides on either side of the brain in the dorsal, posterior, lateral "corners" of the protocerebrum. A peduncle

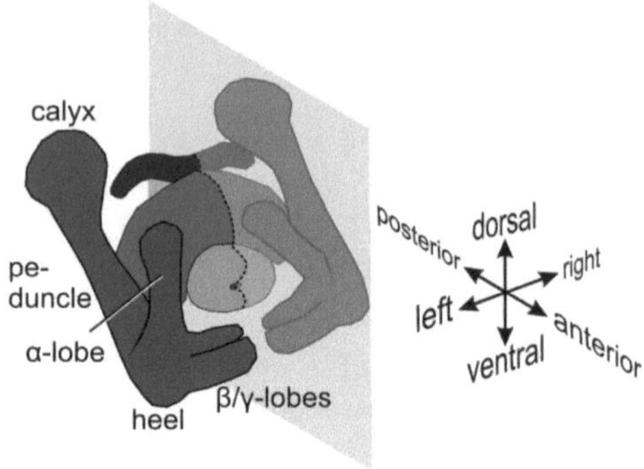


**Fig. 1.11** Localization of the mushroom bodies (MBs) in the brain of *Drosophila melanogaster*. They bracket the CX. MB constituents: Calyx: bulgy top structure; input region. Peduncle: structure running from the calyx diagonally downward. Alpha-lobe: thin, upward pointing structure. Beta/gamma-lobes: structures pointing towards the midline of the brain

runs from each calyx diagonally downward to the ventral anterior lateral "ends" of the protocerebral hemispheres. At these points each MB diverges into an alpha-lobe, which points dorsally upwards, and a pair of beta/gamma lobes pointing towards the midline of the brain. Alpha and beta lobes are further divided in an alpha and alpha prime system and a beta and beta prime system. The beta/gamma lobes of either side just don't "touch" each other at the midline of the brain anterior of the ellipsoid body. The most important constituents of the MB are the approx. 2500 Kenyon cells per side, which run fibers in parallel from the calyx through the peduncle and into the lobes.

## Input/Output Relation and Sensory Modalities

In flies as in most insects, there is a prominent olfactory input from the antennal lobes into the calices. Input from other sensory modalities is anatomically not obvious in *Drosophila* but MB tasks related to vision are nevertheless described for flies. The MBs of honeybees receive—besides olfactory information—prominent visual [36], gustatory, and mechanosensory [37] inputs. These connections likely provide mixed-modality signals that lead to experience-dependent structural changes documented in freely behaving bees (e.g. [38]). In flies and bees the lobe region receives information on sugar reward (via octopaminergic neurons) and electric foot shock (via dopaminergic neurons). There is output of the MBs to pre-motor areas of the brain.

## General Flow of Information

Inside the MBs the flow of information is through the Kenyon cells from the calyx towards the lobes.

#### Chemical Ablation of the MBs

The MBs in *Drosophila* can be conveniently ablated chemically [39]. To this end a cytostatic substance is fed to 1 h old larvae within a time window of 4 h. The neuroblasts of the MBs are the only dividing cells during that time and therefore only those will be killed. The method has been used in various studies on MB function. It is not universally applicable to other neuropils, because those do not usually have a unique time window for their early neuroblasts to divide.

### *1.3.2 Functional Mushroom Body Analyses*

#### The Major Mushroom Body Role in Olfactory Conditioning and Odor Learning

Previously neutral odors can be presented to flies together with electric foot shocks in a classical conditioning experiment. Normal flies will learn to avoid the respective odor in the future (Fig. 1.12). Likewise, a previously neutral odor can be paired with a sugar reward and flies will learn to approach that latter odor in the future (review e.g. [4]). Numerous studies have established that the MBs are necessary for associative olfactory learning and memory in honeybees (review: [40]), locusts [41], and *Drosophila* ([39]; review: [42]). In flies this type of learning is dependent on cyclic AMP and mutants defective in the molecular cascade of learning are known.

Odors are detected by receptor neurons that project from the antennae to the antennal lobes (Fig. 1.12a). Each receptor neuron expresses one type of receptor gene, and those neurons with a common receptor gene converge onto the same glomerulus in the antennal lobe. The glomeruli altogether will show a specific activity pattern for each natural odor, which usually is a blend of different odor molecules. Projection neurons convey the information from the individual glomeruli to the MB calyx and also to a presumptive pre-motor brain region called lateral horn. Within an MB the information is carried from the calyx to the different lobes by Kenyon cells. The activation pattern of the about 150 projection neurons is transformed into an activation pattern of the 2500 mushroom body-intrinsic Kenyon cells per hemisphere. Sparse coding allows activating specific Kenyon cells for each combination of activated projection neurons. Output from the MBs then projects to different target regions including pre-motor areas. The MBs are considered to be a side loop of the olfactory pathway.

The memory trace for the association between odor and shock is thought to be localized within the Kenyon cells. Whenever Kenyon cells are activated by "their" combination of odors and simultaneously with a modulatory reinforcement signal (e.g. sugar or foot shock), output from these activated Kenyon cells onto MB output neurons is suggested to be strengthened ([1, 2]; Fig. 1.12c, d). After memory formation, this strengthened output is thought to mediate conditioned behavior towards

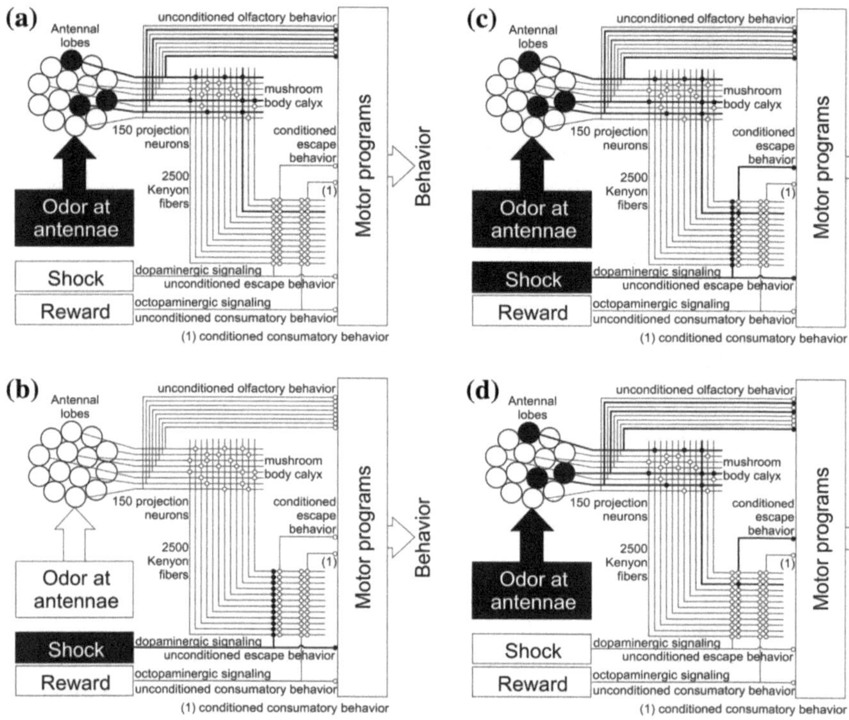


**Fig. 1.12** Model of classical olfactory conditioning in *Drosophila* by Gerber et al. [1], **a** Olfactory pathway. Receptor neurons project from the antennae to the antennal lobes. About 150 projection neurons convey the information to the calyces of the MBs. Each MB consists of about 2500 Kenyon fibers. Synapses between each Kenyon fiber and a small, random subset of projection neurons allow for odor-specific Kenyon fibers. Natural odors are often a blend of hundreds of components of which few are recognized. The projection neurons convey information also to the lateral horn where they might elicit unconditioned behaviors. **b** The reward and punishment pathways. The latter is shown activated. An electric foot shock can elicit unconditioned escape behavior in a direct pathway. The information on a bad experience is also conveyed to the Kenyon fibers in the lobes of the MBs via dopaminergic synapses. Sugar reward information would be conveyed via octopamine in a separate system. **c** During training the olfactory information is paired with foot shock. The output synapses of an active Kenyon fiber (black representing presence of a particular odor) upon pre-motor neurons ("conditioned escape") are strengthened by the simultaneous activity with the dopaminergic system, **d** After training the output synapses of the Kenyon fibers are strong enough to trigger conditioned avoidance without the presence of the shock. The odor alone can trigger avoidance; it has become a predictor for shock

(or away from) the odor without further reinforcement when the odor is encountered again during later testing.

Classical conditioning requires that the coincidence of the odor (previously neutral; conditioned stimulus = CS) and the foot shock (unconditioned stimulus = US) is detected. The odor will become a predictor for future food shocks and triggers the avoidance reaction without a further US (Fig. 1.12a, b). The presence of the odor

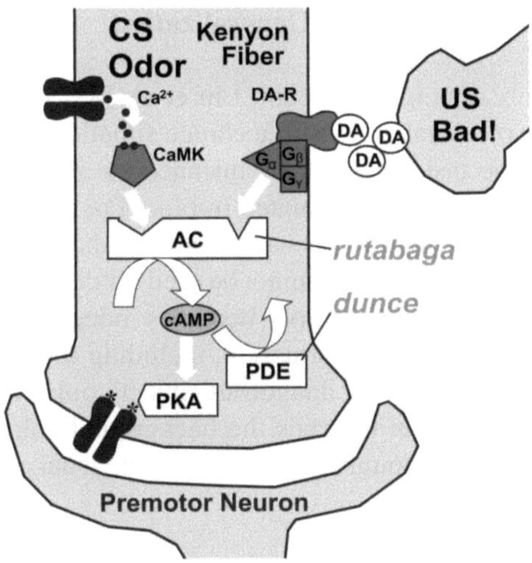


**Fig. 1.13** Model of the cAMP pathway for learning in *Drosophila*. The key player is the adenylyl cyclase (AC; the gene product of *rutabaga*) which synthesizes cAMP. AC is doubly regulated by activated Calcium-Calmodulin (CaMK) and by G$\alpha$. Odor-dependent activity of the Kenyon fiber causes Calcium influx. Calcium ions bind to CaMK. Punishment is signaled by a dopaminergic neuron; dopamine (DA) is released. The dopamine receptor (DA-R) is a G-protein coupled receptor. Upon binding of DA the alpha-unit can activate AC. If both conditions are fulfilled AC will produce over-proportionally many cAMP molecules. cAMP activates protein kinase A (PKA) which in turn can, among other functions, phosphorylate ion channels ($*$). As a result, the output synapse to a premotor neuron becomes stronger. In the future the odor alone will produce strong enough output to activate the premotor neuron which triggers avoidance behavior. cAMP is later catabolized by phosphodiesterase (PDE; the gene product of *dunce*) in order to come back to baseline. The memory can be consolidated into more stable forms

activates one or few particular Kenyon cells; cell activity is accompanied by calcium influx. The foot shock is reported by a neuron that synapses onto many (or all) Kenyon cells (but only few or even just one is activated by the odor). The neurotransmitter dopamine is released. The dopamine receptors at the postsynaptic side at the Kenyon cell are G-protein coupled receptors. Docking of dopamine molecules releases the alpha-subunit of the receptor inside the Kenyon cells. Adenylyl cyclase is the presumptive molecular coincidence detector (Fig. 1.13). Its activity is regulated by calcium (signaling odor presence) and the G-alpha protein (signaling that a shock has been received). If both conditions are fulfilled over-proportionally many cAMP molecules are produced by the adenylyl cyclase. cAMP is the second messenger for further processes which ultimately lead to the strengthening of the output synapses of the active Kenyon cells (e.g. phosphorylation of potassium channels), so that the avoidance reaction can be triggered later by occurrence of the odor alone.

## A Mushroom Body Role in Context Generalization

By means of a life-threatening heat beam, Liu et al. [43] trained stationary flying flies to avoid one pair of visual objects in a choice situation with two different pairs of identical objects. The task is learnt for instance on a blue background, i.e. the surrounding cylindrical arena is illuminated in blue. For the memory test the arena was illuminated in green. Wild-type flies abstract from the background (because it is identical for good and bad objects and cannot be used for discrimination) and learn the shape of dangerous objects; their memory becomes independent of the background color. MB-less flies learn the whole situation, including the background color, and can retrieve the information on the dangerous objects only, when the color of the background in the test situation matches the background color during acquisition. The intact MBs are therefore thought to be involved in visual context generalization.

## A Role of the Mushroom Bodies in Decision Making

Using again a life-threatening heat beam Tang and Guo [44] trained stationary flying flies to avoid a visual pattern in a choice situation with two different pairs of identical objects. One pair of objects was green and had an upright-T shape and the other pair was blue and had an inverted T-shape. In the test situation the colors were switched between the different objects. This measure created a conflicting situation for the flies. Is the predictor of heat the form or the color of the previously punished object? The wild-type flies invariably took their decisions to avoid the previously punished objects based on the color of the objects. Next, flies were trained again on objects with saturated colors but during the test the colors were not only switched but also faded out to variable degrees. Below a certain saturation point of the colors the wild-type flies suddenly and consistently went for the shape instead of the color of the objects to be avoided. There was always a clear decision in all flies of the population.

MB-less flies had no clear decision point at which they would have switched from color to shape. Rather their population average was indecisive at full saturation and then gradually leaned more and more towards shape as the colors got fainter and fainter. The MBs are therefore thought to be involved in decision making.

In a follow up study Zhang et al. [45] showed that this choice behavior consists of an early and a late phase. The first phase requires activation of the dopaminergic system and intact MBs, whereas the later phase is independent of dopamine and MBs. The immunohistological analysis shows that the MBs are densely innervated by dopaminergic axons. Thus, the circuit from the dopamine system to the MBs is crucial for choice behavior in *Drosophila*.

## A Role of the Mushroom Bodies in the Adaptive Termination of Behavior

MB-defective flies have difficulties in giving up on ongoing behavioral activity in an adaptive manner [Zäpf, Mronz and Strauss, in preparation]. For instance, normal but flight-disabled flies that were confined to an elevated platform by means of a surrounding water-filled moat, stop walking at the rim of the platform and turn into a new direction. The same holds true if attractive objects are being presented beyond the water-filled moat. The wild-type flies approached such an inaccessible object but quickly chose a new target when they encountered the moat.

The behavior of MB-defective flies is vastly different: they keep walking on the spot at the rim in front of the attractive but inaccessible object, sometimes for minutes. However, they are able to turn immediately, when the attractive object is removed. Sensing of water (with gustatory receptors in the front legs) and the ability to turn are unimpaired. We hypothesized that the MBs might be involved in the weighting of behavioral options in the light of incoming sensory information. Too high a weight might be given to the ongoing behavior in MB defective flies, whereas alternative behavioral options have too little weight.

We searched for other situations in which MB-less flies would fail in characteristic ways. More examples, where MB-defective flies fail in different behavioral contexts, will allow us to generalize the hypothesis in the above form. For instance, if a small trace of sugar is deposited on the walking surface (not smelling, not visible) wild-type flies mostly stop and eat when they accidentally encounter the spot. Sugar receptors on the front legs allow flies to sense the sugar. In the same situation MB defective flies overrun the sugar. According to the above hypothesis, the ongoing behavior "walking" has a high weight and the highly adaptive alternative option "eat sugar" cannot overwrite the walking.

We have collected more examples, in which the MB-defective flies went on with their previously chosen, then ill-adapted behavior. One example is contributed by gap crossing behavior. Once the decision to climb is taken by the fly at a gap of surmountable width and the fly has thus started climbing, the gap has been broadened belatedly [Thriphan and Strauss, unpublished]. Wild-type flies terminated the ongoing approach earlier than the MB defective flies. Another example is seen in the behavior of MB defective flies in the humidity choice paradigm (Sect. 1.2.2). MB defective flies don't stop at the far end of the dry tube but keep on walking in small circles or on the spot despite the fact that they have reached the sought-after favorable humidity condition. Wild-type flies, in contrast, often rest at the end of the tube.

### A Role of the Mushroom Bodies in Temperature Choice Behavior

Regulation of body temperature in *Drosophila* is the result of heat exchange with the surrounding environment facilitated by the large ratio of surface area to volume of their bodies. Flies must instinctively move to places with an environmental temperature close to their genetically determined desired temperature. Various temperature-gated TRP channels (transient receptor potential family of channels) have been discovered in flies [46, 47] which serve as thermosensors. A study of Hong et al. [48] revealed that MBs and the cyclic AMP-dependent protein kinase A (cAMP-PKA) pathway are essential for controlling temperature-preference behavior. Furthermore, targeted expression of cAMP-PKA pathway components just in the MBs was sufficient to rescue abnormal temperature preference behavior of the corresponding mutants. Preferred temperatures were affected by the level of cAMP and PKA activity in the MBs in various PKA pathway mutants.

### A Role of the Mushroom Bodies in Short-Term Motor Skill Learning

In Sect. 1.2.2 we report that flies can be trained to cross gaps, so that they become more successful and faster. This motor-skill learning is cAMP-dependent and mutants defective in the adenylyl cyclase do not improve their climbing. A neurogenetic partial rescue strategy allowed us to restore adenylyl cyclase function in various small areas of the brain. The short-term component of motor skill learning can be restored in learning mutants, when the wild-type adenylyl cyclase is expressed just in the MBs. This cAMP-dependent plasticity of the MBs is necessary and sufficient to restore the motor skill memory for the range of less than an hour after training. It seems not sufficient to restore the long-term memory which lasts for days and probably life-long ([25]; [Kienitz and Strauss, in preparation]). The long-term component has been restored as described in Sect. 1.2.2.

### Mushroom Bodies and a Memory for Courtship Events

McBride et al. [49] evaluated the role of the *Drosophila* MBs in courtship conditioning, in which rejection experience with mated females causes males to reduce their courtship even toward virgins [50]. MB-ablated males [39] were somewhat able to learn in the courtship paradigm. They resumed courting at naïve levels within 30 min after training, whereas the courtship of control males remained depressed for 1 hour after training. In another courtship conditioning paradigm a long-term memory can be established in wild-type flies, lasting for nine days. In MB-ablated males, memory dissipated completely within one day. These results indicated that the MBs are not required for courtship-related learning and immediate recall of courtship conditioning but are involved in the consolidation of short-term and long-term associative memories.

Courtship and courtship conditioning in *Drosophila* are regulated by multiple sensory inputs, including chemosensation (pheromones) and vision. Globally inhibiting calcium-calmodulin kinase II (CaMKII) activity in *Drosophila* disrupts courtship plasticity while leaving visual and chemosensory perception intact. Volatile and tactile pheromones provide the primary driving force for courtship; memory formation is dependent upon intact MBs and parts of the CX.

Joiner and Griffith [51] used a partial blocking approach to define areas of the brain that require CaMKII for modulation of courtship conditioning in the presence of visual, as well as chemosensory, information. Visual input suppressed the ability of MB- and CX-specific CaMKII inhibition in the dark to disrupt memory formation, indicating that the cellular circuitry underlying this behavior can be remodeled by changing the driving sensory modality. Their findings suggested that the potential for plasticity in courtship behavior is distributed among multiple biochemically and anatomically distinct cellular circuits.

## Sleep Regulation and Memory Consolidation in *Drosophila* Needs Mushroom Bodies

Three recent studies point to a role of the MBs in sleep regulation and memory consolidation. *Drosophila* exhibits all the characteristics of vertebrate sleep [52–55]. Joiner et al. [56] showed that major changes in sleep are achieved by spatial and temporal enhancement of cyclic-AMP-dependent protein kinase (PKA) activity specifically in the adult MBs of *Drosophila*. Other manipulations of the MBs, such as electrical silencing, increasing excitability or chemical ablation, alter sleep as well. These results link sleep regulation to the anatomical locus MBs, known to be involved in learning and memory.

Pitman et al. [57] come to the same result. They performed a screen of GAL4 lines expressing a temperature-sensitive synaptic blocker UAS-$shibire^{ts}$ in a range of discrete neural circuits, and assayed the amount of sleep at temperatures at which the genetically altered flies were normal and at temperatures at which the respective circuits were blocked. They identified three short-sleep lines with shared expression in the MBs. Chemical ablation of the MBs also resulted in reduced sleep. The study highlights once more the central role for the MBs in sleep regulation.

Seugnet et al. [58] ultimately close the chain of arguments towards memory consolidation. Extended wakefulness disrupts acquisition of short-term memories in mammals. However, the underlying molecular mechanisms triggered by extended waking and restored by sleep are unknown. Also the neuronal circuits in mammals that depend on sleep for optimal learning remain unidentified. Seugnet et al. [58] evaluated learning in *Drosophila* with aversive phototaxic suppression. In this task, flies learn to avoid light—that they usually like—by paired presentation with an aversive stimulus (quinine-humidity). The authors find extensive homology in sleep-deprivation-induced learning impairment between flies and humans. Both 6 and 12 h of sleep deprivation are sufficient to impair learning in wild-type flies. Moreover, learning is impaired at the end of the normal, awake day in direct correlation with the time spent awake. Mechanistic studies indicate that this task requires intact MBs and the dopamine D1-like receptors. Importantly, sleep-deprivation-induced learning impairments could be rescued by targeted gene expression of the dopamine D1-like receptor to the MBs. Seugnet et al. conclude that extended wakefulness disrupts learning in *Drosophila* and that their results demonstrate that it is possible to prevent the effects of sleep deprivation by targeting just the MBs within the fly brain.

On the other hand, Liu et al. [59] reported that a single pair of dopaminergic neurons projecting to the fan-shaped body will keep a fly awake, when artificially activated by neurogenetic means. Once more, CX and MBs seem to act antagonistically.

## Mushroom Body Participation in Place Memory in Cockroaches

Experiments of Mizunami et al. [60] demonstrate that the MBs of the cockroach (*Periplaneta americana*) play a crucial role in place memory. Behavioral tests, based on paradigms similar to those originally used to demonstrate place memory in rats (Morris water maze), demonstrate a rapid improvement in the ability of individual cockroaches to locate a hidden target (a cold spot on a hot plate) when its position is marked by distant visual cues. Bilateral lesions of selected areas of the MBs abolish this ability, but leave the ability unimpaired to locate a visible target. The results of Mizunami and colleagues demonstrate that the integrity of the peduncle and medial lobe of a single MB is required for place memory. The results are similar to those obtained by hippocampal studies in rats.

## Sensory Discrimination and the Integration of Sensory Perception with Motor Actions Require the Mushroom Bodies in Cockroaches

Mizunami et al. [61] recorded from neurons in freely moving cockroaches. Their results suggest several hitherto unrecognized functions of the MBs: extrinsic neurons that discriminate between imposed and self-generated sensory stimulation, extrinsic neurons that monitor motor actions, and a third class of extrinsic neurons that predict episodes of locomotion and modulate their activity depending on the turning direction. The relevant neurons sent their processes generally to other areas of the protocerebrum. The results support the idea of multiple roles for the MBs. These include sensory discrimination, the integration of sensory perception with motor actions, and a role in place memory.

## Mushroom Bodies Enhance Initial Motor Activity in *Drosophila*

The CX and the MBs are brain structures in most insect phyla that have been shown to influence aspects of locomotor activity. The CX regulates higher order motor coordination and generally enhances activity whereas MBs have so far been shown to suppress motor activity levels when measured over time intervals ranging from hours to weeks. Serway et al. [62] investigated MB involvement in motor behavior during the first 15 min of walking in a visual fixation task (Buridan's paradigm). They measured different parameters of walking in flies that had MB lesions induced by mutations in one of six different genes and by chemical ablation. All tested flies were later examined histologically to assess their specific MB neuroanatomy.

Mutant strains with MB structural defects were generally less active in walking than wild-type flies during these first 15 min. Most mutants showed additional activity decrements, when the defective MBs were additional ablated chemically. In conclusion, intact MBs up-regulate activity during the initial stages of walking, but suppress activity thereafter. MB influence on decision-making has been shown in a wide range of complex behaviors. Serway et al. suggest that MBs provide appropriate contextual information to motor-output systems in the brain, indirectly fine-tuning walking by modifying the quantity (activity) of behavior.

### Mushroom Bodies Suppress Long-Term Locomotor Activity in *Drosophila*

Martin et al. [63] analyzed locomotor activity of single flies in the time domain of several hours. Flies were walking in narrow tubes equipped with a light-gate in the middle. To assess the influence of the MBs on walking activity, the authors applied three independent, non-invasive methods interfering with MB function: chemical ablation of the MB precursor cells, a mutant affecting Kenyon cell differentiation (*mushroom body miniature*$^{1}$), and the targeted expression of the catalytic subunit of tetanus toxin in subsets of Kenyon cells in order to inactivate synaptic output. All groups of flies with MB defects showed elevated levels of total walking activity. This increase is attributable to the slower and less complete attenuation of activity during the experiment. Walking activity in normal and MB-deficient flies was clustered in active phases (bouts) and resting periods (pauses). Neither the initiation nor the internal structure, but solely the termination of bouts seemed to be affected by the MB defects. The peculiarities of the wild-type bout structure are given in Martin et al. [28].

### Mushroom Bodies Mediate Aspects of Aggressive Behavior

Baier et al. [64] reported the effects of several neurobiological determinants on aggressive behavior in the fruit fly *Drosophila melanogaster*. Male flies become territorial on small food patches (where they wait for females) and show aggressive behavior against incoming males. Females can defend resources against other females. Octopamine, dopamine and the MBs, all profoundly influence the expression of aggressive behavior. Serotonin had no effect. Particularly, if the synaptic output of the MBs was blocked by neurogenetic measures, those males were significantly less aggressive than intact flies.

### A Role of the Mushroom Bodies in Orientation Behavior: Centrophobism and Thigmotaxis

Open field studies showed that *Drosophila* avoids central zones and stays at the periphery, an effect that can be interpreted as centrophobism and/or thigmotaxis. Besson and Martin [65] further characterized this phenomenon and studied the responsible underlying neural mechanisms. In an open field situation the locomotor activity of *Drosophila* flies was recorded by video tracking. Center avoidance was greatly diminished in flies with ablated MBs, suggesting another role of these structures. Furthermore, temperature-controlled blocking of MB substructures with *shibire*$^{ts}$ (which disrupts synaptic transmission) revealed that specifically blocking the gamma-lobes alters centrophobism/thigmotaxis. Blocking the alpha/beta-lobes does not alter the orientation of the flies, suggesting a functional specialization of MB lobes.

## *1.3.3 Towards Modeling of Mushroom Body Functions*

### The Well-Established Mushroom Body Role in Olfactory Learning and Memory and Beyond

In *Drosophila* as well as in other insect species studied in that respect there is a prominent input from the antennal lobes to the MB calices. The MBs of flies serve a clearly proven role in classical odor conditioning (Sect. 1.3.2). By temporary cooling of the MBs, this role had been proven earlier in honeybees [66]. The functional MB model by Gerber et al. ([1, 2]; Fig. 1.12) can explain memory formation and occurrence of conditioned behaviors following learning with either positive or negative reinforcement. This model assumes two separate pathways, one working with octopamine, and the other one with dopamine as neurotransmitters. The roles of both transmitters have been confirmed in larval experiments [67], but a recent report on adult flies places another dopaminergic interneuron between the octopaminergic signaling of sugar reward and the MB lobes [68]. During consolidation, the olfactory memory seems to be shifted between MB-lobe systems [69]. In addition to this well established key role there is a plethora of other functions ascribed to the MBs of insects. Some are related to learning and memory as well (short-term motor skills, Sect. 1.3.2; conditioned courtship suppression; sleep regulation and memory consolidation; place memory). Other functions are of more general nature (termination of behavior; activity control, also with regard to aggression). Ample evidence is accumulating that the MBs are multimodal. Despite the fact that visual neuroanatomical pathways are not known in flies, the outcome of behavioral experiments in *Drosophila* clearly proves MB involvement in visual tasks (visual context generalization; decision making based on visual cues; termination of behavior at landmarks; place memory; orientation strategies). Moreover, temperature-information processing and mechanosensory-information processing are described. This seems a lot of diverse functions for a brain region with comparatively simple internal structure.

### An Excursus on the Hippocampus of Vertebrates

A consideration of the vertebrate hippocampus can remind us of the distributedness of memory functions in the vertebrate brain, but also of the unexpected grouping of diverse memory functions in one particular brain region. The hippocampus is the medial part of the telencephalon and represents the so called archicortex. It integrates information from different sensory systems, which are processed there and sent back to the cortex. The hippocampus is particularly important for declarative-memory consolidation, i.e. the transfer of memory content from a short-term to a long-term memory. Patients with bilaterally removed or destroyed hippocampi are unable to form new long-term memory; they suffer from anterograde amnesia. However, memory formed previously to the loss of the hippocampi stays intact. The hippocampus is seen as a structure which is needed to generate declarative memory, but the content is stored elsewhere in distributed locations of the cortex. As an exception, some aspects of spatial information are stored within the hippocampus itself (c.f. place learning in rats). Consolidation of procedural memory content (e.g. motor skills) works independently of a functioning hippocampus.

The famous patient H. M. lost about two-thirds of his hippocampus, parahippocampal gyrus, and amygdala tissue on both brain sides in a surgery meant to cure his epilepsy [70, 71]. His hippocampi are thought to have remained entirely non-functional, among other reasons because the entire entorhinal cortex—thus the major sensory input to the hippocampus—was destroyed. After the surgery H. M. suffered from severe anterograde amnesia [72]. His working memory and procedural memory were intact, but he could not consolidate new events in his declarative long-term memory. However, his ability to form long-term procedural memories was still intact; thus he was able to learn new motor skills without knowing he did learn them. H. M. seemed to be able to complete tasks that require recall from short-term memory and procedural memory, but not the tasks that required a long-term episodic memory. These facts suggested that recall from these memory systems may be mediated, at least in part, by different areas of the brain. Similarly, H. M. could not estabilish new long-term memories, but could recall long-term memories that existed already before his surgery. These facts suggest that encoding and retrieval of long-term memory information may also be mediated by distinct systems [73]. The medial temporal lobes and with it the hippocampus can be assumed to be a major component involved in the formation of semantic and episodic long-term memories (the medial temporal lobes are described as a convergence zone for episodic memory in Smith and Kosslyn [72]). Further evidence for this assumption has been gained by studies of other patients with respective lesions (e.g. [70, 73]).

New synaptic contacts between nerve cells are formed in the adult hippocampus. This synaptic plasticity is correlated with the creation of new memory content. Moreover, the Gyrus dentatus of the hippocampus is one of the two exceptional regions where new nerve cells are continuously generated in the adult brain. The hippocampus of animals is involved in spatial orientation [74]. Pyramidal cells in the hippocampus proper represent places in the space domain. If neurogenesis is damaged in the hippocampi of experimental animals, they show deficits in spatial learning, not a total loss. Patients with damaged hippocampi can orient in their daily life but they are unable to give directions to others. The hippocampus coordinates memory resulting from different occasions at the same location, and an internal map is formed over space and time.

**A Mushroom Body-Hippocampus Comparison**

In regard to the hippocampus of vertebrates and the MBs of insects there seem to be three different types of memories: some memory seems to be resident in the structures, some memory passes through the structures in order to reach a more durable form elsewhere, and some memory completely bypasses the structures and is not affected by their loss. The olfactory memory is a resident memory in the MBs of insects. In vertebrates, it is the the place memory, which is resident in the hippocampus. Place memory functions of insects have been mapped to the MBs of cockroaches, but no evidence is provided yet for assuming the same in *Drosophila*. Rather, a lot of visual orientation functions are found in the CX of flies. Nevertheless, the background generalization function and the role of the fly MBs in deciding between contradictory visual cues make it clear that the MBs of *Drosophila* do receive and process also visual input. The MBs are a switching station for behaviors, particularly with functions on the termination side. Behavior is initiated also by completely MB-less flies, but the adaptive termination of ongoing behavior is the function that is challenged. It seems that the currently chosen behavioral option keeps too high a priority in the light of sensory information indicating an unforeseen problem.

## 1.4 Global Model of Interactions Between Central Complex and Mushroom Bodies

Based on the current functional studies, the duties of the CX and the MBs seem to be interwoven. The CX is more of a visual neuropil, the MBs more of an olfactory one. The dedication to sensory modalities is, however, not exclusive. The CX is more on the side of initiating behavior and the MB more on the side of adaptive termination. Some forms of learning and memory depend on the CX, others on the MBs. We dare to cast the incomplete information in a global model (Fig. 1.14).

*Action: the "Array of Drives"*

Let us start the journey through the global model at the "Array of Drives", which generates and controls the behavioral activity. Hunger, thirst, need for shelter, control of the body temperature and of other body parameters, as well as the reproductive drive can keep a fly going. In the absence of such needs curiosity might be a drive—new resources can be detected in a relaxed state of mind. Need for sleep can be another drive which can calm down the fly. Drives can be modeled as set-points for certain parameters like glucose level in the hemolymph, minus the actual state of this parameter. If the discrepancy is high, there is an existential drive to meet the demand.

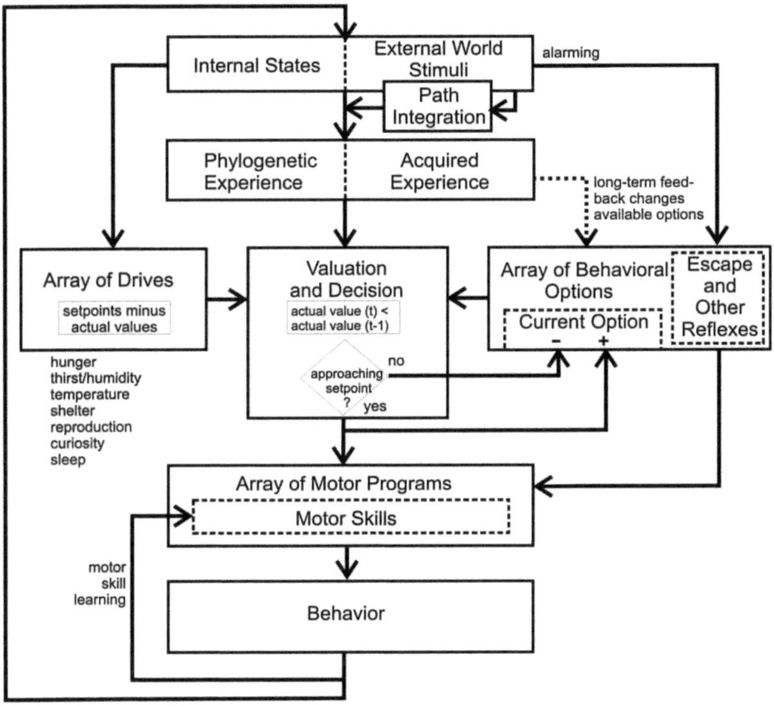


**Fig. 1.14** Global model of CX and MB functions and their interactions. Details are given in the main text

*Option: the "Array of Behavioral Options"*

The fly has a certain behavioral repertoire. Some reflexes might be directly elicited by external stimuli but usually behaviors are chosen in the light of the current needs and the current options present in the environment. Even vital reflexes can be suppressed after respective training. We assume a certain probability for each behavior to occur. The currently chosen behavioral option chosen by the fly gets a higher probability, if it proves useful to come closer to the set-point of a drive. In turn, it gets a lower probability to be chosen again, if the animal gets stuck or is even increasing the difference to the relevant set-point by its current behaviour. A long-term feedback from experience adapts the probability of certain options to be chosen, if they are maladapted, like flight starts, when the wings had been removed. It takes some time for flies to give up entirely on flight starts, if that damage has happened.

*Sensors: "Internal States/External World Stimuli"*

In respect of this global model, stimuli can be highly preprocessed, multidimensional information like object vision, a flow field of motion vision, or single one-dimensional values of all possible sensory modalities, e.g. ambient temperature. *Path integration* can replace actual visual input for some time. *Alarming stimuli* can by-pass time-consuming valuation and decision stages and elicit reflex-like behavior.

*Knowledge Base: "Phylogenetic Experience/Acquired Experience"*

The sensory information passes through look-up tables to see whether there is previous information on that constellation of incoming stimuli. In the fan-shaped body of the CX such a look-up table is seen for visual objects. In the MBs there is a look-up table for previous experience with incoming constellations of odors. Classical conditioning in this global model is contained within the box "acquired experience" because all stimuli, conditioned ones like odors and unconditioned ones like shock and sugar, run through this box. Thus, the MB learning model of Fig. 1.12 as well as learning of object features in the CX fan-shaped body (Fig. 1.8) are all contained there. Note, however, that operant learning and motor-skill learning is modeled elsewhere and will be explained later. The term "phylogenetic experience" subsumes phenomena like congenital food preferences, inborn dislikes such as bitter taste, and the congenital preference of females for the species-specific courtship song of their groom. They might be overwritten by acutely acquired experience.

*Valuation and Decision*

This central box gets input from the drives, ordered in terms of their urgency, the preprocessed sensory information scaled with the previous experience, and the behavioral options ordered in terms of probability to occur. The highest match between all these inputs will be found, and respective premotor commands will be sent out to the array of motor programs. Valuation takes the usefulness of the chosen behavior into account by evaluating whether the animal comes closer to the desired set-point. To this end it can compare the discrepancy between a current state and its set-point at earlier and later times ($t, t-1$). If the behavior proves useful (*yes*), the behavioral option will increase in probability to occur. If not (*no*), the probability of occurrence will be lowered.

*Hysteresis and Operant Learning*

Modeling of the valuation of the usefulness of a chosen behavior can account for two phenomena observed in behavioral experiments, the stability and consistency of behavior for some time and operant learning. Decisions for a certain behavioral options might be narrow, but once a behavior is chosen, the animal usually stays with it for some time. Engagements in certain behaviors come with costs and it would not make sense to switch behaviors easily with every small change in the sensory input. A hysteresis, which automatically increases the probability of further occurrence of the currently chosen behavior, models what we observe also in our own introspection. However, if the chosen behavior turns out to be coarsely maladapted, switching is necessary and possible, just that the threshold is higher.

The valuation loop does also account for operant learning. Operant learning means learning from the consequences of own actions. Depending on the outcome of the valuation at time $t$ as compared to time $t - 1$ the probability for the chosen behavior will be up- or down-regulated.

#### *Stabilization of Object-Approach Behavior*

In order to implement the interdependence between in-run toward an attractive visible object and the run-out after disappearance from sight in the global model, we propose to add a memory for the number of times in which the valuation at times $t$ and $t - 1$ has proven the usefulness of the current ongoing behavior. The probability of the chosen behavior to be chosen again will become higher and higher. The animal can bridge episodes of negative valuation (e.g. because the target is temporarily hidden) without a behavioral change.

#### *Motor-Skill Learning*

In accord with our results on gap-climbing flies, motor programs and the orchestration of motor sub-programs can be improved by motor-skill learning. To this end the outcome of behavior is looped back to the motor programs (The current arrow might be too simple and needs further refinement and future research).

#### *The Real-World Loop*

Behavior is the output of the global model, which changes the sensory input for the model in an iterative fashion.

#### *Localization of Central Functions of the Global Model in the MBs and the CX*

The classical conditioning functions (box "acquired experience") are distributed between the CX (foremost visual functions) and the MBs (foremost olfactory functions). Path integration in the CX ellipsoid body can temporarily replace acute visual input.

Valuation and decision, and particularly the rating of the currently chosen option are seen as a function of the MBs in accord with our finding that MB-less flies cannot adaptively give up ongoing behavior. MB-less flies do, however, initiate behavior. The defects are found more in the termination and correct decision for a certain behavior side, rather than in missing initiation of behavior. We therefore suggest that at least some of the "drives" and part of the "valuation and decision" box are functions of the CX. This would account for the finding, that CX-damaged flies are often inactive. They do react to handling, but fail to act thereafter (cf. [12, 26]). The drive for sleep, in turn, seems likely to be a function of the MBs—MB-less flies sleep less. Alternatively, the fact can be explained by a generally missing termination of behavior in the MB-less flies. When it comes to decision making, the right-left bargaining seems a genuine function of the CX, whereas decisions between sensory modalities and even between two contradicting visual cues ("follow color" or "follow center of gravity") look like functions of the MBs.

Motor skill learning clearly needs intact MBs as far as the short-term component is concerned (and this might be related to valuation and/or consolidation over night) but the long-term component is stored elsewhere in the brain. In insects, motor programs (central pattern generators and the sensory input shaping their output) are seen in the suboesophageal ganglion and in the thoracic ganglia. Rhythmic behavior needs just non-rhythmic go-commands and, of course, directional commands. The first can be replaced by neurotransmitters applied to the cut cervical connective of decapitated flies [75]. Motor programs are therefore not considered to be implemented within the CX. However, the orchestration of simple motor programs in order to synthesize a complex motor sequence, or a behavior, is a typical function of higher brain centers. Particularly, directional commands are being provided by the CX. The ultimate site where motor skills reside is unclear. Our first results on the long-term component point to the CX and we interpret the improvement as an optimized orchestration of motor-program constituents. However, the motor-program constituents out in the periphery might improve as well.

All in all, the investigation of time-tested insect behavior and its neuronal underpinnings with a plethora of available methods will bring us closer to the goal of constructing autonomous mobile robots for the benefit of mankind.

**Acknowledgments** Work on the model was funded by the EU grant no. 216227, SPARK II, call identifier FP7-2007-ICT-1. I am indebted to Paolo Arena for perfectly coordinating this EU project SPARK II and to all the members of the consortium for many fruitful discussions during the meetings. Christian Berg prepared early versions of figures 7–10 and intensively discussed the global model with me.

## References

1. B. Gerber, H. Tanimoto, M. Heisenberg, An engram found? Evaluating the evidence from fruit flies. Curr. Opin. Neurobiol. **14**(6), 737–744 (2004)
2. B. Gerber, H. Tanimoto, M. Heisenberg, Erratum. An engram found? Evaluating the evidence from fruit flies. Curr. Opin. Neurobiol. **15**(4), 494–495 (2005)
3. U. Hanesch, K.F. Fischbach, M. Heisenberg, Neuronal architecture of the central complex in *Drosophila melanogaster*. Cell Tissue Res. **257**(2), 343–366 (1989)
4. X. Liu, R.L. Davis, Insect olfactory memory in time and space. Curr. Opin. Neurobiol. **16**(6), 679–685 (2006)
5. K. Neuser, T. Triphan, M. Mronz, B. Poeck, R. Strauss, Analysis of a spatial orientation memory in *Drosophila*. Nature **453**(7199), 1244–1247 (2008)
6. T. Triphan, B. Poeck, K. Neuser, R. Strauss, Visual targeting of motor actions in climbing *Drosophila*. Curr. Biol. **20**(7), 663–668 (2010)
7. S. Heinze, U. Homberg, Maplike representation of celestial E-vector orientations in the brain of an insect. Science **315**(5814), 995–997 (2007)
8. U. Träger, U. Homberg, Polarization-sensitive descending neurons in the locust: connecting the brain to thoracic ganglia. J. Neurosci. **31**(6), 2238–2247 (2011)
9. U. Homberg, S. Heinze, K. Pfeiffer, M. Kinoshita, B. el Jundi, U. Homberg, S. Heinze, K. Pfeiffer, M. Kinoshita, B. el Jundi, Central neural coding of sky polarization in insects. Philos. Trans. R. Soc. Lond. B Biol. Sci. **366**(1565), 680–687 (2011)

10. U. Homberg, Structure and functions of the central complex in insects, in *Arthropod Brain. Its Evolution, Development, Structure, and Functions* (John Wiley & Sons, New York, 1987), pp. 347–367
11. U. Homberg, The central complex in the brain of the locust: anatomical and physiological characterisation, in *Brain-Perception-Cognition* (Thieme, Stuttgart, 1990), p. 318
12. R. Strauss, U. Hanesch, M. Kinkelin, R. Wolf, M. Heisenberg, *No-bridge* of *Drosophila melanogaster*: portrait of a structural brain mutant of the central complex. J. Neurogenet. **8**(3), 125–155 (1992)
13. P. Callaerts, S. Leng, J. Clements, C. Benassayag, D. Cribbs, Y.Y. Kang, U. Walldorf, K.F. Fischbach, R. Strauss, *Drosophila Pax-6/eyeless* is essential for normal adult brain structure and function. J. Neurobiol. **46**(2), 73–88 (2001)
14. B. Poeck, T. Triphan, K. Neuser, R. Strauss, Locomotor control by the central complex in *Drosophila*—an analysis of the *tay bridge* mutant. Dev. Neurobiol. **68**(8), 1046–1058 (2008)
15. R. Strauss, The central complex and the genetic dissection of locomotor behaviour. Curr. Opin. Neurobiol. **12**(6), 633–638 (2002)
16. J. Pielage, G. Steffes, D.C. Lau, B.A. Parente, S.T. Crews, R. Strauss, C. Klämbt, Novel behavioral and developmental defects associated with *Drosophila single-minded*. Dev. Biol. **249**(2), 283–299 (2002)
17. R. Ernst, M. Heisenberg, The memory template in *Drosophila* pattern vision at the flight simulator. Vision. Res. **39**(23), 3920–3933 (1999)
18. Z. Wang, Y. Pan, W. Li, H. Jiang, L. Chatzimanolis, J. Chang, Z. Gong, L. Liu, Visual pattern memory requires *foraging* function in the central complex of *Drosophila*. Learn. Mem. **15**(3), 133–142 (2008)
19. R. Strauss, J. Pichler, Persistence of orientation toward a temporarily invisible landmark in *Drosophila melanogaster*. J. Comp. Physiol. A: Neuroethol. Sens. Neural Behav. Physiol **182**(4), 411–423 (1998)
20. M.V. Chafee, P.S. Goldman-Rakic, Matching patterns of activity in primate prefrontal area 8a and parietal area 7ip neurons during a spatial working memory task. J. Neurophysiol. **79**(6), 2919–2940 (1998)
21. G. Putz, F. Bertolucci, T. Raabe, T. Zars, M. Heisenberg, The S6KII (*rsk*) gene of *Drosophila melanogaster* differentially affects an operant and a classical learning task. J. Neurosci. **24**(44), 9745–9751 (2004)
22. O. Sayeed, S. Benzer, Behavioral genetics of thermosensation and hygrosensation in *Drosophila*. Proc. Nat. Acad. Sci. **93**(12), 6079–6084 (1996)
23. L. Liu, Y. Li, R. Wang, C. Yin, Q. Dong, H. Hing, C. Kim, M.J. Welsh, *Drosophila* hygrosensation requires the trp channels *water witch* and *nanchung*. Nature **450**(7167), 294–298 (2007)
24. S. Pick, R. Strauss, Goal-driven behavioral adaptations in gap-climbing *Drosophila*. Curr. Biol. **15**(16), 1473–1478 (2005)
25. B. Kienitz, R. Strauss, Need for speed: conditions for the formation of an implicit memory in *Drosophila*, in *32nd Goettingen Neurobiol Conference Neuroforum*, vol. 15, pp. T25–8A (2009)
26. R. Strauss, M. Heisenberg, A higher control center of locomotor behavior in the *Drosophila* brain. J. Neurosci. **13**(5), 1852–1861 (1993)
27. J.R. Martin, T. Raabe, M. Heisenberg, Central complex substructures are required for the maintenance of locomotor activity in *Drosophila melanogaster*. J. Comp. Physiol. A.: Neuroethol. Sens. Neural Behav. Physiol. **185**(3), 277–288 (1999)
28. J.R. Martin, R. Ernst, M. Heisenberg, Temporal pattern of locomotor activity in *Drosophila melanogaster*. J. Comp. Physiol. A.: Neuroethol. Sens. Neural Behav. Physiol. **184**(1), 73–84 (1999)
29. J.R. Martin, P. Faure, R. Ernst, The power law distribution for walking-time intervals correlates with the ellipsoid-body in *Drosophila*. J. Neurogenet. **15**(3–4), 205–219 (2001)
30. U. Hanesch, Der Zentralkomplex von *Drosophila melanogaster*. PhD thesis, Universität Würzburg, (1987)

31. M. Mronz, R. Strauss, Visual motion integration controls attractiveness of objects in walking flies and a mobile robot, in *IEEE/RSJ International Conference on Intelligent Robots and Systems, IROS 2008*, pp. 3559–3564 (2008)
32. G. Liu, H. Seiler, A. Wen, T. Zars, K. Ito, R. Wolf, M. Heisenberg, L. Liu, Distinct memory traces for two visual features in the *Drosophila* brain. Nature **439**(7076), 551–556 (2006)
33. G. Hartmann, R. Wehner, The ant's path integration system: a neural architecture. Biol. Cybern. **73**(6), 483–497 (1995)
34. M. Heisenberg, R. Wolf et al., *Vision in Drosophila: Genetics of Microbehaviour* (Springer, Berlin, 1984)
35. M.F. Wernet, M.M. Velez, D.A. Clark, F. Baumann-Klausener, J.R. Brown, M. Klovstad, T. Labhart, T.R. Clandinin, Genetic dissection reveals two separate retinal substrates for polarization vision in *Drosophila*. Curr. Biol. **22**(1), 12–20 (2011)
36. W. Gronenberg, G.O. López-Riquelme, Multisensory convergence in the mushroom bodies of ants and bees. Acta Biol. Hung. **55**(1), 31–37 (2004)
37. U. Schröter, R. Menzel, A new ascending sensory tract to the calyces of the honeybee mushroom body, the subesophageal-calycal tract. J. Comp. Neurol. **465**(2), 168–178 (2003)
38. S.M. Farris, G.E. Robinson, S.E. Fahrbach, Experience-and age-related outgrowth of intrinsic neurons in the mushroom bodies of the adult worker honeybee. J. Neurosci. **21**(16), 6395–6404 (2001)
39. J.S. de Belle, M. Heisenberg et al., Associative odor learning in *Drosophila* abolished by chemical ablation of mushroom bodies. Science. **263**(5147), 692–694 (1994)
40. M. Giurfa, Behavioral and neural analysis of associative learning in the honeybee: a taste from the magic well. J. Comp. Physiol. A: Neuroethol. Sens. Neural Behav. Physiol. **193**(8), 801–824 (2007)
41. J. Perez-Orive, O. Mazor, G.C. Turner, S. Cassenaer, R.I. Wilson, G. Laurent, Oscillations and sparsening of odor representations in the mushroom body. Science **297**(5580), 359–365 (2002)
42. C. Margulies, T. Tully, J. Dubnau, Deconstructing memory in *Drosophila*. Curr. Biol. **15**(17), 700–713 (2005)
43. L. Liu, R. Wolf, R. Ernst, M. Heisenberg, Context generalization in *Drosophila* visual learning requires the mushroom bodies. Nature **400**(6746), 753–756 (1999)
44. S. Tang, A. Guo, Choice behavior of *Drosophila* facing contradictory visual cues. Science **294**(5546), 1543–1547 (2001)
45. K. Zhang, J.Z. Guo, Y. Peng, W. Xi, A. Guo, Dopamine-mushroom body circuit regulates saliency-based decision-making in *Drosophila*. Sci. Signal. **316**(5833), 1901 (2007)
46. L. Liu, O. Yermolaieva, W.A. Johnson, F.M. Abboud, M.J. Welsh, Identification and function of thermosensory neurons in *Drosophila* larvae. Nat. Neurosci. **6**(3), 267–273 (2003)
47. M. Rosenzweig, K.M. Brennan, T.D. Tayler, P.O. Phelps, A. Patapoutian, P.A. Garrity, The *Drosophila* ortholog of vertebrate TRPA1 regulates thermotaxis. Genes Dev. **19**(4), 419–424 (2005)
48. S.T. Hong, S. Bang, S. Hyun, J. Kang, K. Jeong, D. Paik, J. Chung, J. Kim, cAMP signalling in mushroom bodies modulates temperature preference behaviour in *Drosophila*. Nature **454**(7205), 771–775 (2008)
49. S.M.J. McBride, G. Giuliani, C. Choi, P. Krause, D. Correale, K. Watson, G. Baker, K.K. Siwicki, Mushroom body ablation impairs short-term memory and long-term memory of courtship conditioning in *Drosophila melanogaster*. Neuron **24**(4), 967–977 (1999)
50. R.W. Siegel, J.C. Hall, Conditioned responses in courtship behavior of normal and mutant *Drosophila*. Proc. Nat. Acad. Sci. **76**(7), 3430–3434 (1979)
51. A.J. Mei-ling, L.C. Griffith, Visual input regulates circuit configuration in courtship conditioning of *Drosophila melanogaster*. Learn. Mem. **7**(1), 32–42 (2000)
52. J.C. Hendricks, S.M. Finn, K.A. Panckeri, J. Chavkin, J.A. Williams, A. Sehgal, A.I. Pack, Rest in *Drosophila* is a sleep-like state. Neuron **25**(1), 129–138 (2000)
53. P.J. Shaw, C. Cirelli, R.J. Greenspan, G. Tononi, Correlates of sleep and waking in *Drosophila melanogaster*. Science **287**(5459), 1834–1837 (2000)

54. D.A. Nitz, B. Van Swinderen, G. Tononi, R.J. Greenspan, Electrophysiological correlates of rest and activity in *Drosophila melanogaster*. Curr. Biol. **12**(22), 1934–1940 (2002)
55. B. van Swinderen, R.J. Greenspan, Salience modulates 20–30 Hz brain activity in *Drosophila*. Nat. Neurosci. **6**(6), 579–586 (2003)
56. W.J. Joiner, A. Crocker, B.H. White, A. Sehgal, Sleep in *Drosophila* is regulated by adult mushroom bodies. Nature **441**(7094), 757–760 (2006)
57. J.L. Pitman, J.J. McGill, K.P. Keegan, R. Allada, A dynamic role for the mushroom bodies in promoting sleep in *Drosophila*. Nature **441**(7094), 753–756 (2006)
58. L. Seugnet, Y. Suzuki, L. Vine, L. Gottschalk, P.J. Shaw, D1 receptor activation in the mushroom bodies rescues sleep-loss-induced learning impairments in *Drosophila*. Curr. Biol. **18**(15), 1110–1117 (2008)
59. Q. Liu, S. Liu, L. Kodama, M.R. Driscoll, M.N. Wu, Two dopaminergic neurons signal to the dorsal fan-shaped body to promote wakefulness in *Drosophila*. Curr. Biol. **22**, 2114–2123 (2012)
60. M. Mizunami, J.M. Weibrecht, N.J. Strausfeld, Mushroom bodies of the cockroach: their participation in place memory. J. Comp. Neurol. **402**(4), 520–537 (1998)
61. M. Mizunami, R. Okada, Y. Li, N.J. Strausfeld et al., Mushroom bodies of the cockroach: activity and identities of neurons recorded in freely moving animals. J. Comp. Neurol. **402**(4), 501–519 (1998)
62. C.N. Serway, R.R. Kaufman, R. Strauss, J. Steven de Belle, Mushroom bodies enhance initial motor activity in *Drosophila*. J. Neurogenet. **23**(1–2), 173–184 (2009)
63. J.R. Martin, R. Ernst, M. Heisenberg, Mushroom bodies suppress locomotor activity in *Drosophila melanogaster*. Learn. Mem. **5**(1), 179–191 (1998)
64. A. Baier, B. Wittek, B. Brembs, *Drosophila* as a new model organism for the neurobiology of aggression? J. Exp. Biol. **205**(9), 1233–1240 (2002)
65. M. Besson, J.R. Martin, Centrophobism/thigmotaxis, a new role for the mushroom bodies in *Drosophila*. J. Neurobiol. **62**(3), 386–396 (2004)
66. J. Erber, T.H. Masuhr, R. Menzel, Localization of short-term memory in the brain of the bee, *Apis mellifera*. Physiol. Entomol. **5**(4), 343–358 (1980)
67. C. Schroll, T. Riemensperger, D. Bucher, J. Ehmer, T. Völler, K. Erbguth, B. Gerber, T. Hendel, G. Nagel, E. Buchner et al., Light-induced activation of distinct modulatory neurons triggers appetitive or aversive learning in *Drosophila* larvae. Curr. Biol. **16**(17), 1741–1747 (2006)
68. C.J. Burke, W. Huetteroth, D. Owald, E. Perisse, M.J. Krashes, G. Das, D. Gohl, M. Silies, S. Certel, S. Waddell, Layered reward signalling through octopamine and dopamine in *Drosophila*. Nature **492**(7429):433–437 (2012)
69. M.J. Krashes, A.C. Keene, B. Leung, J.D. Armstrong, S. Waddell, Sequential use of mushroom body neuron subsets during *Drosophila* odor memory processing. Neuron **53**(1), 103–115 (2007)
70. W.B. Scoville, B. Milner, Loss of recent memory after bilateral hippocampal lesions. J. Neurol. Neurosurg. Psychiatry **20**(1), 11–21 (1957)
71. S. Corkin, What's new with the amnesic patient H.M.? Nat. Rev. Neurosci. **3**(2), 153–160 (2002)
72. E.E. Smith, S.M. Kosslyn, *Cognitive Psychology: Mind and Brain*, 1st edn. (Pearson/Prentice Hall, Upper Saddle River, 2007)
73. B. Kolb, I.Q. Whishaw, *Fundamentals of Human Neuropsychology*, 4th edn. (W.H. Freeman, New York, 1996)
74. E.I. Moser, E. Kropff, M.B. Moser, Place cells, grid cells, and the brain's spatial representation system. Annu. Rev. Neurosci. **31**, 69–89 (2008)
75. C. Yellman, H. Tao, B. He, J. Hirsh, Conserved and sexually dimorphic behavioral responses to biogenic amines in decapitated *Drosophila*. Proc. Nat. Acad. Sci. **94**(8), 4131–4136 (1997)

# Chapter 2
# A Computational Model for the Insect Brain

**P. Arena, L. Patanè and P. S. Termini**

**Abstract** As seen in the Chap. 1, the fruit fly *Drosophila melanogaster* is an extremely interesting insect because it shows a wealth of complex behaviors, despite its small brain. Nowadays genetic techniques allow to knock out the function of defined parts or genes in the *Drosophila* brain. Together with specific mutants which show similar defects in those parts or genes, hypothesis about the functions of every single brain part can be drawn. Based upon the results reported in the Chap. 1, a computational model of the fly *Drosophila* has been designed and implemented to emulate the functionalities of the two relevant centres present in insects: the Mushroom Bodies and the Central Complex. Their actions and inter-actions are adapted from the neurobiological prospective to a computational implementation. A complete block scheme is proposed where the proved or conjectured interactions among the identified blocks are depicted. Several simulations results are finally provided to demonstrate the capability of the system both considering specific parts of the complete structure for comparison with insect experiments, and the whole model for more complex simulations.

## 2.1 Introduction

In the bio-inspired robotics field, robots can be used to reproduce animal behavior in order to study their interaction with the environment. Robots help to improve the understanding of animal behavior and animals help to create efficient and robust robotic systems. The study of animal brains leads to new control systems that could

P. Arena (✉) · L. Patanè · P. S. Termini
Department of Electrical, Electronic and Computer Science Engineering,
University of Catania, 95125 Catania, Italy
e-mail: parena@diees.unict.it

L. Patanè
e-mail: lpatane@diees.unict.it

P. Arena and L. Patanè (eds.), *Spatial Temporal Patterns for Action-Oriented Perception in Roving Robots II*, Cognitive Systems Monographs 21,
DOI: 10.1007/978-3-319-02362-5_2, © Springer International Publishing Switzerland 2014

allow robots to be able to orient themselves in complex environments, to take decisions, to accomplish dangerous missions, in order to become completely autonomous. Robotic implementation of biological systems could also lead to the introduction of new models for basic sciences, in particular when investigating the emergent properties of models. Several attempts are present in literature related to algorithms or bio-inspired networks able to mimic the functionalities of parts of the brain. A lot of work has been done in several animal species belonging to mammals, mollusks and insects [1]. Looking into the insect world different research groups around the world are trying to design models which are able to reproduce interesting behaviors shown by insects: cooperation mechanisms in ants [2], navigation strategies in bees [3], looming reflex in locusts [4], homing mechanisms in crickets [5], central pattern generator and obstacle climbing in cockroaches [6, 7], reflex-based locomotion control in the stick insect [8], just to cite some examples. It is evident that the effort is focused on specific peculiarities associated with the different insect species that can be also useful for robotic applications. Nevertheless, a more challenging task consists of trying to model the main functionalities of an insect brain, looking from an higher level, trying to identify the mechanisms involved in the sensing-perception-action loop. The proposed work is focused on the development of an insect brain computational model mainly focused on the *Drosophila melanogaster*, the fruit fly. The insect brain architecture, structured in functional blocks, has been developed in a complete software/hardware framework in order to evaluate the capabilities of this bio-inspired control system on both simulated and real robotic platforms. In order to develop an useful and suitable architecture, the proposed framework is flexible and robust and presents a structure suitable to decouple simulations from control algorithms. The functional separation helps to isolate the application itself from graphic interfaces and the underlying hardware. The main aim is to develop an extensible and general purpose architecture. The insect brain model has been evaluated in scenarios strictly linked to the neurobiological experiments to make a direct comparison. Moreover the available data on wild type flies and mutant brain-defective flies allows to identify the main role of each neural assembly in performing specific tasks like visual orientation, olfactory learning, adaptive termination of behaviours and others. Finally the main guidelines used for the definition of evaluation criteria and the creation of benchmarking scenarios where the system performance can be evaluated, are also reported.

## 2.2 Insect Brain Cognitive Architecture and Learning Issues

In the previous chapter a first model of the interplay between MB and CX was presented. In that model there is not a specific block representing a specific function of the MB or of the CX; the interest was focussed to functional aspects. In the following a model useful in view of a robotic implementation will be first considered. This builds upon a previously designed model [13] and outlines sensory motor pathways with the addition of learning and representation blocks. Subsequently biological aspects will be more and more included leading to the most recent scheme of a complete inset brain computational model, which will be simulated and implemented for robotic experiments.

In this section a preliminary description of the insect brain cognitive architecture developed on the basis of the *Drosophila* experiments is given. An overview of the general architecture is reported and particular attention is devoted to the learning strategies that are envisaged inside the cognitive structure.

### 2.2.1 Various Steps in Modeling the Insect Brain

The actual insect brain computational model is the result of a number of previous versions, that were further refined and updated, once new results, especially on the neurobiological perspective, were available. The first perceptual architecture proposed was designed and organized in various control levels consisting of functional blocks, acting either at the same level, as competitors, or at distinct hierarchical levels showing the capability to learn more complex, experience-based behaviors [9].

The control architecture (as reported in Fig. 2.1) consisted of series of parallel sensory-motor pathways (i.e. basic behaviours) that were triggered and controlled by specific sensory events in a reflexive way, giving the knowledge baseline to the system. Going up in the hierarchical scheme, two relevant centers of the insect brain were considered: the Mushroom Bodies (MBs) and the Central Complex (CX). Taking into account the known facts about these centres, from a biological/neurogenetic point of view and their role in perceptual processes [9–11], some preliminary main functions were initially focussed, to be assessed and refined during the project activities. In particular, a function ascribed to MBs was to have a role, due to their learning capabilities, in the enhancement of causal relations arising among the basic behaviours, by exploiting the temporal correlation between sensory events; information storage and retrieval in the case of the olfaction sense; resolving contradictory cues through the visual sense by imposing continuation or adaptive termination of ongoing behaviour. Relevant functions ascribed to the CX were integration and elaboration of visual information, storing and retrieving information on objects and their position in space, controlling the step length in order to approach or avoid such objects; motor control, landmark orientation and navigation, orientation storage and others.

These learning aspects were treated using causal Hebbian rule in an array of spiking neurons for anticipation [12], on the basis of what already studied in [13], where memory structures based on Recurrent Neural Networks were considered.

At a higher level of the scheme, a representation layer was introduced, able to process sensory information in order to define the final behavior. Here we introduced a lattice of non spiking neurons. This neural lattice shows distinct characteristics of complex dynamical systems. The emerging patterns of neural states take on the meaning of percepts. These ones are then associated to suitable modulations of the basic behaviors. This modulation is performed through an unsupervised learning process which creates associations among sensory stimuli and patterns. In this way,

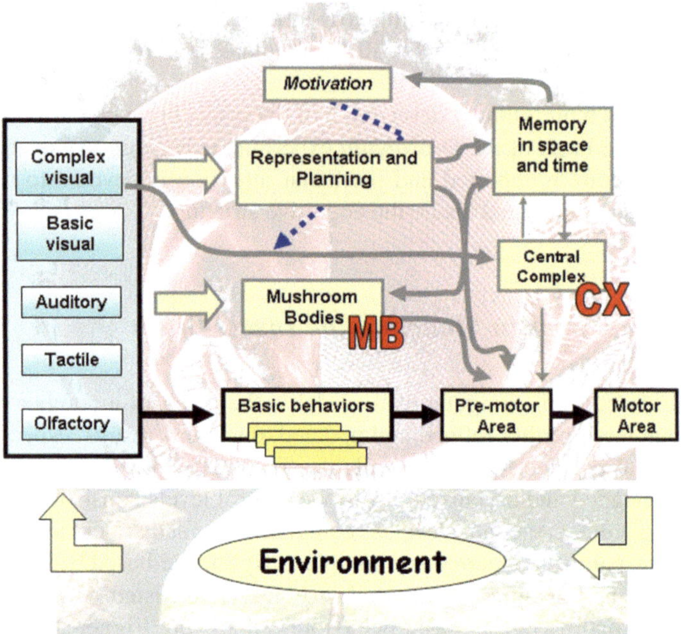


**Fig. 2.1** Functional block diagram of the initial version of the insect brain cognitive architecture. The interaction between the robot and the environment is realized by direct sensory-motor pathways, the *basic behaviors*, which are modulated by the representation layer. MB and CX are relevant centers of the insect brain devoted to temporal correlation, information storage and retrieval, and other functionality summarized in a correlation layer. Finally the high level functions of the representation layer consists of a *preprocessing block*, a *perceptual core*, a *selection network*, while the *Reward function* drives the learning process

at the end of the leaning stage, each pattern represents a particular behavior modulation, while its trained basin of attraction represents the set of all the environment conditions, as recorded through the sensors, leading to the emergence of that particular behavior modulation. The modulation parameters associated with each pattern are learned through a reinforcement learning: here the reinforcement signal is provided by a motivation layer implementing the degree of satisfaction of the robot. This depends on the local satisfaction of the single basic behaviors with the addition of other terms that reflect the robot mission. The presence of additional information into the motivation layer, not used by the basic behaviors can be exploited by the Representation layer in order to increase the robot performance.

Memory was distributed in the whole architecture, but a specific block was also considered (i.e. Memory in space and time in Fig. 2.1). This block develops a contextual layer, like in [14]. Here sequences of successfully emerged patterns can be memorized to be retrieved when needed. The whole architecture was conceived in such a way that the basic behaviors, which are often life-saving sensory-motor pathways, are progressively enriched with emergent capabilities which incrementally increase the animal skills. The main focus was therefore on the application of complex dynamics to obtain a proper, complex, context-learned modulation of the basic skills.

In a first attempt, following the results reported in [13], this process of complex emerging of situation related percepts was an important characteristic of our approach which makes it different from the other control strategies, based on the subsumption architecture proposed by [15]. The latter in fact, uses a high level approach to face with the design of both basic behaviors and the coordination block. In our strategy, complex dynamical systems are successfully used. Both architectures use a behavioral decomposition of the system to exploit parallel computation, although the Subsumption network makes a rigid hierarchy among the basic behaviors: the lower ones cannot influence the upper ones, whereas the latter can act on the former. In our scheme all the basic behaviors are sensory-motor pathways elicited by only one sensory modality and on the same hierarchical level: knowledge is incrementally built upon their modulation, giving importance to one or the other, depending on the context. Under this perspective the proposed architecture resembles the Motor Schemas, introduced by [16]. Turing Patterns in RD-CNN are hosted, in our architecture, within a layer here called *Representation Layer*. This term is here not referred to a place where a predictive model of the body-environment interaction is learned. This is rather a layer where the single-sensory motor modalities, constituted by the parallel sensory motor pathways, are modulated in a feedforward way, taking into account all the incoming sensory stimuli. This leads to the emergence of a contextually self organising activity, focusing at modulating the basic behaviors.

This was the insect brain model inspired by previous results [13] and object of an intense speculation phase. The aim of the research activity was to tightly link the emergent approach to cognition, based on nonlinear complex dynamics, to the knowledge gained from insect neurobiology. Therefore, the initial approach, briefly discussed above, mainly based on emergence and self organization, was modified to take into account the biological perspective. From a deep analysis of the state of the art and direct experiments performed, current knowledge from insect Neurobiology provided precious information on the details of the (mainly) low level information processing (i.e. excluding the representation level). Therefore our effort moved toward the lower level blocks. Once defined these parts, a suitable connection with the higher layers is envisaged. This is clear since the details about how insects gain a structuring of the whole information for decision making and planning is really a challenge, and at the present stage, it is unknown. So to cope with the lack of this information, the high level representation layer could be used to complete the architecture. Following such a path, the functional block scheme in Fig. 2.1 has been modified to include details from Neurobiology that led to a modification on handling the different sensory inputs and the sequence of processing steps involved, adding new details, mostly at the basic sensorimotor pathways and medium level of information processing. Moreover two different learning mechanisms were identified, which are mainly involved in this process: Classical Conditioning through positive and negative reward signals, and Operant Conditioning at the pre-motor area level. Fig. 2.2 shows the relevant elements of the updated insect brain architecture taken

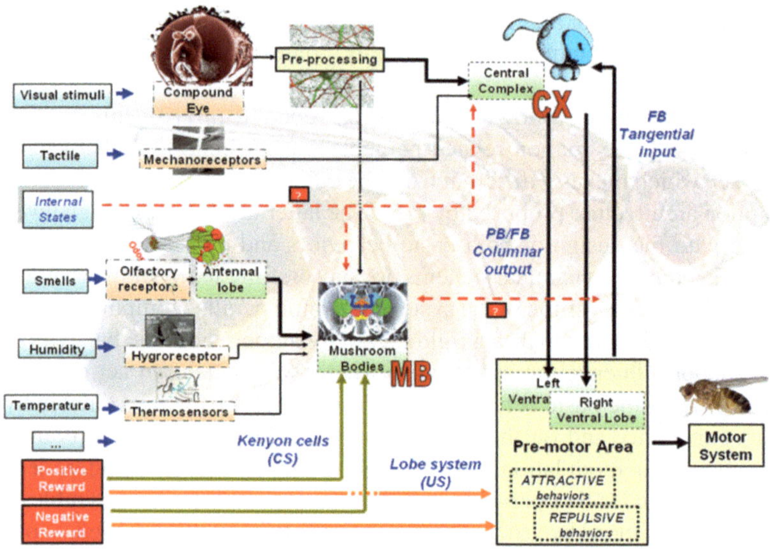


**Fig. 2.2** Revised block diagram of the insect brain architecture. The diagram proposed in Fig. 2.1 is here modified to further match with the biological counterpart. Question marks indicate hypothesized connections that have to be still assessed

into consideration and the interaction among them. The architecture includes a series of sensorial stimuli acquired and preprocessed by the insect, that are successively handled by the two main structures taken into consideration: MBs and CX.

As it can be seen in Fig. 2.2, input sensory modalities are divided into different sensorimotor pathways:

- visual stimuli, through the compound eye and a pre-processing phase are further processed by the CX, which, with all its constituent parts, contributes to vision related functions like orientation (through direct connections to the ventral lobe), object detection, classification and memory.
- Tactile stimuli: there is behavioral evidence for mechanosensory information to be present in the *Drosophila* CX from the legs. Mechanosensation is the next best proven modality to be represented in the CX, besides the visual one.
- Smells and their connections to the MBs: MBs play a relevant role in olfactory processing, through input from the antennal lobe and olfactory receptors. They play also a role in context generalization starting from visual information. As far as the learning aspects are concerned, Classical conditioning mechanisms are constantly used in insects, and MBs seem to be the main center where learning takes place. Two distinct paths for positive and negative rewards exist. Another important learning mechanism occurs in the pre-motor area and it is basically an instrumental learning. Experiments shown how a long term memory can be created and the sleep phase is fundamental to stabilize and improve the learning process; for these reasons a kind of internal model is expected.

- Humidity and temperature conditioning are included into the model but not further directly exploited for the insect brain computational model as themselves. Indeed the information about the role and specific functions of MBs in these processing (like responses to temperature gradients) was used to perform specific experiments to study a memory effect called memotaxis in flies, in view of the robot implementation.
- Other functionalities can be considered taking information from data acquired through neuron recording in freely moving cockroaches and involving MBs processing. Their results suggest several until now unrecognized functions of the MBs: extrinsic neurons that discriminate between imposed and self-generated sensory stimulation, extrinsic neurons that monitor motor actions, and a third class of extrinsic neurons that predict episodes of locomotion and modulate their activity depending on the turning direction. The relevant neurons sent their processes generally ascend to other areas of the protocerebrum. Their results support the idea of multiple roles for the MBs. These include sensory discrimination, the integration of sensory perception with motor actions, and a role in place memory.
- Internal states and motivations are considered in the process but hypotheses and speculations will be done due to the lack of specific neurobiological evidences.

The evolution of the various stages of the model improvement leads to the scheme shown in Fig. 2.3.

In the insect brain block scheme it is possible to distinguish four main sensorial pathways; the olfactory and the visual pathways allow to perceive the environment, whereas gustation and nociception are indispensable to obtain information about the goodness or badness of the current situation. In particular the gustatory sensory modality, placed in the front legs of the fly, is reproduced in robotic experiments through signals coming from light sensors placed in the ventral part of the robots, facing with the ground. This modality is used in experiments like the adaptive termination of behaviours. Nociceptory signals, used for punishment, are reproduced through sound signals (or through the ventrally placed light sensors) and applied in such experiments as visual/odour learning. These sensorial pathways are linked together to make the system able to perform anticipatory actions to improve efficiency in finding rewards and to avoid dangerous situations. In the actual structure, learning is attained using mechanisms based on classical as well as operant conditioning. Olfactory and visual inputs, due to their complexity, are considered as pre-processed at the sensory level. Olfaction, has been studied at the aim to derive the corresponding MB neural models. Regarding the olfactory sensors, since the artificial ones are still too slow and difficult to be efficiently characterized, they were substituted by sound sensors, which are more reliable and able to provide both unconditioned and conditioned inputs to the neural processing network. Soon after the visual pre-processing stage we can find the Central Complex neuropil model, containing all its main components:

PB The Protocerebral Bridge (PB), which, in our model, performs its three main functions (Object Detection, Distance Estimation and Object Position extraction), as drawn by the biological experiments and neuro anatomical evidence;

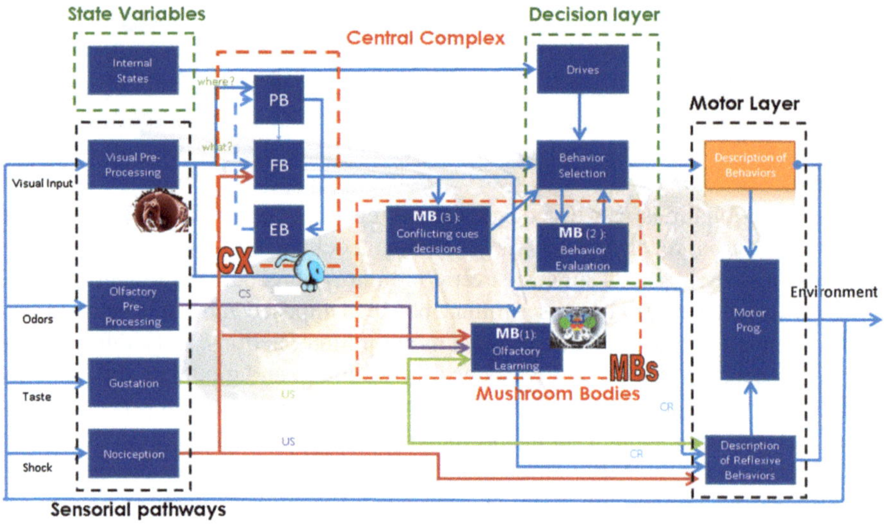


**Fig. 2.3** Block diagram of the insect brain model

FB the Fan-shaped Body (FB), which performs two main functions: feature extraction (color, orientation, size, center of gravity, wideness, height) and feature evaluation/learning (the robot collects features and is able to associate those features to punishment or neutral situations.

EB the Ellipsoid Body (EB), where the robot spatial and the newly discovered decision making memory is formed and contained.

The other fundamental neuropil of the insect brain is constituted by the Mushroom Bodies (MBs). MBs were found to influence many different aspect of the insect brain plasticity. The main function of MBs is olfactory learning: this was implemented in our architecture through a hebbian or STDP based learning scheme in spiking networks. The other function experimentally found in MBs is behaviour evaluation, mainly acting at the decision level. For this reason this MB functionality, here called MB2 (Fig. 2.3) is included into the decision layer and implemented as a separate block with respect to the olfactory learning block. Another addressed function is decision making: this function was discovered working with MB defective flies which were unable to make a sharp choice among two different contradictory visual features (color and shape) in front of the fading of the preferred one (color). This is a function that, involving visual learning, cannot be ascribed to the conventional functionality of MBs (olfactory representation and leaning). So this function was modeled as a separated block (MB3) and placed at the decision layer (Fig. 2.3) [17]. Moreover, from the block-size perspective, direct connections among FB and MB cannot be directly drawn for the lack of experimental evidence in the fly. So it is hypothesized that particular visual information, like color saturation reaches the MBs indirectly through other brain parts (like for example the Lateral Horn) and gives it the possibility to concurrently act on the Behaviour Selection block at the level of decision making.

A series of internal states are monitored through a set of virtual proprioceptive sensors; these internal states undergo a continuous interaction with the ongoing external state of the agent, recorded through the exteroceptive sensors. Internal states are chosen according to the applications prepared, discussed within each experimental scenario. An internal state (like hunger or "need for charging", need to sleep, etc.) is supposed to be directly related to drives which are typically reference control signals for the following Behaviour Selection Network (BSN) block (like desired battery level, zero home distance, etc.). In order to satisfy its drives, the robot has to choose a precise behaviour from a pre-defined set of available behaviours, each one oriented to satisfy one or several contemporary drives. Up to this stage, the BSN is implemented through a spiking network with dynamic synapses, leaving opened the possibility to learn other behaviours better satisfying the strongest drive. This functionality within the BSN takes place at the highest layer in the insect brain architecture. Till now, there are not yet specific experiments that can demonstrate the existence of such a network in the *Drosophila* brain; therefore the hypothesized artificial BSN was maintained to represent the highest level control functionality. The BSN was endowed, at this stage, with auto-excitatory synapses to avoid a continuous switching among the selected behaviours.

The other block residing at the decision layer is Behavior Evaluation. Experiments on the MB less flies show that this function is ascribed to MBs, even if apparently separated by the common MB functionality. So also this block was modeled separately with respect to the main MB block and so called MB2, as also mentioned above. This block evaluates the capability of the selected behaviour to satisfy the active drive, represented by a given setpoint to be reached. As soon as a given behaviour is initiated (behaviour initiation is ascribed as a specific CX role) the MB2 block starts an increasing inhibitory function on the ongoing behaviour in order to completely inhibit this one if the drive is not satisfied within a certain time window. In this case another behaviour wins the competition and is selected.

The Motor layer contains the following blocks: The Description of Reflexive Behaviours describes the fixed actions that allow the robot to take the right direction in the case of punishment. Here additional functions are included, considering the fact that a fly, repetitively punished, can reach a "no-motion state": i.e. the insect is frozen for a certain amount of time.

The Description of Behavior block describes the available behaviours that the robot can follow. The type and number of the possible behaviours the robot can exhibit depends on the robot applications. As an example implemented is the targeting behaviour. This behaviour, when selected in the BSN, causes a series of actions focussed at moving the robot towards the visual target that elicited that behaviour, while maintaining it at the centre of the visual scene.

The Motor Programs block contains all the possible elementary actions the robot can perform. They are supposed, up to now, to be pre-programmed unless a wide space for hosting learning strategies exists, which is currently under investigation. This block is strictly dependent on the robotic architecture to be used. It contains a series of control signals for the wheels/legs in order to realize the desired advancement, steering or rotation. In particular, dealing with legged robots, the central Pattern Generator paradigm was taken into account. This approach was recently accompanied to some powerful theoretical conditions which a-priori guarantee global exponential convergence to any imposed gait that the structure is asked to show [18].

As it can be derived from the analysis of the new scheme, almost all of the blocks enhanced are truly biologically driven, from experiments on flies or other insects. The last item outlined, i.e the path internal states—drives—behavior selection—behavior evaluation (the really high level functionalities) up to now does not find any specific experimentally driven model. The behavior evaluation is addressed to the MBs, and indeed such areas are recognized to have a role in decision making, resolving contradictory cues, imposing adaptive termination or continuation of ongoing behaviors, but how this is linked to the choice of the behaviors and to the internal states and drives, is unknown. This information flow can really be included into the *Representation layer* and could be modeled using the Reaction-diffusion approach discussed at the highest layer in the former scheme of Fig. 2.1. In this part of the insect brain architecture could well find place all the spatial-temporal dynamics leading to the experience based control of ongoing behaviors.

## 2.3 Memory and Learning Mechanisms in Nature

Artificial agents (i.e. simulated and real robots) need learning algorithms to construct a knowledge and to improve their basic capabilities. In Nature, learning mechanisms are part of living beings and act at different form and level of complexity. According to a general definition, "*learning* is an adaptive change in behavior caused by experience" [19].

It is important to notice that learning mechanisms, in whichever form, need a memory structure to correctly work. Memory is the storage and recall of previous experiences. *Memory* is necessary for learning; it is the mechanism whereby an experience is incorporated into the organism, so that it can later be used to bring about adaptive changes in behavior [19].

There are a number of different types of learning and memory. Table 2.1 lists the main categories. In other words, learning is one of the most important mental function present in humans, animals and artificial cognitive systems. It relies on the acquisition and processing of different types of knowledge supported by perceived information. It leads to the development of new capacities, skills, values, understanding.

### *2.3.1 Memory and Learning in MBs and CX*

In relation with the Insect Brain Model, memory elements and learning mechanisms are distributed on the whole architecture. The instruments used to unravel

**Table 2.1** Main categories of learning and memory [19]

| Types of learning | Types of memory |
|---|---|
| *Simple* | Immediate |
| Habituation | Short-term |
| Sensitization | Long-term |
| *Associative* | Specific |
| Passive (classical) | |
| Operant (instrumental) | |
| One-trial (aversion) | |
| *Complex* | |
| Imprinting | |
| Latent | |
| Vicarious | |

**Table 2.2** Memory and learning in insect brain

| Experiments | Functionality | STM | LTM | Memory/Learning |
|---|---|---|---|---|
| Exploration | Increase of mean free path | ? | ? | Working memory |
| Olfactory-based navigation | Olfactory | MBs | MBs | Classical conditioning |
| Visual-based navigation | Visual learning | FB (CX) | ? | Operant conditioning |
| Detour paradigm | Path integration | EB (CX) | – | Working memory |
| Heatbox learning | Orientation memory | ? | – | Operant conditioning |
| Gap climbing | Motor learning | MBs | CX | Operant learning |

the information flow that characterizes this complex structure are focused biological experiments that can be used to distinguish involved blocks and functionalities. A summary of the results obtained is reported in Table 2.2 where a series of experiments are used to identify which parts of the insect brain are involved in performing specific behaviors. In particular the table includes the locations of the short term (STM) and long term (LTM) components of the memory system involved and the learning mechanism used.

The identification of the role of each center in the cognitive process that characterizes the fly is not always easy to obtain. An example could be the analysis of the exploration phase. Through experiments a centrophobic behavior has been identified: the fly adopts an exploration strategy that probably includes the increasing of its mean free path. The resulting behavior in a closed arena consists in reaching and following the external walls. This kind of behavior needs a working memory but up to now the location of this element is unknown.

As far as the olfactory learning is concerned, the fly is able to navigate into an environment following a smell. Both the short term and long term memory can be located in the MBs and a simple but efficient classical conditioning is performed to associate a meaning to specific smells that can be either rewarded or punished.

Similarly to the olfactory learning where the MBs are the structure responsible in the fly, visual processing and the corresponding memory and learning structures is completely associated to the Central Complex. The fly is able to learn both through classical and operant conditioning to classify objects, extracting from the segmented image a series of characteristic features. The *fan-shaped body* (FB) is the structure of the CX devoted to the STM: the fly associates a meaning to objects, depending on its past experience.

Another important neural structure belonging to the CX is the *ellipsoid body* (EB), that is responsible for orientation memory.

In the detour experiment (addressing the capability to aim to a formerly seen target, even after being luring away) the parameters characteristic for path integration have been identified. The capability to retain, recall and integrate positional information about the target into guiding behavior has been summarized under the term spatial working memory and is ascribed to the EB.

Finally as concern the motor learning, relevant information were obtained using the gap climbing experiment where the fly is forced to climb several consecutive gaps. The strategy used to overcome the obstacle can be improved through operant learning both in the short term (the STM was located in the MBs) and in the long term after a sleeping phase (the LTM was located in the CX). The modeling of the mechanisms used for motor learning is complex task that up to now represents an open issue and will be investigated.

As previously discussed, important functions can be referred to the MBs and CX: all the experiments carried on up to now reveal that these two areas are not directly connected, although many indirect connections are present. Therefore olfactory and visual stimuli could be treated independently but, finally, they converge to the pre-motor area. The final decision will be taken but we do not know where. Some experiments envisaged that the CX decides what to do but the MBs modulate the intensity of the response.

## 2.4 Description of the Computational Architecture

The main parts of the insect brain have been modeled and integrated into a computational architecture. This global computational model inspired by the *Drosophila* brain is presented in Fig. 2.3. It has been designed in order to be directly linked to a robotic implementation [20, 21]. The main parts of the whole architecture are described in the following.

### *2.4.1 Sensorial Pathways and Internal States*

In the model, it is possible to distinguish four main sensorial pathways; the olfactory and the visual pathways allow to percept the environment while the gustation and the nociception are indispensable to obtain information about the goodness or badness of the current situation. The interaction among the sensorial pathways allow the emergence of anticipative actions to improve efficiency in finding rewards and to avoid dangerous situations. Learning mechanisms based on classical and operant conditioning are used in the architecture [22]. Olfactory and visual inputs, due to their complexity, are pre-processed to be easily handled by the spiking networks used for the learning processes. The internal states of the system are also important and can be monitored through a set of virtual proprioceptive sensors. These states are chosen according to the application; for example, autonomous navigation-based tasks need an accurate monitoring of the batteries level of the robot.

### 2.4.2 Drives and Behavior Selection Network

External stimuli and internal states are essential information for living beings to survive in unstructured environments. An internal state is supposed to be directly related to drives like hunger, thirst, the will to sleep that are used by animals to adapt the behavioral responses. In a robotic implementation, drives will be chosen following the robot applications: a classical example is the need for power supply. In order to satisfy its needs, the robot has to choose a behavior from a pre-defined number of available behaviors. Behavior is meant like a sequence of programmed actions. Each behavior is oriented to satisfy one or more drives. The aim is to make the robot able to choose the right behavior that can satisfy the strongest drives. Even if there are not specific experiments that can demonstrate the existence of such a network in the *Drosophila* brain, an artificial Behavior Selection Network (BSN) was envisaged and implemented. The BSN was thought as a two-layers neural network, in which each unit is an Izhikevich Class I spiking neuron [23], having the following equation:

$$\begin{cases} \dot{v} = 0.04v^2 + 5v + 140 - I \\ \dot{u} = 0.02(-0.1v - u) \end{cases} \tag{2.1}$$

if $v \leq 0.03$, then $v \leftarrow -0.055$ and $u \leftarrow u + 6$

where $v$ is the membrane potential of the neuron, $u$ is a recovery variable and $I$ is the synaptic current. A typical example of a BSN structure is shown in Fig. 2.4.

The number of neurons in the first layer matches the number of drives the robot has to satisfy whereas, in the second layer, each neuron corresponds to the available behaviors. Every drive introduces a current, that is then converted in a spike-rate by the corresponding first layer neuron. The weight of the synapses connecting the first and the second layer neurons are chosen according to the capacity of each behavior to satisfy each drive. Synaptic weights $W_{ij}$ represent the importance of drive $i$ for the behavior $j$. Synaptic efficiencies are fixed: no learning is considered at this step. The second layer is a Winner-Takes-All (WTA) network; during every simulation step the neurons in the second layer are competing and only one neuron can win the competition: the behavior represented by the winning neuron is the selected behavior for the next robot step. To avoid a continuous switching among

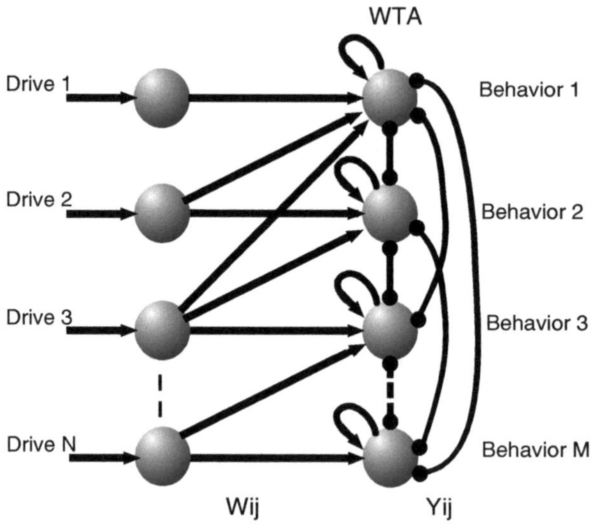


**Fig. 2.4** Example of a spiking network used to simulate the behavior selection functionalities. Drives are represented by input currents. Each drive can excite more than one behavior. Synaptic efficiencies between the input layer and the WTA layer represent the influence that each drive has in each behavior. Only the most excited behavior can win the competition and can be selected

the selected behaviors, a self-excitatory synapse has been introduced in each neuron of the second layer of the BSN. In this way, if a behavior is been selected during a simulation step, the probability to be selected again is increased during the next step. Synaptic weights $Y_{ij}$, i$\neq$j, represent the inhibitory synapses between neuron i and j in the WTA layer. Synaptic weights $Y_{ii}$ represent the self-excitatory synapses of neuron i in the WTA layer. The last point to clarify is how to transform drives in an input current. Considering for instance the drive "recharge", the robot analogue to "sleep", strongly connected to an internal sensor that measures voltage in batteries. It is possible to implement a transfer function that takes as input the battery level and gives as output a numerical evaluation of the "sleep" drive. Other methods for behavior selection have been used in literature, in particular for sequence learning [16]. The proposed approach could be modified in order to implement sequence learning, even if, up to now, there are no biological evidences about the capabilities of *Drosophila* in learning sequences of behaviors.

## *2.4.3 Central Complex Model*

### 2.4.3.1 Protocerebral Bridge Model

Object detection and distance estimation are functions related to the PB in fruit flies. Mronz and Strauss proposed a simple model based on parallax motion [24] that can be used to model these functions and a hardware implementation for an autonomous roving robot has been proposed in [25]. However, it is possible to use a generalized PB model, realized by a cascade of three simple blocks:

- *Object Detection Block.* This block takes input from the visual system and is used to detect the presence of an object.
- *Distance Estimation Block.* When an object has been detected, this block estimates its distance from the robot. In real flies, distance is estimated using a parallax motion approach.
- *Object Position Block.* It is possible to reproduce fly behavior assuming as interesting objects those standing in the compartments ranging from the frontal direction to $\pm 100°$ in the two lateral sides and repulsive ones those standing in the compartments from $\pm 100°$ to $\pm 160°$ on the rear part of the robot. Objects at angles exceeding $160°$ cannot be seen.

#### 2.4.3.2 Fan-Shaped Body Model

The fan-shaped body model has been designed as a cascade of two sub-blocks: a feature extraction and a feature evaluation element.

- *Feature Extraction.* Once an object has been detected, the Feature Extraction Block classifies it by using a series of features. As underlined in focused experiments with flies [26], the following features can be considered:
  - *Color.* Using a HSV representation, it is assumed to consider only the Hue value.
  - *Orientation.* Orientation is meant as the angle between the vertical direction and an axis that represents the direction in which an object is mainly distributed.
  - *Size.* Size is meant as the portion of total visual area of the robot occupied by the object, normalized with respect to the distance from the robot.
  - *Center of Gravity.* This feature is given by the height of the center of gravity normalized with respect to the vertical dimension of the visual area and the distance from the robot.
  - *Wideness.* Wideness is meant as the maximal horizontal extension of the object, normalized with respect to the total horizontal dimension of the visual area and the distance from the robot.
  - *Height.* Height is meant as the maximal vertical extension of the object, normalized with respect to the total horizontal dimension of the visual area and the distance from the robot.
- *Feature Evaluation.* The robot is able to associate object features to punishment or neutral situations. Every feature has a *Punishment Value*: if this value exceeds a threshold, the robot escapes every time it meets an object with that feature. The Punishment Value of a feature decreases if the robot is not punished when that feature is encountered. When the robot meets an object, it evaluates its Escaping Value: this is a weighted sum of the Punishment Values of the features of that object. When the Escaping Value is high enough, the robot escapes from the object, even

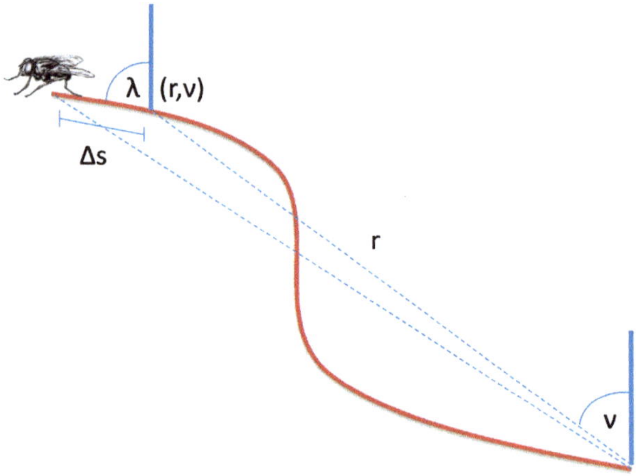


**Fig. 2.5** Path Integration scheme. The values of $r$ and $\nu$ represent the position of the robot from the object. Every robot step $r$ and $\nu$ are updated according to the last robot movements, in direction $\lambda$ by a path increment $\Delta s$

if not punished. This is the simplest way to implement a classifier. Other, more performing and sophisticated algorithms, either bio-inspired, or more information theory-bases, like the Neural Gas [27], could also be taken into account to improve the system plasticity. A neural model has been finally used to implement visual learning capabilities in flies [28].

#### 2.4.3.3 Ellipsoid Body Model

Neuser [29] described the role of the *Drosophila* ellipsoid body in the visual short term memory. That functional analysis leads to the implementation of a model able to create a spatial memory in a robot. By using polar coordinates to code the robot position in the environment, it is possible to design neural architectures inspired by the ant's path integration [30]. However other solutions could be based on a mathematical implementation of a polar path-integration algorithm and this kind of approach (easier and more robust) has been taken into consideration. A scheme describing the path integration mechanism is shown in Fig. 2.5. Supposing $\Delta s \ll r$, defining $\lambda$ as the direction of the current robot movement and $\delta = \lambda - \nu$, the approximation of the current robot position is recursively given by:

$$\begin{cases} r_{n+1} = r_n + \Delta r_n = r_n + \Delta s_n cos(\delta_n) \\ \nu_{n+1} = \nu_n + \Delta \nu_n = \nu_n + \Delta s_n sin(\delta_n / r_n) \end{cases} \tag{2.2}$$

where $\Delta s_n$ is the length of the robot step of index $n$ and $r$ and $\nu$ are the coordinates that represent the position of the robot with respect to the object that is supposed to be the origin of the polar coordinate system.

## 2.4.4 *Mushroom Bodies Model*

The MBs are a key structure of the insect brain. In particular, two main functions are related to this structure. Mushroom Bodies are primarily involved in olfactory learning [31, 32] and in a more complex function that will be called *behavior evaluation*. Because experiments are not able to demonstrate the connection between the two functions, two uncoupled models were implemented.

### 2.4.4.1 Olfactory Learning Model

A two layer spiking neural network was designed and implemented to model the olfactory learning function. The Spike Timing Dependent Plasticity (STDP) has been applied as learning algorithm [33, 34]. This algorithm can reproduce Hebbian learning in biological neural networks. The algorithm works on the synaptic weights, modifying them according to the temporal sequence of spikes occurring. The algorithm is represented by the following formula:

$$\begin{cases} \Delta W = A_{+} \exp(\Delta t/\tau_{+}), & if\, \Delta t < 0 \\ \Delta W = -A_{-} \exp(\Delta t/\tau_{-}), & if\, \Delta t \geq 0 \end{cases} \quad (2.3)$$

where $\Delta t$ is the time delay between pre and post synaptic spikes. In this way the synapse is reinforced if the pre-synaptic spike happens before the post-synaptic one, it is weakened in the opposite situation. Parameters $\tau_{+}$ and $\tau_{-}$ represent the slope of exponential functions, while positive constants $A_{+}$ and $A_{-}$ represent the maximal variations of the synaptic weight.

Each neuron is modeled by an Izhikevich Class I neural model [23]. A scheme of the neural model is shown in Fig. 2.6. The Shock (Punishment) Neuron takes as an input a current proportional to the value of the robot punishment, while the Good (Reward) Neuron takes as input a current proportional to the reward. In experiments with *Drosophila*, the punishment could be represented by an electrical shock, while the reward is represented by sugar. Each of the remaining neurons of the first layer takes as an input a current proportional to the odors the robot can detect in the environment. Each odor has a corresponding receptor and a neuron that converts the current in a spiking-rate if the current is high enough above the threshold for the Class I Izhikevich model. In a real robotic implementation odors can be substituted with other sensorial inputs, according to the application. Specific neural network composed by Izhikevich neurons and STDP learning were already implemented to realize approaching or escaping behaviors [12, 35]. In the present implementation Synapses between the Shock and the Reward Neuron and the output layer have a fixed value. Outputs of the second layer neurons are connected to a Motor Program block. The robot escapes from the actual position if the Escape Neuron is firing, while it begins an approaching algorithm if the Approach Neuron is firing. Synapses between unconditioned stimuli (i.e. shock and reward) and motor neurons are fixed

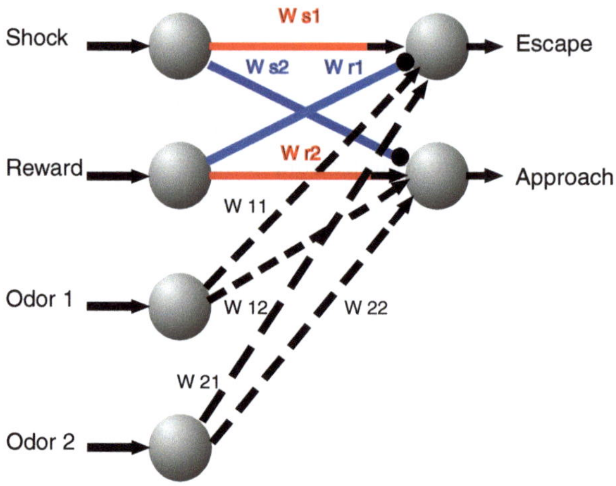


**Fig. 2.6** Olfactory Learning Model. *Solid line* (*dashed*) connections correspond to fixed (plastic) synapses; *arrows* (*bullet*) correspond to excitatory (inhibitory) connections. The model here presented can be easily extended to the desired number of odors

and represent the inherited knowledge whereas connections between conditioned stimuli (i.e. odors) and the motor system are subject to learning, according to the STDP rule. If not reinforced, the efficiency of a synapse decays with time.

#### 2.4.4.2 Behavior Evaluation Model

When a behavior is selected, the robot defines a setpoint to be reached. A setpoint is meant as a desired value for a vector of physical quantity linked to the definition of the drives: to satisfy its needs, the robot has to minimize the error between this setpoint and its actual state. For example, let us assume that the robot has a low battery voltage and that a charging-station is present in the environment. In this situation the robot could choose to go to the base station whereas the desired battery level would be the setpoint. If the selected behavior is not able to allow the robot to satisfy its needs, that behavior has to be inhibited: in the opposite situation, if the selected behavior leads to satisfy its needs, that behavior has to be excited. The Behavior Evaluation block inhibits or excites the actual behavior depending on the amount of time already spent and the level of success in reaching the setpoint. Inhibition or stimulation is easily implemented sending a current to the neuron of the WTA layer in the BSN associated to the ongoing behavior. Moreover some additional plasticity could be implemented into the Behavior Selection Network through the Behavior Evaluation model. In particular, the synapses between the neuron related to the selected behavior and drives that represent the setpoint could be reinforced (or weakened) if the last selected behavior has been able (or not) to reach the last setpoint. However, there are no biological evidences about this point.

### *2.4.5 Motor Programs and Description of Behaviors*

The Motor Program block describes all the possible elementary actions that the robot can perform. Motor learning is not considered up to now, although it is envisaged to be investigated and added in the near future.

### *2.4.6 Reflexive Behaviors*

When the robot is punished in some way it needs to escape as fast as possible from the object responsible for the shock. The *Description of Reflexive Behaviors* is a simple high level block that allows the robot to take the right direction in the case it is punished.

### *2.4.7 Complex Behaviors*

The *Description of Behavior* block is a high level part of the complete model that describes the available behaviors that the robot can follow. The choice of the possible behaviors the robot can exhibit depends on the robot applications. Applying a searching strategy to find a charging station could be an example of a typical behavior. The description of each behavior, however, could depend on the robotic structure and the embedded sensorial system.

## 2.5 Simulation Description

The implementation of drives and behaviors on a real robot strongly depends on the field of application in which that robot is involved. The simulation of the general model of an insect brain requires a simulated environment, where behaviors and drives, which the robot has to satisfy, need to be defined. The focus is to simulate the model in a context that can present analogies with the *Drosophila* real experimental set-up, in order to obtain the experimental validation of the model. Details about the implementation of the complete model are reported in the following.

### *2.5.1 Drives*

A brief description about the drives chosen in simulations and their analogies with real fruit flies is presented here.

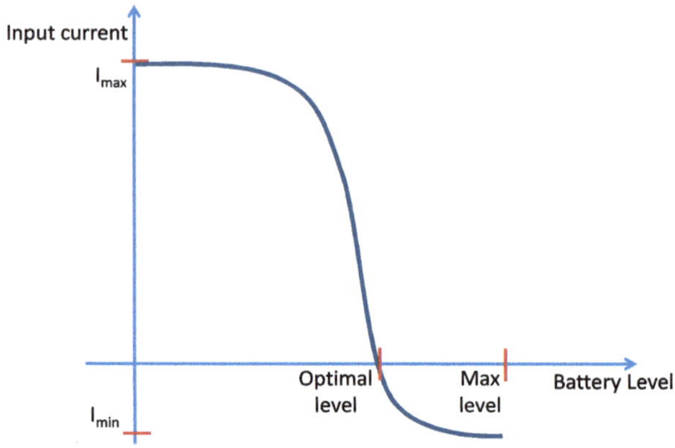


**Fig. 2.7** Example of the transfer function used to model the sleep drive. A low battery level leads to a high value of the current related to the *Sleep* drive. The optimal battery level is represented by the point in which the drive *Sleep* is equal to zero. If the battery level exceeds the optimal point, the drive became negative, in order to inhibit a possible dangerous battery charging

- *Sleep*—The drive Sleep is assumed to be the need for a robot to charge its batteries. In real fruit flies sleep is indispensable for learning [36]. In common robots we can quantify the drive sleep using a function of the battery level:

$$I_{drive} = K_{drive} \tanh(\Lambda - \chi) + \psi \tag{2.4}$$

where $K_{drive}$, $\Lambda$, $\chi$ and $\psi$ are the parameters of the function. Through these parameters it is possible to set the maximum and the minimum value of the current and the optimal battery level. An example is shown in Fig. 2.7. The optimal battery level is represented by the point in which the drive Sleep is equal to zero. If the battery level exceeds the optimal point, the drive became negative, in order to inhibit a battery charging. In order to simulate the battery level, it is convenient to implement a virtual sensor. The output of such sensor is the estimated battery level. The battery level must decrease each step in which the robot is far from the charging station, and reaches the Max Battery Level after a given time spent in the charging station area. A sleep drive is indispensable for every time it is necessary to have a completely autonomous robot, which must be able to find power supply sources and use them to move for a long time.
- *Hunger*—The need of food can be reproduced putting inside the environment objects or landmarks that the robot should periodically find and/or visit. The drive Hunger could be thought as proportional to the time the robot left the object. This drive is indispensable to obtain a behavior that can match with reality but also to force the robot to find objects that can be periodically useful. The drives Hunger and Sleep have some similarities; their differences will be remarked according to the application.

- *Shelter*—When in danger, a fly looks for a safe place. A fly in open spaces often has the tendency to protect itself, typically aiming to approach and follow walls. Shelter can be related to the distances of the robot from the walls.
- *Curiosity*—The drive Curiosity allows a fly to search for other resources when the other drives are satisfied. Curiosity can be quantified with a constant value. From a robotic point of view, curiosity leads the robot to explore the environment and to acquire information about the detected objects.

### 2.5.2 Behaviors

To make the robot able to satisfy its drives, the following behaviors have been implemented. They constitute the output of the Behavior Selection Network:

- *Exploration*. During an exploration behavior the robot tries to find new resources. In flies, the environment exploration is characterized by an increase of the mean-free path algorithm [37]. As in real flies, during an exploration behavior it is possible to distinguish two behaviors [38]:
  - Sitter larvae behavior: short path length and tight turning angles.
  - Rover larvae behavior: long path length and wide turning angles.

  Exploration can be thought as a default behavior: the robot could choose this behavior when no particular drives are enabled. Usually curiosity is the drive that mainly influences the choice of an exploration behavior. The implementation of the exploration behavior requires also the management of the obstacle avoidance and object detection. Moreover, the robot must be able to update its position at each step; in our case, the ellipsoid body model is involved.
- *Homing*. The homing behavior consists in returning to the charging station, where the simulation is started. Of course, the position of the Home must be known and updated every step. An obstacle avoidance algorithm has to be implemented during the homing behavior. Homing behavior is needed to have an autonomous robot, able to charge its battery before its autonomy is compromised.
- *Landmark Recalling and Achievement*. During the navigation the robot meets objects: if some objects are associated to food, the robot must remember their position in order to reach them when it is needed. The biological plausibility of this behavior is evident. The robot must use the path integration system to update its position from each interesting object.
- *Centrophobism*. A centrophobic behavior has been found in flies [39]. From a biological point of view, centrophobism in flies could be a consequence of the increase of mean-free path in the exploration behavior. A fly uses centrophobic behavior to protect itself in dangerous environments. Shelter is the drive that mainly influences the choice of a centrophobic behavior.

### 2.5.3 The Robot and the Simulator

The robot used in the first experiments is a Pioneer P3-AT differential-drive roving robot. The platform operates as a server in a client-server environment; the onboard PC is used to host the control architecture.

MobileSim is the software used for simulating the Pioneer P3-AT roving robot in a virtual 2D environment. This simulation environment has been used to evaluate the performance of the proposed control system.

### 2.5.4 Implementation of Odors, Punishments and Rewards in the Simulator

In order to implement olfactory classical conditioning it is necessary for the robot to have sensors that can detect odors and that can monitor rewards or punishments given to the robot. In a simulation environment it is convenient to implement virtual sensors. For instance, if an object releases an odor called Odor1, it is convenient to assume the output of the olfactory sensor as a Gaussian function of the distance $d$ from the robot to that object:

$$f_{od}(d) = K_{od} e^{-d/\tau_{od}} \tag{2.5}$$

where $K_{od}$ is a constant gain and $\tau_{od}$ represents the decay of the sensor output when the robot goes away from the object.

It is possible to use a similar strategy to determine the output of a punishment sensor and the output of a reward sensor:

$$f_{pun}(d) = K_{pun} e^{-d/\tau_{pun}} \tag{2.6}$$

$$f_{rew}(d) = K_{rew} e^{-d/\tau_{rew}} \tag{2.7}$$

The values of the constants can be determined in order to obtain a tighter Gaussian function for the output of the punishment and reward sensors: in this way, if the robot is approaching the object, it will first detect the odor and then it will be rewarded or punished if that object is not neutral.

## 2.6 Simulation Results

This section presents the experiments made in order to perform a first evaluation for each model of the general computational architecture of the *Drosophila melanogaster* brain.

Table 2.3 Summary of the characteristic of the objects in the MBs model test

| Object | Odor | P/N/R |
|---|---|---|
| Object A | Odor 1 | Punishment |
| Object B | Odor 2 | Reward |
| Object C | Odor 2 | Punishment |
| Object D | Odor 2 | Neutral |

### 2.6.1 Mushroom Bodies and Olfactory Learning

The following simulation shows how the MBs model for odor learning works. The simulation of the model has been done using the Euler integration method with a constant integration time of 20 ms. The synapses time constant is 800 milliseconds and synaptic weights are initialized to the value of 0.05. A priori known information is codified in the fixed synaptic weights that have been initialized to the value of 10 (excitatory) and -3 (inhibitory). A decay rate has been introduced: every 100 simulation steps all synaptic weights are decreased by 1 % of their value. The implemented network is the same as shown in Fig. 2.6. This simulation was performed to verify the capability of the MBs model to make the right associations between odors and rewards or punishments in a complex environment. The robot is introduced into a square arena, $10 \times 10$ m, in which four objects are present. There is an odor spreading out from each object in the environment. In particular, Eq. 2.5 has been used. Two different odors are associated to these objects and a reward or a punishment is given to the robot when one of the objects is reached, following the association reported in Table 2.3.

Exploring the environment, the robot has to learn that there is a strong association is between the Odor 1 and the punishment: in a testing phase, the robot will be able to escape when detecting that odor, before the shock occurs. The behavior of the network neurons during the simulation is shown in Fig. 2.8 while Fig. 2.9 presents the trend of the synaptic weights during the simulation. The network evolves for 100 simulation steps for each robot action.

At the end of the simulation the robot explored the environment completely and it is able to make the right association. Other experiments were performed, obtaining similar results.

### 2.6.2 Protocerebral Bridge and Fan-Shaped Body

Through a functional analysis of *Drosophila* protocerebral bridge and fan-shaped body, it is possible to suppose that object detection and distance estimation are mainly performed by the PB, while the FB is related to feature extraction and classification. In the following experiments, the properties of the PB and FB models and the capabilities of the robot in terms of visual learning are presented.

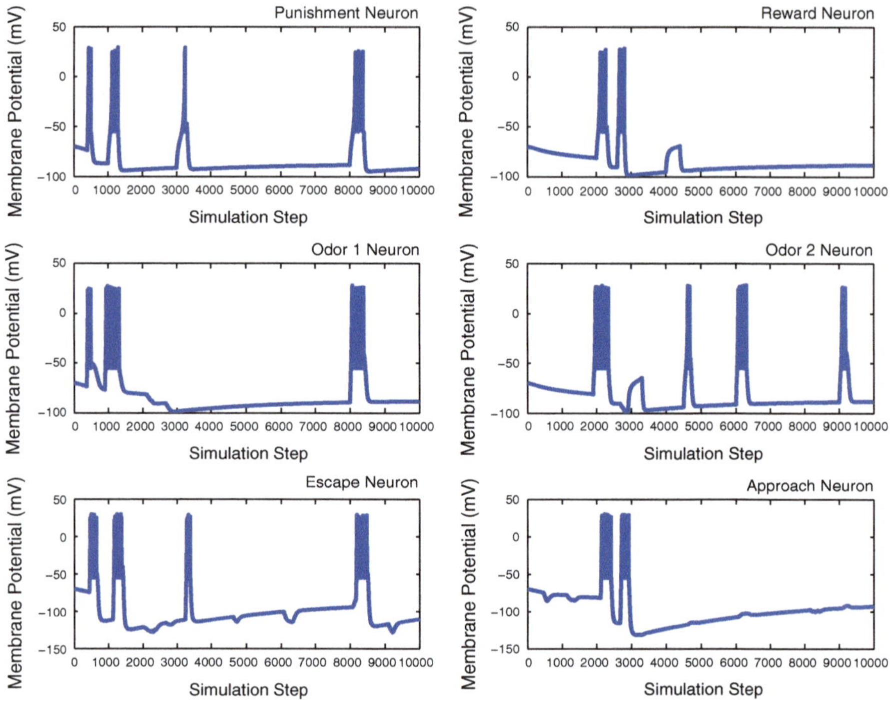


**Fig. 2.8** Simulation results of the olfactory learning model: behavior of the neuron membrane potential during simulation. The implemented network is shown in Fig. 2.6

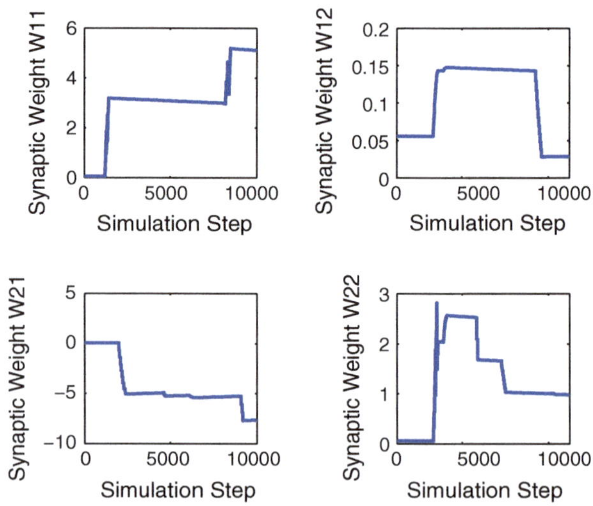


**Fig. 2.9** Simulation results of the olfactory learning model: trend of the synaptic weights during the simulation. The implemented network and the parameter meaning are illustrated in Fig. 2.6

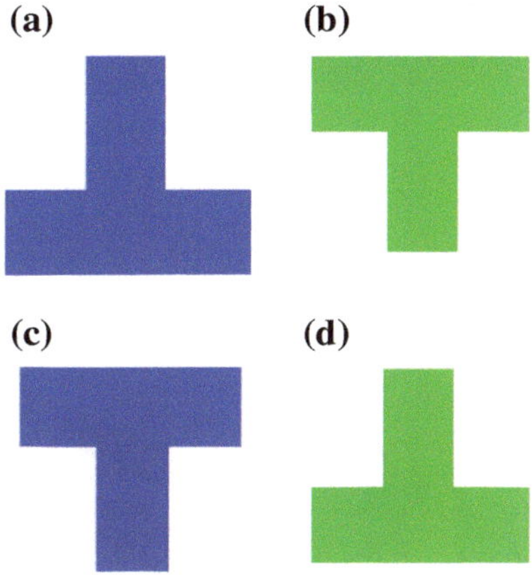


**Fig. 2.10** Objects used in the fan-shaped body model simulation. **a** *blue* inverted T-shape; **b** *green* upright T-shape; **c** *blue* upright T-shape; **d** green inverted T-shape

In particular, the proposed simulation is inspired by the experiment designed by Liu and collaborators on real flies [40] about visual learning and object recognition. The robot has to explore a square arena, ($10 \times 10$ m), in which four objects are present. Even if these objects are different, they can have some similar features. The objects used in this simulation are shown in Fig. 2.10. Every time the robot meets an object, it tries to recognize that object, extracting features and comparing them with the stored ones. If the robot meets an object for the first time, it extracts and stores the new features. It has been assumed to consider six features: color (in the Hue Saturation Brightness representation, here only the Hue value is considered), orientation, size, center of gravity position, wideness and height. The PB model has been set so that the robot is able to detect objects in a range of 4 m. Objects associated to a punishment shock the robot if its distance from these objects is less than 2.7 m. In this experiment the color "green" is a bad feature: the robot will be punished every time it tries to approach a green object. The robot has to learn to avoid green objects. The arena and the simulation results are shown in Fig. 2.11. At the beginning of the simulation, the robot tries to approach every object standing in its visual range. If punished, the robot increases the punishment value of the features of the approached object. If an object is neutral, the punishment value of the features associated to that object decreases. If the escaping value of an object reaches a threshold, the robot will escape when that object is detected. In this simulation the robot learns correctly to avoid green objects. Figure 2.12 shows also the Punishment Value of the bad feature (green color) and the Punishment Value of a neutral feature, the wideness, that is the same for all the objects. In order to implement a hysteretic response, when the Punishment Value exceeds 2, it is simply raised to 7. In this way the robot will remember this bad feature association for a long time, even if the learning is not reinforced. If the robot detects an object, the Punishment Value of all the features that do not belong to that object will remain the same. A time-dependent decay rate could also be introduced.

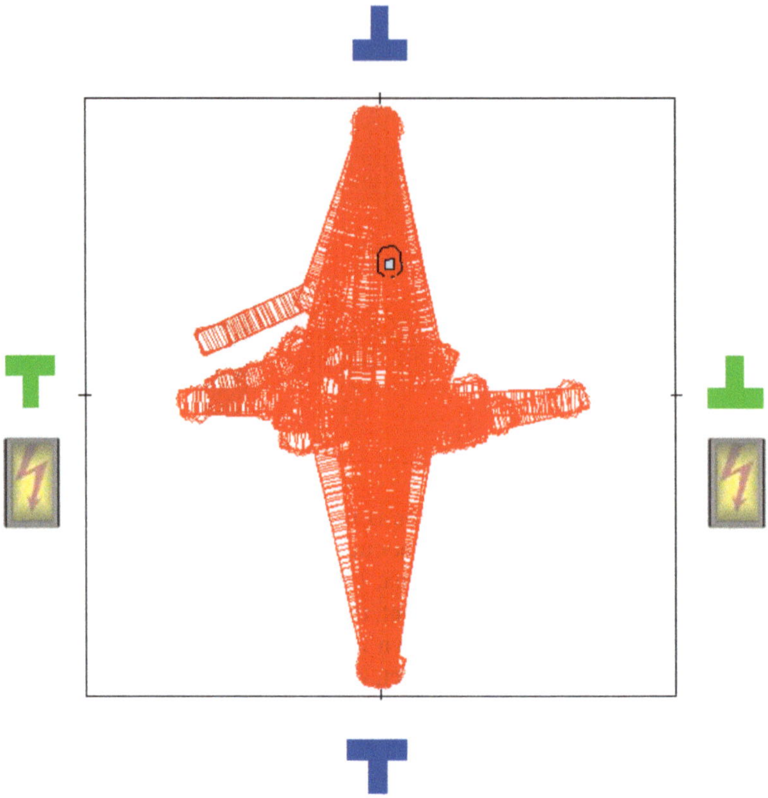

**Fig. 2.11** Robot trajectories obtained during the testing of the fan-shaped body model. After being punished enough times, the robot is able to isolate the dangerous feature (the *green color*) and escapes when a *green* object is detected

In a second experiment, the robot has to learn to avoid each "T" object. Color is now neutral for the robot. Even if the shape is not a feature, a T is different from an inverted T because of the different center of gravity. This experiment leads to the same conclusion of the first experiment; the robot is able to recognize bad features and to avoid them.

## *2.6.3 Ellipsoid Body*

In real fruit flies, the ellipsoid body is necessary for a visual short-term memory and orientation. In the following simulation, the behavior of the EB model while the robot is moving around the environment is evaluated. In the simulated environment an odometry error has been introduced, to make the results more realistic. In this first

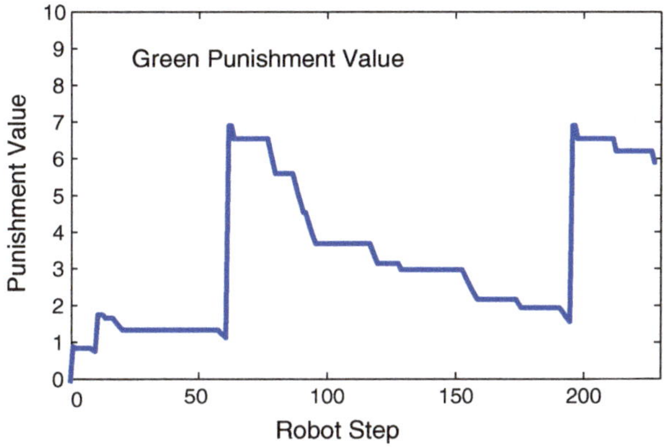


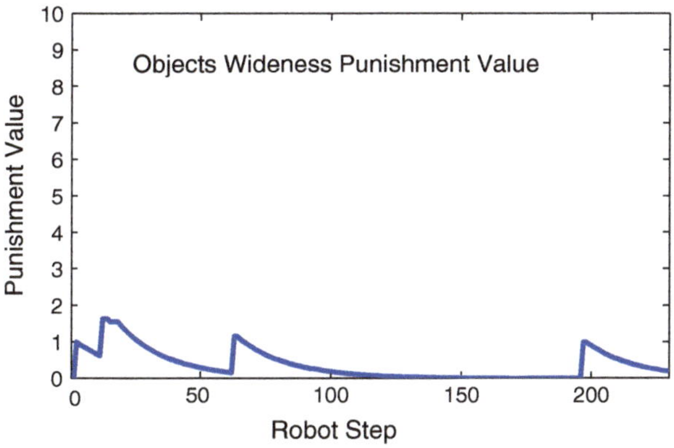


**Fig. 2.12** Comparison between the punishment value of the dangerous feature and the punishment value of a neutral feature, the wideness. Using the punishment value algorithm the robot is able to discriminate the dangerous feature. The decreasing of the punishing value of the dangerous feature is due to the steps in which the robot can detect a green object but is not so near to be punished

experiment we want to show how the ellipsoid body model works: the robot must be able to update its position while moving into a square arena ($8 \times 8$ m). The robot starts from the Home position and moves randomly into the arena: its capability to update its relative position with the Home is analyzed. Of course the coordinates stored into the robot memory will be different from the real ones, because of the odometry errors and the approximation of the path integration method. Figure 2.13 show an example of trajectory and the response of the ellipsoid body. The same test has been repeated many times, in order to make a better analysis of the model.

In order to test the capability of the model in real situations, it is convenient to simulate the robot behavior and the EB response in more complex arenas. In the following experiment the robot has to explore a large arena, in which several objects are present. The robot starts from the Home and initially it moves randomly: this

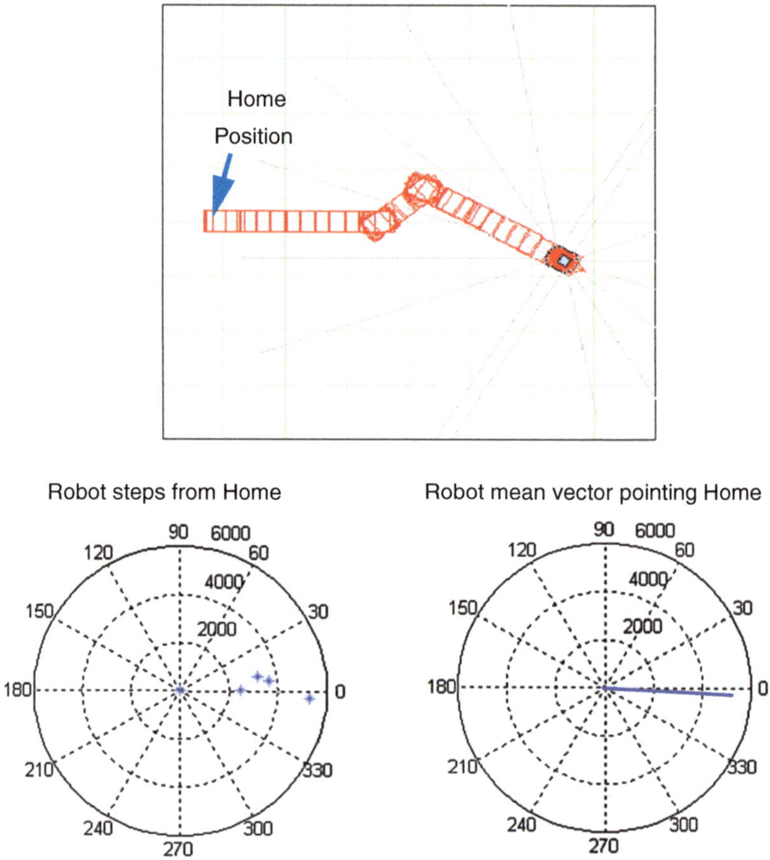


**Fig. 2.13** Information stored in the EB. The relative position of the robot with respect to the Home is represented in polar coordinates and it is indicated in millimeters (distance) and degrees (angular position). In this case the current robot position is $r = 5730\,\text{mm}$, $\nu = -12°$

behavior is created to simulate a typical escaping reaction of real flies when newly introduced into an arena.

After that, the robot starts an Exploration behavior. If the robot meets objects it is able to learn about their danger or neutrality, thanks to the MBs model. During the exploration, the robot updates its position from the Home. An obstacle avoidance mechanism was also implemented. During this experiment two behaviors are available: Exploration and Homing. The level of the battery decreases while the robot explores the arena. A virtual battery sensor has been implemented. If the level of the battery is too low, the BSN switches the selected behavior to the Homing Behavior. If the stored position is correct, the robot must be able to return to the Home position. Obstacle avoidance is used also during the Homing behavior. Simulation results are shown in Fig. 2.14. The robot starts to move and, after the escaping reaction imple-

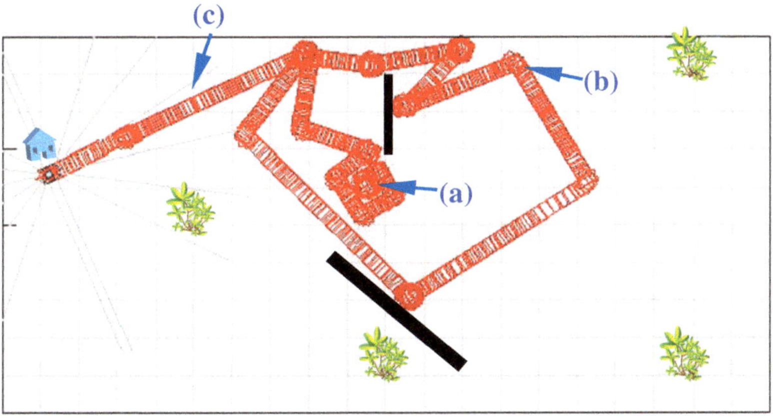


**Fig. 2.14** The robot starts to move and, after the escaping reaction implemented to match the biological experiments with real flies, it begins an exploration phase (**a**). The escaping reaction from the Home position to position (**a**) is not outlined for clarity reasons. After fifteen exploration steps, the battery level is low and the robot starts its homing behavior (**b**). Using an obstacle avoidance algorithm, the robot is able to return to the Home (**c**). In order to have a more complex simulation, some objects have been also introduced into the arena

mented to match the biological experiments with real flies, it begins an exploration (a). The escaping reaction from the Home position to position (a) is not outlined for clarity reasons. After fifteen exploration steps, the battery level is low and the robot starts its homing behavior (b). Using an obstacle avoidance algorithm, the robot is able to return to the Home (c).

### *2.6.4 Behavior Selection*

In order to allow the robot to choose the more suitable behavior, the Behavior Selection Network (BSN) has been implemented. The BSN has been tested and its properties have been analyzed. In a real implementation of the model the drives are the inputs of the first layer of the network. In the following simulations drives have been simulated in order to study the response of the BSN in different possible situations. This experiment shows how the Behavior Selection Network works. It has been assumed to have four behaviors and four drives, and to represent these drives with four input currents.

In this first example a perfect symmetry in the complete network has been supposed: $W_{ij} = 1.5$; $W_{ii} = 10$; $Y_{ij} = -3$; $Y_{ii} = 3$ (see Fig. 2.4 for the network topology). A random Gaussian noise has been added in the input currents ($\sigma$ =2). Fig. 2.15 presents the behavior of the neurons of the network. When a second layer (WTA layer) neuron is firing faster than the others, the respective behavior is selected.

The network has been simulated for ten thousand simulation steps, with an integration step of 20 milliseconds. During a short transient, all the WTA neurons are firing: this situation is due to the response of the synapses between WTA neurons. After this transitory period, only one neuron can win the competition.

## 2.7 Real Life Scenario Application

In the Sect. 2.6.4 a model of the main parts of a fly brain computational model has been introduced. Herewith the capability of the model to solve realistic tasks is shown. Modifying the behavior repertoire but maintaining the conceptual structure of the general model we can obtain a versatile robot that is able to learn about the environment, to make choices and to face potentially dangerous situations. The experiment presented in this section is only one example of the real applications of the insect brain model, and it could be easily modified or generalized [41, 42].

### *2.7.1 Description of the Experiment*

Let us imagine to have a critical situation in which, after a disaster (earthquake, fire) it is necessary to rescue people trapped in a place. Often situations like this are very dangerous both for survivors and people who try to help them. Now let us image to have a smart robot able to explore the environment and which can learn, recognize people and remember their position. Such a robot could menage a critical situation acquiring the information needed to solve it. In the present experiment an environment that can represent a place after a disaster has been implemented into a robot simulator. The robot has to explore the environment, find some good objects that it is able to recognize, remember their position and learn about all kinds of dangers present in the environment. At the end of the exploration, the robot must escape from the environment and give all the information useful for humans to know the position of the survivors and organize a safe rescue. In order to solve this problem, the behavior repertoire of the robot has been limited to two possible behaviors, exploration (rover type) and homing. In the same way, two drives are considered, Curiosity and Sleep, the latter indispensable for the robot to understand when to leave the environment and return home; for this, a virtual battery level sensor is used. The MBs model was also simplified: only the olfactory learning model will be considered. The synaptic weights, the synapses time constant and the integration step are the same of the previous simulations. Every robot step of the robot includes only one hundred simulation steps of the MBs and BSN neural networks.

The arena implemented for the simulation and the results are shown in Fig. 2.16. The Home represents the starting point for the robot exploration and the point the robot has to reach at the end of the simulation.

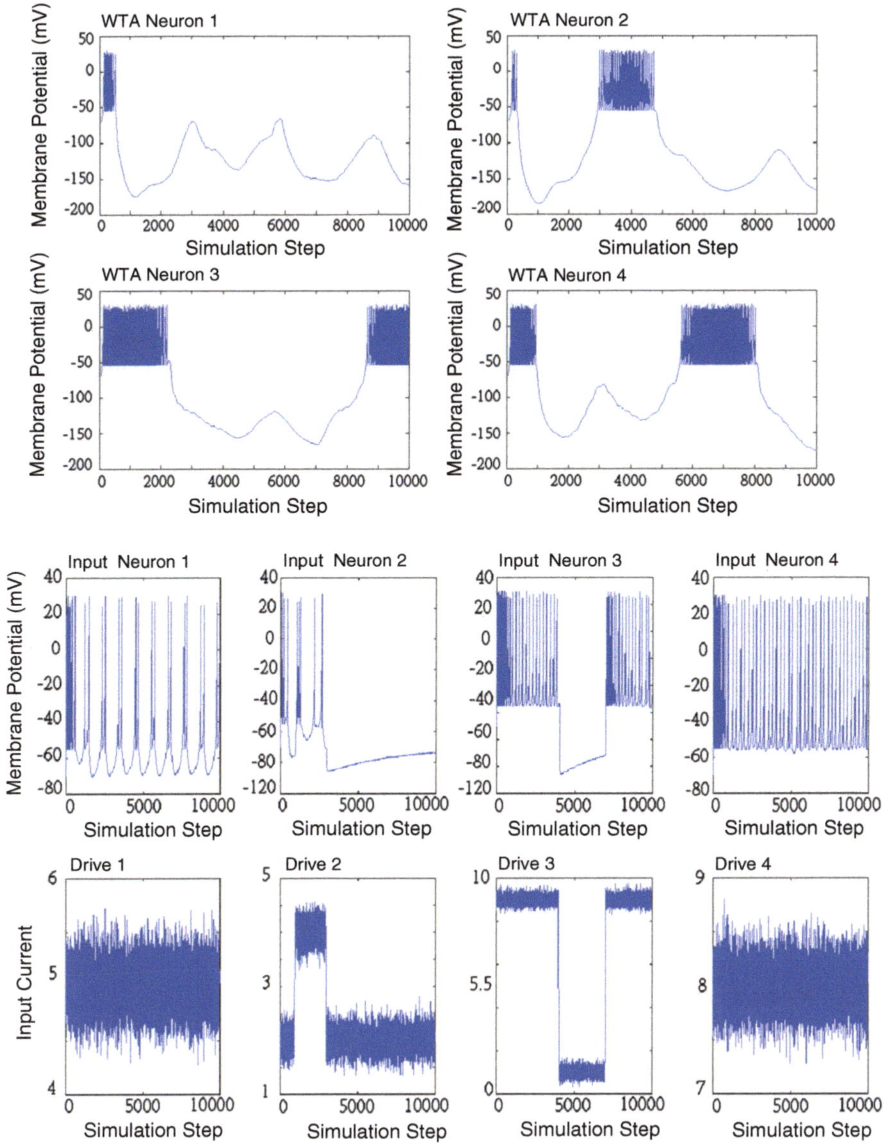


**Fig. 2.15** Results of the simulation of the BSN. After a short transient in which all the WTA neurons are firing, only one neuron can win the competition. The transient is a consequences of the time response of the synapses between WTA neurons. Variation of the drives could also lead to new transient, in which the WTA neurons compete. A low value of the auto-excitatory synapses weights in the WTA layer can cause a continuous switching of the selected behavior, while a too high value leads to a conservative behavior selection

$S_1$ and $S_2$ represent the position of the targets: let us assume the robot considers them as interesting objects and, after an approach, it is able to recognize them. Let us assume that the targets are a blue T-shaped object and a blue inverted T-shaped object. Identical objects $Obj_1$, $Obj_2$ and $Obj_3$, are also considered. The robot cannot see them, but can sense them thanks to another sensorial system (i.e. olfactory). The robot is punished every time it tries to approach them. In the environment, two other objects are present, a green upright T-shaped object and a green inverted T-shaped object. The robot can detect them with the visual system. The robot is punished only when it tries to approach the first one, while the second one is neutral.

After a long exploration, the robot must be able to detect the targets, learn to avoid as soon as possible the objects $Obj_1$, $Obj_2$ and $Obj_3$, understand that the green upright T-shaped object is dangerous and finally reach the Home and give the position of the targets at the end of the exploration. Mushroom Bodies model will be used for the learning involving $Obj_1$, $Obj_2$ and $Obj_3$; the protocerebral bridge model will be used for the detection of the objects and the fan-shaped body model for the visual learning; the ellipsoid body model is indispensable for homing and remembering the position of the targets. For this simulation, the capabilities of real flies have been extended, for instance, improving the performances of the EB that is now able to store multiple target information in a long time memory. This is an example of how the elementary functions of the *Drosophila* brain that allows the insect to face with its world can be easily extended in a modular way to make a robot able to fulfill more complex tasks, not affordable for the real fly. The Behavior Selection Network is useful to select the homing behavior if the battery level is too low. The parameters of the model have been set so that the robot can sense odors if its distance is lower than 3 m away from the nearest odor source, while it is punished if its distance from that source is $<1$ m. In the same way, the visual system of the robot can detect objects if they are closer than 2.5 m. It is punished if an object is closer than 1.5 m. The arena used for the simulation is 28 m long and 15 m wide.

### 2.7.2 Results and Discussion

In this section experimental results obtained in a typical simulation are shown, discussing step by step the behavior of the robot. Only the most relevant robot steps are depicted in Fig. 2.16, whereas Fig. 2.17 shows the MBs model response during the whole simulation.

At step 1 the robot starts the simulation from the Home position. At the second step the robot enters the arena and begins an exploration behavior. The ellipsoid body model updates the position of the robot. Neurons of the MBs model are not stimulated and they lie in their silent state. At the following step (step 5, not shown), the robot uses the increase of the mean free path algorithm. The EB model updates the position of the robot. During the exploration, the robot must find objects and sense odors. At step 9, the robot senses $Odor_1$, but it is not punished, because it is not close enough to $Obj_1$. In the following step the robot continues its exploration following

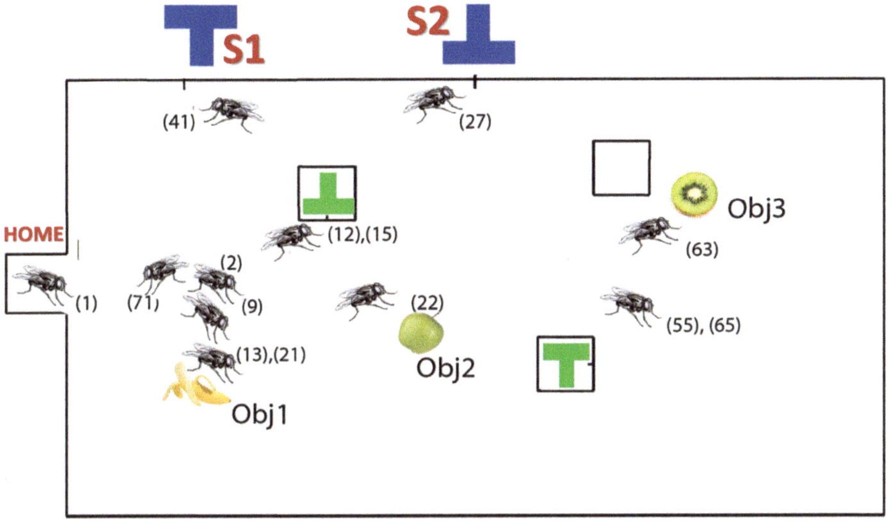


**Fig. 2.16** Most relevant robot steps of the proposed simulation. After the exploration of the environment the robot returns to the Home and gives the position of the target $S_1$ and $S_2$. Moreover, information about the dangers in the environment are stored in the FB and the MBs model

the increase of the mean free path algorithm, while the EB model updates the position it has stored. At step 11 (not shown), the robot detects the green T-shaped object. The FB model extracts features from this object and the robot tries to approach it. While the robot is approaching the new object, it is punished (step 12). After being punished, the robot escapes from the green upright T-shaped object (step 13). It has to be notice an unexpected situation: the robot sensed $Odor_1$ and was punished after two subsequent steps, due to the punishing visual input, and not for the odor. So, even if not planned in this way, the robot has made an association between $Odor_1$ and Punishment. This situation is plausible and it is a natural consequence of the correlation based on STDP learning.

As a consequence, the association between $Odor_1$ and the need to escape is reinforced. While it is escaping, the robot again detects $Obj_1$, senses $Odor_1$, is punished and escapes again in the opposite direction (step 14, not shown), reaching once more the green inverted T-shaped object (step 15). The robot is then punished for the third time. At step 16 (not shown), the robot is escaping again. At step 20 and 21 the robot is sensing $Odor_1$ again, without being punished. It is very interesting to analyze how the MBs model responds to this contradictory situation. Studying the firing of each neuron of the MBs model, it is possible to see that at a first time the robot was punished immediately after sensing $Odor_1$, while at a second time it senses $Odor_1$ but it is not punished. In this way, at a first time the robot made an association between punishment and $Odor_1$, but at a second time this association was weakened. However, the synaptic weight between the $Odor_1$ neuron and the Escape neuron of the olfactory learning model was not high enough to make the

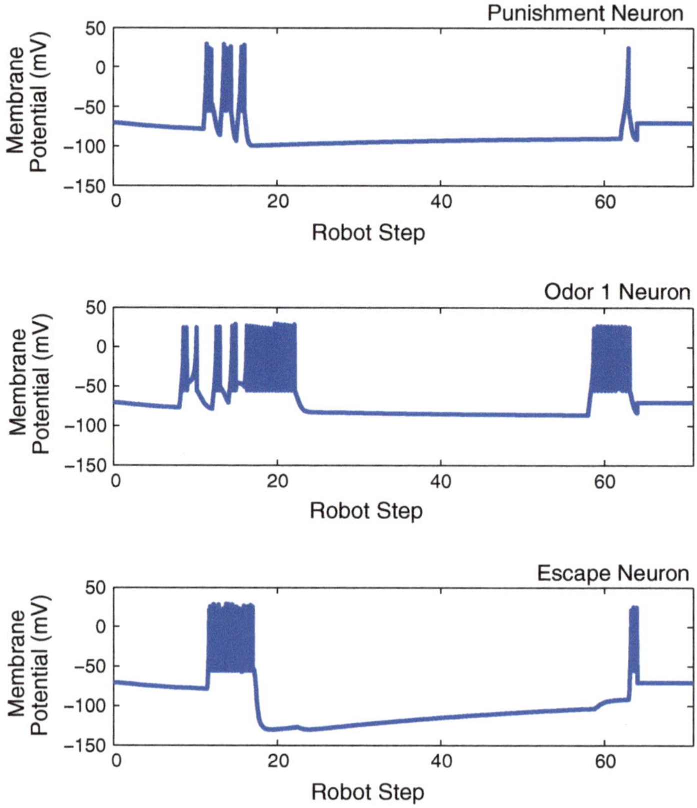


**Fig. 2.17** Mushroom Bodies model response during the simulation

robot escape when sensing again $Odor_1$, without being punished. Now the robot continues its exploration of the arena. At step 22 the robot is near $Obj_2$, it is sensing $Odor_1$ again but it is not close enough to be punished. The association between punishment and $Odor_1$ must decrease again. While exploring, the robot detects the first target (step 27). The fan-shaped body analogue extracts the features of the object, the robot recognizes the target and tries to approach it. The EB model stored the position of the robot. The target $S_2$ is now reachable in the future. The robot leaves the object and begins another exploration.

After many steps, the robot detects and reaches the target $S_1$ and stores its position (step 41). After leaving the second target, the robot begins another long exploration. At step 55, the robot is into the area of detection of the green T-shaped object, but in this case the PB model leads the robot to consider that object repulsive because it is standing in the rear of the robot, therefore the robot leaves the object. The robot continues its exploration and, detecting $Obj_3$, the robot senses $Odor_1$ again at step 59. At step 63, the robot is close enough to $Obj_3$ to be punished. Because of the position of the robot, the punishment is not so strong, but the robot is sensing $Odor_1$

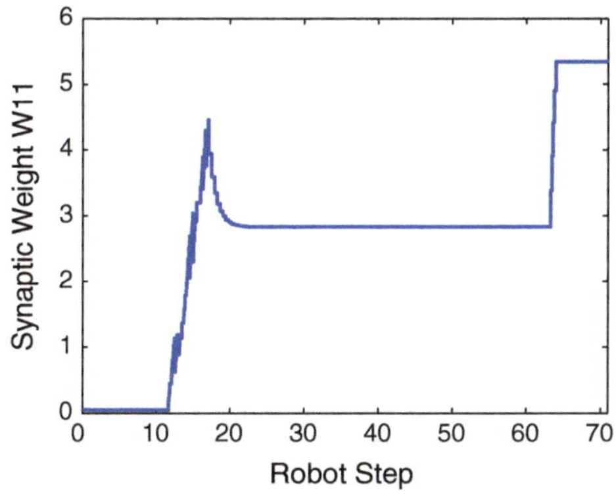


**Fig. 2.18** Trend of the synaptic weight of the synapses between the $Odor_1$ receptor neuron and the Escape neuron, in the pre-motor area. The synaptic weights of the MBs olfactory learning model are subject to STDP learning. The higher the value of the weight is, the faster the robot will escape if punished while sensing that odor. If the weight exceed a certain threshold, the robot sensing that odor will escape even if not punished at all. For clarifications about the parameters refers to Fig. 2.6

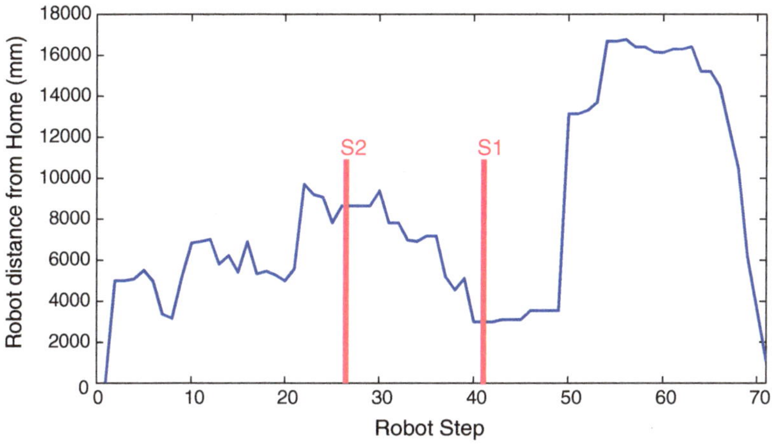


**Fig. 2.19** During the simulation, the EB model estimates the distance of the robot from the Home position. Errors in the position are due to the simulated odometry error and to the path integration method approximations

and it is recalling the association with punishment: even if the Punishment neuron only spikes once, the robot escapes.

Analyzing MBs response and the synaptic weights at step 64, it is evident how the robot reinforced the association between $Odor_1$ and Punishment, as shown in Fig. 2.18. Learning allowed the robot to escape fast, without strong punishment. After escaping, at step 65, the robot meets again the green T-shaped object.

While the robot tries to approach it, the low output level of the virtual battery sensor determines the behavior and initiates the homing procedure. The EB model is involved to remember the Home position. The response of the EB model at step 65 is shown in Fig. 2.19. At steps 67, 69 (not shown) and 71 the robot tries to return to the Home position. At the end of the simulation, the robot can communicate the approximated position of the targets. Moreover, the robot is aware of the association between an odor and a danger. Nevertheless, in this simulation, the robot was not able to certainly associate a visual feature with reward or punishment, because it has been punished only once while approaching a landmark.

## 2.8 Conclusions

The concept of cognitive abilities is commonly associated to humans and animals like mammals, birds and others. Nevertheless, in the last years several research groups have intensified the studies on insects that posses a much simpler brain structure even if they are able to show interesting memory and learning capabilities. In this chapter, some results toward the design and implementation of a model of the insect brain inspired by the *Drosophila melanogaster* have been presented. Particular attention was paid to the main neural centers the Mushroom Bodies and the Central Complex. In this chapter the parts of the model have been presented and simulation results are reported. In the following Part III of the book a Software/Hardware framework, where the complete architecture could be tested and evaluated by using both simulated and real robots, is presented.

## References

1. B. Webb, T.R. Consi, *Biorobotics: Methods and Applications* (AAAI Press/MIT Press, Menlo Park, 2001)
2. M. Dorigo, T. Stutzle, *Ant Colony Optimization* (MIT Press, Cambridge, 2004)
3. M.V. Srinivasan, S. Zhang, M. Altwein, J. Tautz, Honeybee navigation: nature and calibration of the odometer. Science **287**(5454), 851–853 (2000)
4. T.W. Secomb F.G. Barth, J.A.C. Humphrey, Locust's looming detectors for robot sensors. in *Sensors and sensing in biology and engineering*, ed. by F. Rind, R. Santer, J. Blanchard, P. Verschure (springerwien, newyork, 2003)
5. B. Webb, T. Scutt, A simple latency dependent spiking neuron model of cricket phonotaxis. Biol. Cybern. **82**(3), 247–269 (2000)
6. P. Arena, L. Fortuna, M. Frasca, L. Patané, M. Pavone, Realization of a CNN-driven cockroach-inspired robot, in *International Conference on Circuits and Systems (ISCAS)*, 2006
7. J.T. Watson, R.E. Ritzmann, S.N. Zill, A.J. Pollack, Control of obstacle climbing in the cockroach, Blaberus discoidalis. I. Kinematics. J Comp Physiol A **188**, 39–53 (2002)
8. H. Cruse, T. Kindermann, M. Schumm, J. Dean, J. Schmitz, Walknet a biologically inspired network to control six-legged walking. Neural Networks **11**, 1435–1447 (1998)
9. B. Webb, J. Wessnitzer, Multimodal sensory integration in insects—towards insect brain control architectures. Bioinspiration Biomimetics **1**, 63 (2006)
10. U. Homberg, Structure and functions of the central complex in insects. in *Arthropod Brain, Its Evolution, Development, Structure and Functions*, vol. 347–367, ed. by A.P Gupta, (Wiley, New York, 1987)

11. G.O. Lopez-Riquelme, W. Gronenberg, Multisensory convergence in the mushroom bodies of ants and bees. Acta. Biol. Hung. **55**, 31–37 (2004)
12. P. Arena, L. Fortuna, M. Frasca, L. Patané, Learning anticipation via spiking networks: application to navigation control. IEEE Trans. Neural Networks **20**(2), 202–216 (2009)
13. P. Arena, L. Patané, (eds.), Spatial temporal patterns for action-oriented perception in roving robots. in *Cognitive Systems Monographs*, vol. 1 (Springer, Berlin, 2009)
14. P.F.M.J. Verschure, T. Voegtlin, R.J. Douglas, Environmentally mediated synergy between perception and behaviour in mobile robots. Nature **425**, 620–624 (2003)
15. R. Brooks, New approaches to robotics. Science **253**, 1227–1232 (1991)
16. R.C. Arkin, *Behavior-Based Robotics* (MIT Press, Cambridge, 1998)
17. P. Arena, L. Patané, P.S. Termini, Decision making processes in the fruit fly: a computational model. in *Frontiers in Artificial Intelligence and Applications—Proceedings of the 21st Italian Workshop on Neural Nets*, vol. 234 (Seville, Spain, 2011), pp. 284–291
18. E. Arena, P. Arena, L. Patané, Efficient hexapodal locomotion control based on flow-invariant subspaces. in *18th World Congress of the International Federation of Automatic Control (IFAC)*, Milan, Italy, 2011
19. G.M. Shepherd, *Neurobiology* (Oxford University, New York, 1997)
20. P. Arena, C. Berg, L. Patané, R. Strauss, P.S. Termini, An insect brain computational model inspired by Drosophila melanogaster: architecture description, *WCCI 2010 IEEE World Congress on Computational Intelligence* (Barcelona, Spain, 2010), pp. 831–837
21. P. Arena, M. Cosentino, L. Patané, A. Vitanza, Sparkrs4cs: a software/hardware framework for cognitive architectures, *5th SPIE's International Symposium on Microtechnologies* (Czech Republic, Prague, 2011), pp. 1–12
22. P. Arena, L. Patané, Simple sensors provide inputs for cognitive robots. Instrum. Measur. Mag. **12**(3), 13–20 (2009)
23. E.M. Izhikevich, Simple model of spiking neurons. IEEE Trans. Neural Networks **14**(6), 1569–1572 (2003)
24. R. Strauss, M. Mronz, Visual motion integration controls attractiveness of objects in walking flies and a mobile robot, *International Conference on Intelligent Robots and Systems* (Nice, France, 2008), pp. 3559–3564
25. L. Alba, P. Arena, S. De Fiore, L. Patane, R. Strauss, G. Vagliasindi, Implementation of a Drosophila-inspired model on the eye-ris platform. in *Proceedings of the IEEE International Conference of CNNA* (Berkley, 2010)
26. G. Liu, H. Seiler, A. Wen, T. Zars, K. Ito, R. Wolf, M. Heisenberg, L. Liu, Distinct memory traces for two visual features in the Drosophila brain. Nature **439**, 551–556 (2006)
27. H. Hoffmann, R. Moller, An extension of neural gas to local pca. Neurocomputing **62**, 305–326 (2004)
28. P. Arena, S. De Fiore, L. Patané, P.S. Termini, R. Strauss, Visual learning in Drosophila: application on a roving robot and comparisons, *5th SPIE's International Symposium on Microtechnologies* (Czech Republic, Prague, 2011), pp. 1–12
29. K. Neuser, T. Triphan, M. Mronz, B. Poeck, R. Strauss, Analysis of a spatial orientation memory in Drosophila. Nature **453**, 1244–1247 (2008)
30. R. Wehner, G. Hartmann, The ant's path integration system: a neural architecture. Biol. Cybern. **73**, 483–497 (1995)
31. R.L. Davis, X. Liu, Insect olfactory memory in time and space. Curr. Opin. Neurobiol. **6**, 679–685 (2006)
32. J. Dubnau, C. Margulies, T. Tully, Deconstructing memory in Drosophila. Curr. Biol. **15**, 700 (2005)
33. L.F. Abbott, S. Song, K.D. Miller, Competitive hebbian learning through spike-timing-dependent plasticity. Nat. Neurosci. **3**, 919–926 (2000)
34. L.F. Abbott, S. Song, Cortical development and remapping through spike timing-dependent plasticity. Neuron **32**, 339–350 (2001)
35. P. Arena, S. De Fiore, L. Patané, M. Pollino, C. Ventura, Stdp-based behavior learning on tribot robot. in *Proceedings of IEEE/RSJ International Conference SPIE* (2009)

36. L. Vine, L. Gottschalk, P.J. Shaw, L. Seugnet, Y. Suzuki, D1 receptor activation in the mushroom bodies rescues sleep-loss-induced learning impairments in Drosophila. Curr. Biol. **18**, 1110–1117 (2008)
37. R. Biesinger, K.G. Gotz, Centrophobism in *Drosophila melanogaster*. Physiological approach to search and search control. J. Comp. Physiol. A **156**, 329–337 (1987)
38. M.B. Sokolowski, S. De Belle, Heredity of rover/sitter: alternative foraging strategies of *Drosophila melanogaster* larvae. Heredity **5**(9), 73–83 (1987)
39. J.R. Martin M. Besson, Centrophobism/thigmotaxis, a new role for the mushroom bodies in Drosophila. J. Neurobiol. **62**, 386–394 (2005)
40. A. Wen, T. Zars, K. Ito, R. Wolf, M. Heisenberg, L. Liu, G. Liu, H. Seiler, Distinct memory traces for two visual features in the Drosophila brain. Nature **439**, 551–556 (2006)
41. P. Arena, L. Patané, P.S. Termini, A. Vitanza, R. Strauss, Software/hardware issues in modelling insect brain architecture. in *4th International Conference on Intelligent Robotics and Applications (ICIRA)*, Aachen (Germany), 2011
42. P. Arena, L. Patané, P.S. Termini, An insect brain computational model inspired by *Drosophila melanogaster*: simulation results, *WCCI 2010 IEEE World Congress on Computational Intelligence* (Barcelona, Spain, 2010), pp. 838–845

# Part II
# Complex Dynamics for Internal Representation and Locomotion Control

# Chapter 3
# Compact Internal Representation of Dynamic Environments: Simple Memory Structures for Complex Situations

**J. A. Villacorta-Atienza , M. G. Velarde and V. A. Makarov**

**Abstract** In this chapter the novel concept of Compact Internal Representation (CIR) is introduced as a generalization of the internal representation extensively used in literature as a base for cognition and consciousness. CIR is suitable to represent dynamic environments and their potential interactions with the agent as static (time-independent) structures, suitable to be stored, managed, compared and recovered by memory. In this work the application of CIR as the cognitive core of moving autonomous artificial agents is presented in the context of collision avoidance against dynamical obstacles. The structure that emerges, even if not directly related to the insect neurobiology, is quite simple and could enhance the capabilities of the computational model already presented in the previous chapter, in view of its robotic implementation.

## 3.1 Introduction

The real world is a continuously changing environment whose complexity often invalidates rigid reactive behaviors. Animals during long evolution developed mechanisms that enable predicting the future and a purposed based selection of behaviors. Growing experimental evidence (for review see, e.g. [1, 2]) suggests that these mechanisms rely on the ability of generating an Internal Representation (IR) of the body and external environment and on the parallel virtual "simulation" of multiple alternatives. In the context of an agent (either animal or robot) moving in an arena, IR can

J. A. Villacorta-Atienza (✉) · M. G. Velarde · V. A. Makarov
Instituto Pluridisciplinar, Universidad omplutense de Madrid, Madrid, Spain
e-mail: joseavillacorta@pluri.ucm.es

M. G. Velarde
e-mail: mgvelarde@pluri.ucm.es

V. A. Makarov
e-mail: vmakarov@mat.ucm.es

P. Arena and L. Patanè (eds.), *Spatial Temporal Patterns for Action-Oriented Perception in Roving Robots II*, Cognitive Systems Monographs 21,
DOI: 10.1007/978-3-319-02362-5_3,  

be defined as an abstract, purpose-based spatiotemporal construction mimicking crucial features of the environment and describing the possible (including forthcoming) interactions between its elements and the agent.

Four main types of neurons involved in construction of the abstract IR of spatial environments have been reported. In the pioneering work O'Keefe and Dostrovsky [3] proved the existence of a hippocampal neuronal population whose activity correlates with the animal's position in the environment. Moreover, they showed that this activity does not depend on the animal's orientation. When a rat moves in a known environment and reaches a specific spatial location the corresponding place cell increases its firing rate. The rate returns to the base line when the animal moves away from the place [3, 4]. The complementary information, also independent of the animal's position, is provided by head-direction cells mainly found in the anterior thalamus, presubiculum, and lateral mammillary bodies [5, 6]. These neurons respond according to the direction pointing by the animal's head. More recently a new type of neurons, so-called grid cells, has been discovered in the dorsocaudal medial entorhinal cortex, providing a structured IR of the surrounding space [7, 8]. A grid cell fires at specific animal's locations given by vertexes of a triangular or hexagonal grid "covering" the environment. The orientation of the grid field may be controlled by the head-direction cells. It has been also shown that there exists a population of neurons in the entorhinal cortex responding to the presence of obstacles (i.e., object borders) [2]. The activity of the border cells is related to the orientation of obstacles and it correlates with the activity of grid and head-direction cells [2, 9]. This suggests that border cells contribute to the creation of a reference frame for grid and position cells.

The mentioned above studies put experimental ground for the representation of surrounding space and static situations, however, dynamic (i.e., essentially time dependent) situations claim a spatiotemporal IR capable of dealing with time-changing environments [10] or even with the representation of the outer world in more abstract terms such as intentions, context, empathy, etc. IR of the time dependent situations must be based on two critical elements: i) the world mental model including both the agent and its environment; and ii) the principle of causality.

The first component is widely accepted (see, e.g., [11]) and can be viewed as a description (not necessarily complete) of geometrical and physical laws governing the observed dynamical processes. For example, a free falling object accelerates with 9.81 m/s$^2$ in the earth. Experiments with astronauts catching a thrown ball under reduced gravity revealed that the peak of anticipatory muscle activation occurs earlier than the real impact of the ball [12]. After a few days the astronauts adapted to the new gravity conditions. This finding leads to two inferences: the brain has an internal mental model which "calculates" the ball trajectory and this model can be tuned by learning. Many other psychophysiological studies show that humans tend to represent observed and imagined (e.g., through reading) situations as dynamic structures and anticipate forthcoming changes.

The second IR component, the principle of causality, means that there is no phenomenon that does not give rise to certain consequences and has not been caused by other phenomena. Thus an intelligent agent cannot restrict its actions to what is immediately perceived. Instead, it should synchronize in time and space its behavior with other objects or agents. Thus the causality should also be presented in mental simulations. As we shall show below the use of this principle allows separating the "past" from "future" in the mental time and mapping moving objects into disjointed space-time regions.

The existence of neuroanatomical substrates responsible for the high-level time-dependent functions mentioned above has been supported by the finding of mirror neurons [13]. These cells respond both when a subject makes a particular action and when he sees how another individual performs the same action. However, a more intriguing is that mirror neurons can react on the intention to perform a purpose-based action. When a monkey sees the initial phase of the action of extending a hand to grasp something (cause), specific neurons fire even if the final consequence, i.e., grasping (effect), is not directly perceived [14]. Thus up-to-date neuroanatomical and electrophysiological studies converge to the hypothesis that the brain has specific spatially distributed neural networks (presumable in the hippocampus and neocortex) responsible for IR of static as well as dynamic situ- ations, which may offer multiple, in fact countless, ways of performing behaviors. The latter means that IR is not reducible to a search for a best solution, since the solution fitness may be subjective or/and goal dependent. For example, a football player can turn aside from or catch a thrown ball depending on which side he plays on. Thus IR should be considered as a universal spatiotemporal description of the environment, useful for flexible decision making.

Recent theoretical research on IR (although the first record goes back to 1943 [15]) elucidated its main advantage: it allows detaching the behavior from direct environmental control by, e.g., inhibition of the motor execution [16–20]. This enables responses of the organism to features of the world that are not directly present, but whose causes already exist, and hence to plan ahead smartly. Besides, IR can be evoked mentally (without direct sensory input) to evaluate potential solutions. It is widely believed that IR is a prerequisite of a cognitive behavior [17]. Particularly, the IR concept has been used to control a multisegmented manipulator with redundant degrees of freedom [21]. Specific feed-forward networks and Recurrent Neural Networks (RNNs) have been used as holistic models of geometrical structures like bodies with arms and legs (see, e.g., [20, 22] and references therein). It has been shown that such networks can operate as forward models, inverse models or any mixed combination. Later in a series of papers [22–25] it has been argued that IR based on RNNs assembled using simple building blocks can implement working memory. Namely, it has been shown that RNNs can store and then reproduce some specific static and dynamic situations. We have shown that a network of n neurons can learn and then replicate a sequence of up to n different frames [25]. Another promising application of the IR approach is the construction of sensory-motor maps that implement external anticipation of the sensory information. For example, it has been shown that echo state networks can successively learn forward models for a blind navigation in a square shaped static environment [26]. Toussaint [27] argued that standard sensory-motor maps lack a proper neural representation and proposed the use of neural field models. Then motor activations induce anticipatory shifts of the activity peak in the sensory-motor map, which in turn can be used more naturally for planning and navigation in a maze.

In the present work we propose an Internal Representation Neural Network (IRNN) capable of dealing with significantly dynamic situations, i.e., when a mobile agent has to cope with an environment also containing moving objects (obstacles). In such a situation the quasistatic approach or sequential behaviors, as proposed by Kuhn and Cruse [18] for language production, may not work properly. Instead we formulate the model and study its behaviors in the framework of the dynamical systems theory. Thus the mental simulations of behaviors occur through operations feasible in the phase space of the IRNN and their results also belong to this space. Hence an IR (as a result of simulation) is an element of the phase space whose desired properties are: (a) mental calculations should converge to the same (or similar) IR under small perturbations and (b) IR should admit a compact description ready to be stored in (static) memory and allowing operations like classification, grouping, averaging, etc. We show that the IRNN satisfies both properties and the generated compact IRs (CIRs) of dynamic situations can be associated with the simplest form of attractors, i.e., time-invariant stable steady states. The latter at first glance can appear to be contradictory, but this apparent puzzle is resolved by an appropriate use of the principle of causality in the mental time. Finally, the strategies to be chosen by for the agent are defined by the position of the steady state in the multidimensional phase space, i.e., the CIRs can be unfolded into the real space.

In order to illustrate the capabilities of this internal representation, the present work is focused on the simplest of these decisions making process, namely the choice of a possible trajectory to the target avoiding obstacles in base of diverse criteria, for example less time spending, more safety, less energy consumption, etc. A particular case of this problem is the path planning problem, which has been for decades a frequently visited topic in robot research. The path planning is defined as the search for the shortest path between two points avoiding collisions with the obstacles present in the environment [28]. Many methods have been developed to tackle this problem [29–33] but at present one of the most fruitful approaches for the path planning is based on the potential fields, defined as potential functions covering the perceived environment, whose shape defines the paths following the negative gradient. Numerous potential fields have been proposed inspired by physical or biological analogies [28, 34–38], motivated by the avoidance of local minima [39, 40], based on the application of mathematical methods suitable to optimize the computations [41, 42], leading with time-changing environments [43], etc. Nevertheless in the context of the internal representation the optimum path is only one possible choice, based on one particular motivation. For that reason we make use of a potential field as a secondary component of a general system which first establishes the relations between spatiotemporal evolution of agent and environment, and after generates all possible paths. Nonetheless the potential of the proposed Compact Internal Representation out of te path planning problem will be shown with different decision making concerning interaction with dynamic environments.

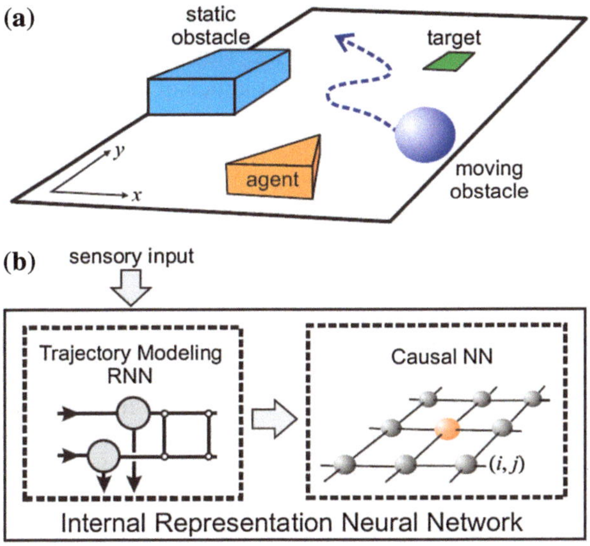


**Fig. 3.1** **a** Sketch of an arena containing agent, target, and static and moving obstacles. The agent perceives all objects in the arena and creates an internal spatiotemporal representation of itself and of the environment with the goal of reaching the target avoiding collisions with the obstacles. **b** General block-scheme architecture of the IRNN consisting of two coupled subnetworks: (**i**) *Trajectory Modeling* and (**ii**) *Causal neural networks*. The TM-RNN receives sensory information, simulates trajectories of all objects (except the agent) and maps them into the Causal Neural Network, which creates the CIR of the dynamic situation

## 3.2 General Architecture of IRNN

Let us consider an agent moving in a time-dependent environment with the objective to reach a target (Fig. 3.1a). For the sake of simplicity (but see Discussion) the target is considered immobile and can be an object or a specific place or even an area in the arena. We assume that the target emits or reflects a sensory signal, e.g., sound or light, which is perceived by the agent (i.e., the target is "visible", although its perceiving may not be 100.

Figure 3.1b shows the general architecture of the agent's IRNN. It consists of two coupled subnetworks: Trajectory Modeling RNN (TM-RNN) and Causal Neural Network. The TM-RNN receives a sensory information either directly from the sensory system or from another brain area (e.g., memory). In the latter case the IRNN takes part in completely or partially mental (i.e., without direct observation of the environment) simulations of dynamic situations. The TM-RNN should be trained by real trajectories. Once the training is deemed finished the network becomes capable of simulating arbitrary trajectories of objects using as an input only their initial conditions (e.g., position and velocity) in a similar way as it has been observed in experiments with astronauts under reduced gravity [12].

To create a suitable IR the agent has to synchronize its possible motions with those given by the TM-RNN for the obstacles. This is achieved in the Causal Neural Network that is a two dimensional $(N \times M)$ lattice of locally coupled neurons that geometrically reproduces the arena, i.e., position of a neuron in the lattice described by the pair of indexes $(i, j)$ corresponds to the actual $(x, y)$-coordinates in the arena (Fig. 3.1). Such mapping is the simplest but not unique. For example, to gain better realism the mapping may be logarithmic, i.e., the distance in the real space scales logarithmically into the neuronal lattice (see, e.g., [7] for grid cells). The assumption of complete sensory knowledge of the environment suggests the egocentric reference frame for the Causal Neural Network, i.e., the agent maps itself into the origin of the neuronal lattice (red bigger neuron in Fig. 3.1b). Such reference frame has been argued to be used by insects [44], although more complex navigational systems have hierarchical organization depending on the context. For instance, honeybees switch between the egocentric and allocentric reference frames depending on whether they are flying along an unknown or a familiar route, respectively [45].

## 3.3 Compact Internal Representation for Navigation in Complex Environments

The IR of a perceived situation is a task-oriented abstraction of the environment created for the agent to fulfill the task. Static environments have been extensively considered in the literature of autonomous navigation, however real complexity appears in dynamic situations, when the environment changes with time. In this work we show how CIR provides the agent with solutions to this problem through a virtual exploration of the perceived arena. During such mental exploration the agent extracts the possible interactions with the (dynamic and static) obstacles and targets, whose trajectories are predicted from their initial conditions. These critical events (collisions and reaches) will be projected into the Causal Neural Network, together with the possible pathways to navigate, creating the CIR.

Let us illustrate the concept of Compact Internal Representation (Fig. 3.2**a**) by considering an arena where a single obstacle (vertical bar) moves, for the shake of simplicity, with constant velocity. The agent (red circle) generates a mental exploration by virtual agents that follow all possible pathways from its initial position; such trajectories are "locally optimal", i.e. they are straight lines when possible. Figure. 3.2**b**, shows the numerical implementation of this process. The set of infinite virtual agents, propagating at the same velocity, forms a wavefront that progressively covers the represented arena. Such wavefront separates the three-dimensional space (2D space + time) into disjoint temporal regions represented in the 2D space. The past corresponds to the area enclosed by the wavefront, containing those points visited by the virtual agents (in gray), whereas the future is represented by the outer part (in yellow), with the locations to be virtually visited by the agent. The wavefront describes the present, so when it interacts with the mental representation of the moving bar, a possible collision between the obstacle and the agent appears (pink area). As long as the wavefront propagates it encloses the set of possible collisions, which

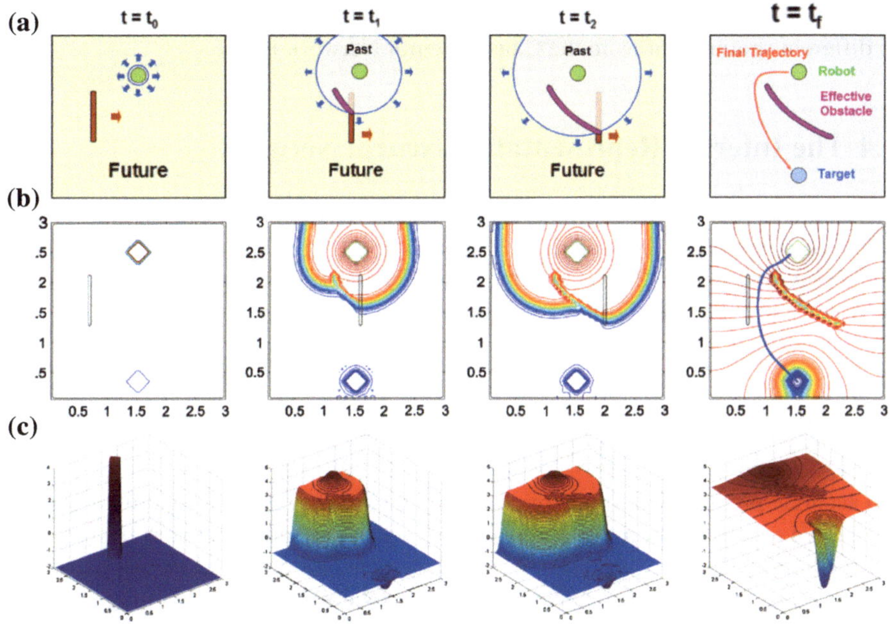


**Fig. 3.2** CIR of a dynamic situation. **a** Illustration of the concept of Compact Internal Representation. A moving bar in a 2D arena must be avoided by the agent, which mentally explores the environment by following simultaneously all possible pathways. This is simulated by a propagating wavefront separating the space in past and future. When the wavefront finds the dynamic obstacle an effective static obstacle (*pink*) is created. The avoidance of this effective obstacle ensures the avoidance of the moving bar. **b** Computational simulation of the CIR creation process of the previous situation. The last panel shows the CIR superimposed to the initial perceived situation. **c** Evolution of the diffusion field creating the CIR. The x and y axes correspond to the spatial 2D arena, the z axis corresponds to the variable r defining the dynamic of the Causal Neural Network. It can be observed the coexistence of the two dynamic regimes, the active diffusion creating the propagating wavefront, and the passive diffusion, leading to the pathways required for navigation

belongs to the past. The principle of causality ensures that the past cannot affect the future, so these locations will remain unalterable, forming a static effective obstacle.

Once the mental exploration has finished, the dynamic situation has been reduced to a static representation, where the perceived moving objects (a single obstacle in this case) are replaced by their corresponding effective (static) objects (which can be both obstacles and targets). In addition the evolution of the wavefront defines a (infinite) set of trajectories that, while avoid the static effective obstacle, allow the agent to reach the target avoiding the moving obstacle. This final static representation of a dynamic situation is the CIR. The two processes involved in its creation are then:

1. Wave regime. The propagation of the wavefront separates the virtual future from the virtual past. Effective obstacles are formed in the Causal Neural Network.
2. Diffusion regime. The evolution of the Causal Neural Network with effective (static) obstacles shapes the CIR.

Notice that both regimes occur simultaneously in the virtual mental time, but belong to different spatial regions in the Causal Neural Network lattice.

## 3.4 The Internal Representation Neural Network

The Compact Internal Representation is created by the Internal Representation Neural Network, consisting of two interacting subnetworks (Fig. 3.1b), one for the prediction of the environment evolution and the other for the generation of the CIR. In this section we provide their formal implementation.

### *3.4.1 Trajectory Modeling Neural Network*

An accurate prediction of trajectories of moving objects results from previous learning [17], i.e., observation of moving objects tunes the neural network designed for the prediction. Recently we have discussed elements for a general memory structure [46], and proposed a Recurrent Neural Network (RNN) capable of learning and replicating some specific dynamic situations. The trajectory of a moving object can also be viewed as an example of such a dynamic situation. Thus the general RNN-based concept can be applied to the learning and modeling trajectories of moving objects.

Let us consider the trajectory of an object in 2D as described mathematically by two parametric equations, one for each dimension. The parameter is the time and according to Newton's laws, the equations are polynomials whose coefficients are functions of initial positions, velocities, and accelerations. Although this information is acquired by the agent's sensors during the initial stages of the object's movement, the question is, how can this movement equations be implemented in and dynamically performed by a neural network? As shown in [47] the RNNs are able to acquire this capacity by learning, adapting the connectivity among this neurons to match its output with the presentation of different physical trajectories. Once the learning is complete the RNN reproduces dynamically the trajectory corresponding to a given initial conditions (initial position, velocity, and acceleration), permitting an accurate prediction of object's movement.

### *3.4.2 Causal Neural Network*

The Causal Neural Network is based on a set of coupled modified FitzHugh-Nagumo neurons, able to elicit autowaves (waves that propagate keeping their shape and velocity) to generate the wavefront, and exhibiting simultaneously passive diffusion: Mathematically:

$$\begin{aligned} \dot{r}_{ij} &= q_{ij}\left(H(r_{ij})\left[f(r_{ij}) - v_{ij}\right] + d\Delta r_{ij} - r_{ij}p_{ij}\right) \\ \dot{v}_{ij} &= (r_{ij} - 7v_{ij} - 2)/25 \end{aligned} \tag{3.1}$$

where $v_{ij}$ is the so-called recovery variable; $f(r)$ is a cubic nonlinearity, which for numerical simulations we set to $f(r) = (-r^3 + 4r^2 - 2r - 2)/7$; and $H$ is the regime controlling (Heaviside step) function:

$$H(r) = \begin{cases} 1, & \text{if } r \le r_{\mathrm{h}} \\ 0, & \text{otherwise} \end{cases}$$

where $r_{\mathrm{h}}$ is the threshold separating the wave and diffusion regimes.

The binary variable $q_{ij}(\mathrm{t}) \in \{0, 1\}$ in (3.1) describes the influence of effective obstacles (green area in Fig. 3.2b) on the Causal Neural Network dynamics. This inhibitory term mimicks the possible effect that border cells may exert over the activity of grid cells [2]. The TM-RNN provides the obstacle position, thus causing firing of the corresponding artificial border cell, which in turn inhibits the response of the Causal Neural Network ("grid") cell $r_{ij}$. When the "border" cell is silent (no effective obstacle) $r_{ij}$ evolves according to the Causal Neural Network intrinsic dynamics.

At $\mathrm{t} = 0$ no effective obstacle exists and $q_{ij}(0) = 1\ \forall(i, j)$. For $\mathrm{t} > 0$ a concentric circular wave (sketched in Fig. 3.2a) is generated. Once the wavefront catches up an obstacle (mobile or immobile) it slips around. Cells, where the wavefront "touches" an obstacle at $\mathrm{t} = \mathrm{t}_0$ (Fig. 3.2b) become "frozen", $q_{ij}(\mathrm{t} \ge \mathrm{t}_0) = 0$. As a consequence, for the frozen cells we have

$$r_{ij}(\mathrm{t}) = r_{ij}(\mathrm{t}_0), \quad \text{for } \mathrm{t} \ge \mathrm{t}_0 \tag{3.2}$$

It can be shown that the unit's intrinsic dynamics (for $d = 0$, $p = 0$, $q = 1$, and $H = 1$) is bistable, with two stable steady states at $r_{\mathrm{d}} = 0$ and $r_{\mathrm{u}} = 3$. This yields multi-stability and even spatial chaos for a low enough coupling strength $d < d_{\mathrm{cr}}$ [48, 49]. The upstate $r_{\mathrm{u}}$ has much bigger basin of attraction than the downstate $r_{\mathrm{d}}$. For a strong coupling $d > d_{\mathrm{cr}}$ by fixing just a single neuron in the upstate we create a wave that will switch all neurons from the downstate to the upstate. Hence to obtain a wavefront in our problem we select a high enough inter-neuronal coupling $d > d_{\mathrm{cr}}$.

The propagating wave switches neurons to the upstate and hence $H = 0$ and also $q_{ij} = const$ behind the wavefront. Thus (3.1) also exhibits the gradient property for $\mathrm{t} > \mathrm{t}^*$, although its transient process is not gradient. Thus trajectories in the phase space $\Psi = \mathbb{R}_+^{NM} \times \mathbb{R}^{NM}$ of the Causal Neural Network (3.1) tend to one of the stable steady states that defines the CIR for a given dynamic situation.

We notice that once the wavefront reaches the target image in the Causal Neural Network at $\mathrm{t} = \mathrm{t}_{\mathrm{tr}}$ the calculations can be stopped. Then by construction there exists at least one path starting from the agent position and ending at the target. Thus we get a first suitable approximation to the IR. Running the Causal Neural Network further for $\mathrm{t} > \mathrm{t}_{\mathrm{tr}}$ improves the shaping of the $r_{ij}$ pattern in $\{\mathbb{Z}^2, \mathbb{R}_+\}$, which leads to an exponentially saturating optimization of paths.

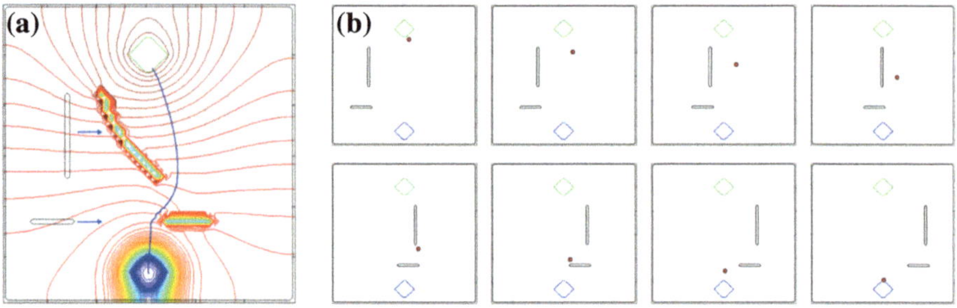


**Fig. 3.3** CIR-based navigation avoiding two obstacles moving in a straight line. **a** two bars (*vertical* and *horizontal*) move from left to right. The corresponding CIR is superimposed to the perceived initial situation (*black* silhouettes for obstacles, *green* for agent's initial location, and *blue* for target). *Blue* line represents the shortest navigation trajectory followed by the agent for reaching the target avoiding the moving obstacles. **b** Snapshots showing the navigation in the dynamic situation, following the pathway depicted in A. The agent is represented by the *red* circle

## 3.5 CIRs of Different Dynamic Situations

Let us now test the performance of the IRNN in different simulated environments.[1]

### 3.5.1 Avoiding Simple Moving Obstacles

We begin with a simple (but no trivial) dynamic situation where two moving bars move from left to right following straight lines (Fig. 3.3). The agent, initially located at the top of the arena (in green), must reach a target at the bottom (in blue). After observing the object during the first three time steps (needed for the estimation of its initial velocity and acceleration) the TM-RNN predicts its future trajectory. This calculation is fed into the Causal Neural Network where the CIR of this dynamic situation is created.

Figure 3.3a shows superimposed the perceived dynamic situation (black silhouettes for the obstacles and blue arrows for the velocities) and the corresponding CIR, with the shortest navigation trajectory depicted in blue. Figure 3.3b shows a sequence of snapshots where the agent navigates in the dynamic situation following such trajectory. The agent can therefore move safely among the moving obstacles simply by avoiding the static effective obstacles.

[1] Corresponding videos can be found at http://www.mat.ucm.es/~vmakarov/IRNN.html.

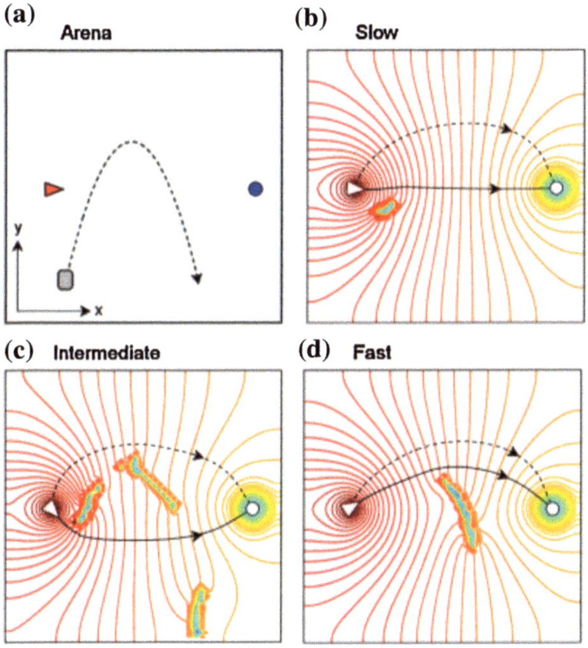


**Fig. 3.4** Gaps in effective obstacles. **a** Dynamic situation. Obstacle (*grey rectangle*) moves along a parabolic trajectory (*dashed curve*). **b–d** CIRs of the situation significantly depend on the obstacle velocity

## *3.5.2 Nonlinear Motion and Gaps in Effective Obstacles*

Let us now illustrate how relatively simple trajectories of a moving obstacle can lead to nontrivial CIRs.

We consider an obstacle moving along a parabolic trajectory (Fig. 3.4a), which can simulate, for example, a free fall of an object launched with initial "horizontal", $v_x(0)$, and "vertical", $v_y(0)$, velocities. For convenience the trajectory shape is fixed and we examine three cases, which differ by the obstacle velocity: (i) slow, (ii) intermediate, and (iii) fast motions.

As we have shown above, parabolic trajectories can also be predicted by the TM-RNN. In the case of slow motion, the obstacle image coincides with the wavefront in time and space in the first (ascending) part of the trajectory only, where a small effective obstacle appears (Fig. 3.4b). Then the created CIR enables straight-line path to the target (solid line). We note that although this path crosses the obstacle trajectory in $(x, y)$-plane these intersections do not occur in the complete $(x, y, t)$-spacetime. Besides the shortest the agent can select safer paths going above the obstacle trajectory (Fig. 3.4b, dashed curves).

Now we increase the obstacle velocity but conserving its trajectory. This significantly changes the spatiotemporal domains where the wavefront coincides with

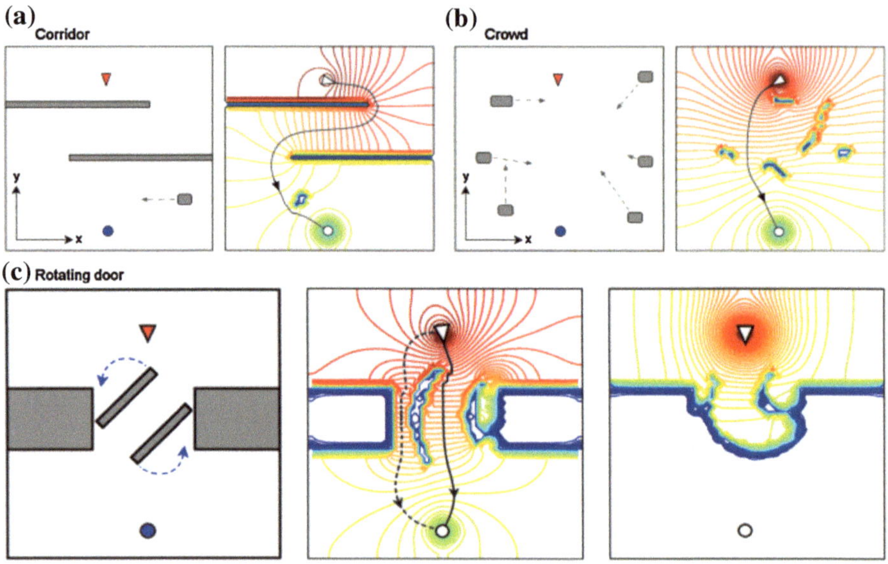


**Fig. 3.5** Examples of CIRs of realistic dynamic situations. *Triangles* and *circles* mark the initial agent and target positions, respectively. *Grey* objects are obstacles. *Solid curves* show the shortest paths to the target. **a** Passing through a corridor and avoiding a moving obstacle. **b** Moving in a crowd environment. **c** Passing through a rotating door

the obstacle image. We get three spatially separated domains of frozen cells in the Causal Neural Network (Fig. 3.4c). Thus a single moving obstacle can cause several *spatially separated* effective obstacles in the Causal Neural Network space. Then the diffusion finds all the gaps, and the final CIR offers to the agent a possibility to pass through the gaps. The black solid curve in Fig. 3.4c shows the shortest path to the target that goes below the first two effective obstacles and above the third one. Obviously, there still exist safer paths passing over the whole obstacle trajectory (Fig. 3.4c, dashed curves).

Finally, Fig. 3.4d shows the CIR for the case of a fast obstacle, whose effect on the wavefront leads to a single effective obstacle in the final (descending) part of the obstacle trajectory. Thus the shortest path to the target goes just a little bit above this effective obstacle.

### *3.5.3 Realistic Environments*

Realistic dynamic situations can be quite complex. For illustration we selected three particular situations.

Figure 3.5a (left panel) shows an example where the agent on the way to the target should go through a corridor delimited by two elongated immobile obstacles and then avoid a moving obstacle. The CIR obtained for this situation and the shortest path to the target are shown in Fig. 3.5a (right panel).

Figure 3.5b (left panel) illustrates a dynamic situation with multiple (six) moving obstacles whose simultaneous avoidance is not trivial even for living beings. The IRNN successfully solves this problem by generating a CIR with five effective obstacles (Fig. 3.5b, right panel). The shortest path passes between two of them.

Finally, Fig. 3.5c (left panel) shows a dynamic situation with a rotating (door like) obstacle. To reach the target the agent should synchronize its movement with the door. For prediction of the door position we use a slightly modified TM-RNN similar to those described in Sect. 3.4.1 and based on the modeling of periodic processes [25]. Depending on the rotation velocity this dynamic situation can admit (Fig. 3.5c, middle panel) or not (Fig. 3.5c, right panel) solutions. In the latter case the effective obstacle occupies all the space between the walls, thus precluding the pass. In the former case the CIR enables only passage synchronized with the door rotation, i.e., the agent, when going through, always moves between two parts composing the door. This particularly forces the agent to make a short excursion to the left in such a way that its movement synchronizes with the door rotation (dashed trajectory in Fig. 3.5c, middle panel).

## 3.6 CIR as Substrate of Decision Making Processes

The power of Compact Internal Representation presented in this work has been illustrated by means of the widely-explored problem of the path planning. Nonetheless the capabilities of CIR go further from the choice of an optimum path joining two point of the agent's environment. In the path planning approach the decisions provided by CIR is represented by paths but in more general circumstances this decisions could involve different strategies not necessary related with the construction of a route. A clear example is shown in Fig. 3.5c right, where the velocity of the rotating door makes impossible to reach the target; in this case the obvious strategy to be made from the agent will not be "what path should I choose?" but "Should I try to pass?".

A different situation is illustrated in Fig. 3.6, where the agent is initially equidistant to two targets with two moving obstacles describing parabolic trajectories with several velocities (as in Figs. 3.4b, c). The present case models a biologically-inspired situation where survival impels the agent to make a 'fast' decision choosing one of the targets (modeling for example the agent's nests) where it does not matter the path to reach it. In these circumstances the decision could be supported by the measure of the distance between agent and obstacles when a straight line joining the agent's position and each target is followed. The chosen target will be those whose corresponding distance is largest.

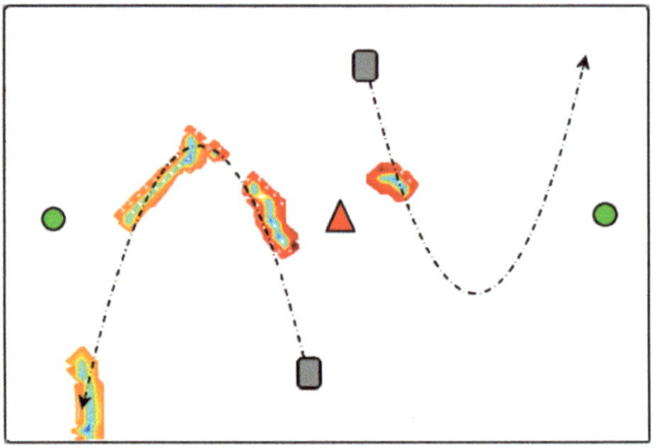

**Fig. 3.6** Fast-decision scenario. This decision does not involve a selection of a path but a selection of a target, choosing for example that in which the obstacles present in the straight trajectory joining it with the target are furthest away for the agent

## 3.7 Discussion

In "I of the vortex" R. Llinas [50] put forward the hypothesis that "the mindness state evolved to allow predictive interactions between mobile creatures and their environment". He claimed that the capacity to predict and anticipate motions is most likely the ultimate brain function. Recently the concept of internal representation (IR) of static and dynamic situations has been strongly supported by experimental and theoretical studies. It has been shown that IR indeed offers new perspectives for the decision making and it is said to be a prerequisite for a cognitive behavior [17]. A big challenge in the IR research is to understand how the brain makes a compact effective description of complex, time-evolving situations, in such a way that they can later be stored in (presumably static) long term memory and then retrieved on purpose. In this paper we have proposed an IR Neural Network (IRNN) that can simulate in the phase space real world *dynamic* (i.e., time-dependent) situations and generate their time-invariant compact IRs (CIRs). We have shown that each "mental" simulation converges to a stable steady state, i.e., a single point in the network phase space describes the given dynamic situation. Besides direct on-line strategy planing for immediate motor execution, CIRs as points in a mutlidimensional space can be compared, clustered, averaged, etc. by introducing an appropriate distance measure. Libraries of stereotypic CIRs can significantly speedup and make automatic the previously learned behaviors converting them into "unconscious states". It is also noteworthy that the IRNN can work as a part of "autonomous thinking", i.e., instead of sensory information some mental situations can be supplied to the IRNN for evaluation.

The IRNN is composed of two coupled subnetworks (Fig. 3.1b). The Trajectory Modeling Recurrent Neural Network (TM-RNN) predicts trajectories of objects moving in the arena. Noticeably, its inner structure does not depend on the features of particular movements, hence once the TM-RNN has been trained, it is ready to predict any new trajectory (belonging to a certain class). The output of the TM-RNN is fed into the Causal Neural Network (Causal Neural Network) organized as a two dimensional lattice. Here we have used the simplest one-to-one mapping between the real $\mathbb{R}^2$ and the lattice $\mathbb{Z}^2$ spaces, though more complex mappings, e.g., giving preference to nearest space, are also possible. The mapping allows unfolding the steady states of the IRNN, i.e., CIRs into real space for, e.g., execution of motor actions.

Spatiotemporal synchronization of the agent movements with external elements makes up the Causal Neural Network dynamics, which goes through two concurrent processes (Fig. 3.2). (i) The wave regime simulates parallel mental exploration of the environment. The lattice interior enclosed by the wavefront corresponds to the agent's virtual past, while the exterior belongs to its future. Then the principle of causality (i.e., the virtual future can't influence the virtual past) allows collapsing the time-dependent structure of the dynamic situation into a static pattern, i.e., it makes the IR compact and thus biologically plausible. The propagating front ("mental present") freezes the dynamics of neurons mapping into locations of space-time coincidence of the agent and obstacles in the mental time (what we called formation of effective obstacles). (ii) The diffusion-like process shapes the states of all neurons that have not been "frozen", such that the initial excitation relaxes to a stationary 2D pattern. Thereby obtained global (steady) state of the network provides the CIR of a given dynamic situations.

CIR is an abstract construction of "what can be done", i.e., it is not reducible to a search for a best trajectory, since the solution fitness may be subjective or/and goal dependent (see, e.g., Fig. 3.5c). Nevertheless, one of the straightforward applications of CIRs is universal path planing, i.e., how to reach the target without collisions and taking into account the context (e.g., on the way to the target the agent should pass nearby the table) and motivation (e.g., the target has to be reached as fast as possible). We notice that the additional constraints and criteria are posed by other agent's brain areas, whereas the IRNN offers a free choice among all feasible alternative paths to the target. Using several simulated arenas, from a simple static situation to diverse realistic environments (Figs. 3.3, 3.4 and 3.5), we have shown that the IRNN indeed provides CIRs of complex spatiotemporal interactions between the agent and the environment. We note that a CIR can always be created. However actually realizable solutions (paths to the target) may not exist (Fig. 3.5c, right panel).

When developing the IRNN we intentionally made several simplifying assumptions on the agent and environment to disentangle influence of the main composing elements. For example, a single immobile target has been assumed; however, inclusion of several mobile targets is straightforward; see [51] for details. Then the IRNN will allow interception of targets or selection among them. Another restriction applies to the agent, which we modeled at the very basic level. However, more sophisticated agent models, e.g., including the spatial dimensions or constraints, can also be incorporated to the IRNN, which will lead to higher realism in the modeling. Besides, the IRNN can cope with several agents cooperating or/and competing for some goals.

For example, in the simplest cooperation two agents can share their actual positions and run Causal Neural Networks synchronously, thus exploring much faster the environment and deciding who should do one or another task.

Finally, the generality of the IRNN approach makes it susceptible for biophysical verification. Indeed, some critical components we have introduced here already are based on biological findings (e.g., border and grid cells). Other parts, like cells "effectively frozen" under mental simulations of dynamic situations, have been postulated. Their theoretical existence is crucial for CIRs of dynamic situations. Thus the existence of border cells, firing in the presence of static obstacles, allows us to hypothesize the existence of a neuronal population that detects possible space-time coincidences of the animal and moving obstacles. The activity of these conjectured cells should simultaneously depend on the locations and velocities of the animal and obstacles. We expect that these neurons may receive an inhibitory disynaptic inputs from grid and border cells.

## References


1. E.I. Moser, M.B. Moser, A metric for space. Hippocampus **12**(18), 1142–115 (2008)
2. F. Savelli, D. Yoganarasimha, J.J. Knierim, Influence of boundary removal on the spatial representations of the medial entorhinal cortex. Hippocampus **12**(18), 1270–1282 (2008)
3. J. O'Keefe, J. Dostrovsky, The hippocampus as a spatial map. Preliminary evidence from unit activity in the freely-moving rat. Brain Res. **1**(34), 171–175 (1971)
4. C. Barry, C. Lever, R. Hayman, T. Hartley, S. Burton, J. O'Keefe, K. Jeffery, N. Burgess, The boundary vector cell model of place cell firing and spatial memory. Rev. Neurosci. **17**(1–2), 71–97 (2006)
5. J.S. Taube, R.U. Muller, J.B. Ranck, Head-direction cells recorded from the postsubiculum in freely moving rats. i. description and quantitative analysis. J. Neurosci. **10**, 420–435 (1990)
6. J.S. Taube, R.U. Muller, J.B. Ranck, Head-direction cells recorded from the postsubiculum in freely moving rats. ii. effects of environmental manipulations. J. Neurosci. **10**, 436–447 (1990)
7. T. Hafting, M. Fyhn, S. Molden, M.B. Moser, E.I. Moser, Microstructure of a spatial map in the entorhinal cortex. Nature **436**(7052), 801–806 (2004)
8. E.I. Moser, E. Kropff, M.B. Moser, Place cells, grid cells, and the brain's spatial representation system. Ann. Rev. Neurosci. **31**, 69–89 (2008)
9. T. Solstad, C.N. Boccara, E. Kropff, M.B. Moser, E.I. Moser, Representation of geometric borders in the entorhinal cortex. Science **5909**(322), 1865–1868 (2008)
10. J. Sharma, V. Dragoi, J.B. Tenenbaum, E.K. Miller, M. Sur, V1 neurons signal acquisition of an internal representation of stimulus location. Science **300**, 1758–1763 (2003)
11. O. Holland, R. Goodman, Robots with internal models–a route to machine consciousness? J. Conscious. Stud. **10**, 77–109 (2003)
12. J. McIntyre, M. Zago, A. Berthoz, F. Lacquaniti, Does the brain model newton's laws? Nat. Neurosci. **4**, 693–694 (2001)
13. G. Rizzolatti, L. Fogassi, V. Gallese, Neurophysiological mechanisms underlying the understanding and imitation of action. Nat. Rev. Neurosci. **2**, 661–670 (2001)
14. M.A. Umilta, E. Kohler, V. Gallese, L. Fogassi, L. Fadiga, C. Keysers, G. Rizzolatti, I know what you are doing: a neurophysiological study. Neuron **32**, 91–101 (2001)
15. K. Craik, The nature of explanation (Cambridge University (in press), Cambridge, 1943).
16. G. Hesslow, Conscious thought as simulation of behaviour and perception. Trends Cogn. Sci. **6**, 242–247 (2002)

17. H. Cruse, The evolution of cognition a hypothesis. Cogn. Sci. **27**, 135–155 (2003)
18. S. Kuhn, H. Cruse, Static mental representations in recurrent neural networks for the control of dynamic behavioural sequences. Conn. Sci. **17**, 343–360 (2005)
19. M.J. Aitkenhead, A.J.S McDonald, The state of play in machine/environment interactions. Artif Intell Rev. **25**, 247–276 (2006)
20. V. Mohan, P. Morasso, Towards reasoning and coordinating action in the mental space. Int. J. Neural. Syst. **17**, 329–341 (2007)
21. U. Steinkuhler, H. Cruse, A holistic model for an internal representation to control the movement of a manipulator with redundant degrees of freedom. Biol. Cybern. **79**, 457–466 (1998)
22. S. Kuhn, H. Cruse, Modelling memory functions with recurrent neural networks consisting of input compensation units: Ii. dynamic situations. Biol. Cybern. **96**, 471–486 (2007)
23. S. Kuhn, W.J. Beyn, H. Cruse, Modelling memory functions with recurrent neural networks consisting of input compensation units: I. static situations. Biol. Cybern. **96**, 455–470 (2007)
24. H. Cruse, D. Hubner, Selforganizing memory: active learning of landmarks used for navigation. Biol. Cybern. **99**, 219–236 (2008)
25. V.A. Makarov, Y. Song, M.G. Velarde, D. Hubner, H. Cruse, Elements for a general memory structure: properties of recurrent neural networks used to form situation models. Biol. Cybern. **98**, 371–395 (2008)
26. H. Svensson, A. Morse, T. Ziemke, Representation as internal simulation: a minimalistic robotic model. In Proceedings CogSci'09, pp 2890–2895, (2009).
27. M. Toussaint, A sensorimotor map: modulating lateral interactions for anticipation and planning. Neural. Comput. **18**, 1132–1155 (2006)
28. D. Keymeulen, J. Decuyper, The fluid dynamics applied to mobile robot motion: the stream field method. In Proceedings of the IEEE International Conference on Robotics and Automation **1**, 378–385 (1994)
29. S. Sundar, Z. Shiller, Optimal obstacle avoidance based on the hamilton-jacobi-bellman equation. IEEE Trans. Robot. Autom. **2**(13), 305–310 (1997)
30. A.P. Munuzuri, L.O. Chua, Shortest-path-finder algorithm in a two-dimensional array of nonlinear electronic circuits. Int. J. Bifurcat. Chaos. **12**(8), 2493–2501 (1998)
31. S. Waydo, R.M. Murray, Vehicle motion planning using streamfunctions. **2**, 2484–2491 (2003)
32. A. Adamatzky, P. Arena, A. Basile, R. Carmona-Galan, B.D.L Costello, L. Fortuna, M. Frasca, A. Rodriguez-Vazquez, Reaction-diffusion navigation robot control from chemical to vlsi analogic processors. IEEE Trans. Circuits. Syst. 51(5), 926–938, (2004).
33. R. Barzamini, M.B. Motamedinejad, J. Jouzdani, *A* (Khosravi, A new fuzzypathplanning for multiple robots, 2006), pp. 295–300
34. G.K. Schmidt, *K* (Azarm, Mobile robot navigation in a dynamic world using an unsteady diffusion equation strategy, 1992), pp. 642–647
35. C. Louste, A. Liegeois, Near optimal robust path planning for mobile robots: the viscous fluid method with friction. J. Intell. Robot. Syst. **1–2**(27), 99–112 (2000)
36. A.A. Masoud, S.A. Masoud, M.M. Bayoumi, Robot navigation using a pressure generated mechanical stress field: the biharmonic potential field approach. pp. 124–129, (1994).
37. Y.F. Wang, G.S. Chirikjian, A new potential field method for robot path, planning. pp. 977–982, 2000.
38. L. Guan-Chun, L. Wei-Wen, Dynamic mobile robot navigation using potential field based immune network. Systemics, Cybern. Inf. 2(5), 43–50 (2006).
39. V.I. Utkin, H. Hashimoto, Robot obstacle avoiding in n-dimensional space using planar harmonic artificial potential fields. J. Dyn. Syst, Meas. Contr. 119, 160–166 (1997).
40. J.B. Burns, C.I. Connolly, R. Weiss, Path planning using laplaces equation. In IEEE International Conference on Robotics and Automation, pp. 2101–2106, 1990.
41. L.E. Kavraki, Computation of configuration-space obstacles using the fast fourier transform. IEEE Trans. Robot. Autom. **3**(11), 408–413 (1995)
42. O. Uzol, I. Yavrucuk, *N* (Sezer-Uzol, Collaborative target tracking for swarming mavs using potential fields and panel methods, 2008)

43. Q. Cao, Y. Huang, J. Zhou, An evolutionary artificial potential field algorithm for dynamic path planning of mobile robot. **1**, 3331–3336 (2006)
44. T.S. Collett, J. Zeil, Spatial representation in animals, chapter Places and landmarks: an arthropod perspective. pp 18–53. Oxford University (in Press), 1998.
45. R. Menzel, R. Brandt, A. Gumbert, B. Komischke, J. Kunze, Two spatial memories for honeybee navigation. Proceedings R. Soc. Lond. B **267**, 961–968 (2000)
46. B.C. Berg, Random walks in biology (Princeton University, Princeton, 1993)(in Press).
47. V.A. Makarov, J.A. Villacorta-Atienza, in *Recurrent Neural Networks for Temporal Data Processing*, ed. by H. Cardot, INTECH. Compact internal representation as a functional basis for protocognitive exploration of dynamic environments (2011)
48. V.I. Nekorkin, V.A. Makarov, Spatial chaos in a chain of coupled bistable oscillators. Phys. Rev. Lett. **74**, 4819–4822 (1995)
49. V.I. Nekorkin, V.A. Makarov, V.B. Kazantsev, M.G. Velarde, Spatial disorder and pattern formation in lattices of coupled elements. Physica D **100**, 330–342 (1997)
50. R.R. Llinas, I of the vortex: from neurons to self. MIT, 2001.
51. J.A. Villacorta-Atienza, V.A. Makarov, Neural network architecture for cognitive navigation in dynamic environments, IEEE Transactions on Neural Networks and Learning Systems. doi:10.1109/TNNLS.2013.2271645 (2013)

# Chapter 4
# CPG for Motor Control

**E. Arena, P. Arena and L. Patanè**

**Abstract** In this Chapter the main results related to locomotion control are reported. Theoretical results referring to Contraction theory are exploited to design reaction diffusion dynamical systems able to work as stable Central Pattern Generators for multipodal robotic structures. A series of strategies are also discussed referring to the control of migration through different locomotion patterns as well as to the steering control for trajectory planning. Relevant parameters are also outlined in view of their modulation for a low low-level feedback control. Both theoretical and experimental results are reported to verify the suitability of the approach.

## 4.1 Introduction

As reported in the first two chapters, the insect brain model architecture cannot disregard the way in which the agent interacts with the environment, i,e, its ability to move. Locomotion requires motor organs and suitable nervous circuits to control them. Looking at the robotic implementation, the ways in which motion will be practically realized will depend on the particular application and also on the complexity of the particular scenario considered. Typically, demonstrations on robots will deal both with rovers on wheels, hybrid prototypes and legged robots. The motion control complexity will grow from the first to the last category. In particular, for legged machines, the traditional motion control approaches will have to give way to more complex methodologies: the single rotational degree of freedom of a moving wheel is substituted by the concurrent actuation the different links of a leg in specific phase

E. Arena · P. Arena (✉) · L. Patanè
Department of Electrical, Electronic and Computer Science Engineering,
University of Catania, 95125 Catania, Italy
e-mail: parena@dieei.unict.it

L. Patanè
e-mail: lpatane@dieei.unict.it

P. Arena and L. Patanè (eds.), *Spatial Temporal Patterns for Action-Oriented Perception in Roving Robots II*, Cognitive Systems Monographs 21,
DOI: 10.1007/978-3-319-02362-5_4, © Springer International Publishing Switzerland 2014

relations among one another. Moreover, the overall gait is generated via a suitable control of all the legs of the moving structure. In this chapter one of the most important paradigms of biologically inspired locomotion control, namely the Central Pattern Generator (CPG) will be adopted. Many experiments have demonstrated, along the last decades, the presence of a centralized generator of neural activity for controlling locomotion in animals (from invertebrates to vertebrates) [1]; this allows them to adopt rhythmic, stereotyped patterns of movement, called gaits.

The CPG is a functional unit containing the key mechanisms needed to generate motion rhythm: it essentially provides the feedforward signals needed for locomotion, even in the absence of sensory feedback and high-level control. Examples of successful applications of the CPG are the biomimetic walking robots [2], underwater robots [3], hexapod robots [4–6], quadrupeds [7] and bipeds [8]. Several researches were inspired by the seminal works on bio-inspired locomotion, like [9]. A mathematical approach to efficiently handle such complex concurrent control of multiple actuator is reported in [10, 11]: CPGs are viewed as networks of coupled nonlinear systems mutually connected with parameters which can be modulated so as to give rise to different locomotion gaits, often exploiting the symmetry of animal gait. In the dynamical system approach, the term CPG refers to the network controlling the whole motor system, i.e., it comprises both single leg controllers and connections between them. The control of the overall structure is thus conceived as the task of reaching a given phase-shift synchronization regime [12]. This means that the overall dynamics will generate a traveling wave pattern through all the motor neurons, imposing a specific gait to the controlled robotic structure. Even if wave pattern is seen more apparently in snakes or lampreys and in their inspired robots [13, 14], looking carefully at the locomotion gaits, also hexapods possess different types of coordination schemes among the legs, that can be seen as particular wave patterns allowed by the plastic nature of the neural control network [15]. These wave-like patterns are mostly visible in insects at their larva stage, where motion is produced by peristaltic waves of contraction of the body wall musculature [16]. The CPG contains the essential neural mechanisms for generating coordinated rhythmic motor patterns. Even if can work autonomously, it is able to receive command signals from the higher motor centers. These neural structures deal with high level behavior : orientation control, gait transition, attitude control and others.

The authors have been involved in the design of CPGs and in their realization by means of Cellular Neural Networks [17], going down into the implementation issue with VLSI CPG analog networks on chip [5]. The approach was essentially based on the exploitation of the Reaction-diffusion paradigm, where traveling waves were generated in arrays of locally connected circuits, which were efficiently mirrored into self-organized actuation commands to the joints in a legged robotic structure. The present chapter summaries some recent results on a new formulation of the locomotion control issue, in light of modern stability analysis tools. In fact locomotion stability can be seen as a control issue toward specific phase locked trajectories in arrays of nonlinear oscillators. Within this framework it is useful to exploit the *Contraction Theory*, recently introduced in [18], and later on extended to Partial Contraction [19].This efficiently handles with exponential convergence to behaviors or even to specific relations among the state variables, instead of trajectories. This approach was extended [20] and applied in [12] to generate models for CPG-based locomotion. Building upon these theoretical results, in this Chapter some useful extensions are demonstrated on undirected diffusive tree graphs, represented as special CNN structures, where convergence to phase locked oscillatory dynamics is obtained. This provides the possibility to define, within the network, a specific phase relation among the oscillators which reflects a specific locomotion gait. In particular, the partial contraction theory allows to derive, for a given undirected diffusive tree graph, a unit control gain value that does not depend on the specific phase imposed, but represents a sufficient condition for any desired phase-shift among the neural oscillators, i.e. for any flow invariant subspace, which can be modulated on-the-fly, having guaranteed the exponential convergence to the new imposed phase shift.

Starting once again from the results reported in [12], we apply a feedback law that modulates the frequency of a reduced number of oscillators in the network. This induces a phase shift among the other oscillators, which is used for gait control [21]. We exploit the effect of a frequency modulation in terms of phase and amplitude variation in the overall network to obtain efficient steering patterns which strictly resemble those ones described in insects. The results proposed are presented in a dynamic simulation environment, and in some cases also by experimental results on a hexapod robot prototype : the Minihex. This is controlled through a custom board built around a digital low cost microcontroller, which allows a simple, efficient and flexible test bed. Once a stable architecture will be reached, an analog, digitally programmable VLSI chip, of the type already available in our laboratory [5], is envisaged to be used.

Finally, a strategy for controlling the trajectory in a legged machine via suitable turning methods is discussed, which takes into account both a frequency and a phase control on the robot legs [22]. The control methodology aims at implementing the steering strategy using a minimal number of control variables, which are related to the phase displacements among the oscillators of the femur and tibia joint of the robot legs. Experimental results carried out in a dynamic simulation environment are also reported. The Chapter is organised as follows: in the first Sect. 4.2 some basic information will be given on Cellular Nonlinear Networks and in particular on the adopted cell and connections. Then the approach based on partial contraction analysis will be discussed referring directly to our application to locomotion control in a six legged machine. Furthermore the control issues related to frequency for gait and steering control will be presented.

## 4.2 The CNN cell and Network

CNNs are arrays of identical, nonlinear and simple computing elements whose main characteristic is local interaction [23]. Therefore the CNN paradigm is well suited to describe locally interconnected dynamics in a lattice-like structure. The simplest cell, embedded into the firstly introduced CNN, consisted in a first order simple circuit, showing an output saturation-like non linearity. The state of each cell is locally connected to its neighbors by means of the so-called cloning templates, containing the parameters modulating the effect of the input, state and output of the neighboring cells on the considered one. The local connectivity feature made the CNN a suitable architecture for an analog VLSI implementation, obtaining real time computation devices composed of large scale circuits, and several VLSI prototypes were introduced. Even if the primary application field was related to real time image processing, soon after their introduction in 1988, a number of different other applications were found, among which the emulation of neural-like spatial temporal dynamics, obtained by changing the type of the CNN cells and also the kind of connections. For the application discussed here, the CNN cell used as the building block of the CNN array has a number of interesting features: it is able, for the set of parameters used below, to show a stable limit cycle behavior, which can range from a harmonic like flow, to a slow-fast dynamics, typical of neural oscillations with resetting. It can be thus used to represent a wide variety of oscillatory conditions. The basic cell is reported below whereas the connections are introduced in Sect. 4.2.1, under the form of coupling matrices, that can be easily arranged as the classical CNN cloning templates. The CNN-CPG generic cell $i$ is the following a second-order nonlinear system:

$$\dot{x}_{i,1} = -x_{i,1} + (1+\mu)y_{i,1} - sy_{i,2} + i_1, \tag{4.1}$$

$$\dot{x}_{i,2} = -x_{i,2} + (1+\mu)y_{i,2} + sy_{i,1} + i_2 \tag{4.2}$$

$$y_i = \frac{1}{2} \cdot (|x_i + 1| - |x_i - 1|) \tag{4.3}$$

The parameters used to allow the cell oscillate with a stable limit cycle, are: $\mu = 0.7$, $\varepsilon = 0$, $s = 1$, $i_1 = 0$, $i_2 = 0$, as demonstrated in [17]. When needed for theoretical analysis, the piecewise linear nonlinearity in Eq. (4.3) will be substituted by its smooth approximation. $y_i = tanh(x_i)$. In the following, CNNs will be used as CPGs for legged robots. In particular, a special case of a graph will be adopted, which avoids loops into the network connection, still retaining local connectivity among the cells. Moreover a diffusion-like connectivity function will be formalized, leading the the paradigm of undirected diffusive graphs, where reaction-diffusion dynamics can be formulated, possessing special stability characteristics, as outlined in the following.

### *4.2.1 Contraction and Partial Contraction in a Tree Graph CPG Network*

In our analysis, we'll refer to CNNs, considered as dynamical graphs and in this subsection a brief overview of the main results and tools derived from Contraction and Partial Contraction theory used in the following treatment will be given. For details the reader is referred to [18]. Let us consider a network organised in an

**Fig. 4.1** A simple undirected diffusive network of three cells

undirected[1] diffusive tree graph, where each neuron is characterised by its position $i$ within the graph. The network is composed of:
$n_n$: total number of neurons in the network;
$n_c$: order of the single neuron (in our case $n_c = 2$ as in Eq. (4.3);
$n_t = n_n \cdot n_c$: total number of state variables in the network.
The whole dynamics can be described as:

$$\dot{x}_i = f(x_i, t) + k \sum_{i \neq j; j \in N_r} (R_{ij} x_j - x_i) \tag{4.4}$$

where:
$f(x_i, t)$ represents the dynamic of the ith uncoupled oscillator, in our case Eq. (4.3); the summation is restricted to all immediately neighboring oscillators $j$; $k$ is the feedback gain, whereas the argument of the sum represents the feedback error between the state variables of the ith cell and the corresponding state variables of cell $j$ after those ones being phase-shifted with respect to cell $i$ via the rotational matrix $R_{ij}$. Under these conditions, the two oscillators are in reciprocal equilibrium only if the error term is zero, which implies *phase-shift synchronization*. Using the traditional formulation of a reaction-diffusion system, the whole system becomes:

$$\dot{x} = f(x, t) - k \cdot L \cdot x \tag{4.5}$$

with: $x = [x_1, \cdots, x_{nt}]^T$, $f(x, t) = [f(x_1, t), , f(x_{nn}, t)]^T$, $L \in \mathbb{R}^{n_t x n_t}$. In details, the laplacian $L$ is a matrix built-up of blocks $L_{ij} \in \mathbb{R}^{n_c x n_c}$ which are:

$$\begin{cases} L_{ij} = -R(\phi_{ij}), & \text{if cells i and j are connected with a phase shift } \phi_{ij} \\ L_{ji} = -R(-\phi_{ij}) & \text{for undirected balanced networks, as our case} \\ L_{ij} = L_{ji} = 0, & \text{if no connection does exist between nodes i and j} \\ L_{ii} = d_i I_{nc}, \forall i = 1..n_n, & d_i\text{: unweighted degree of node } i\text{th.} \end{cases}$$

For instance, supposing to consider a system with chain-topology shown in Fig. 4.1, the L matrix is defined as:

$$L = \begin{bmatrix} I & -R(\phi_{12}) & 0 \\ -R(-\phi_{12}) & 2I & -R(\phi_{23}) \\ 0 & -R(-\phi_{23}) & I \end{bmatrix} \tag{4.6}$$

[1] The term undirected means that for each connection from the node i to node j, the same connection is duplicated in the opposite direction, from j to i.

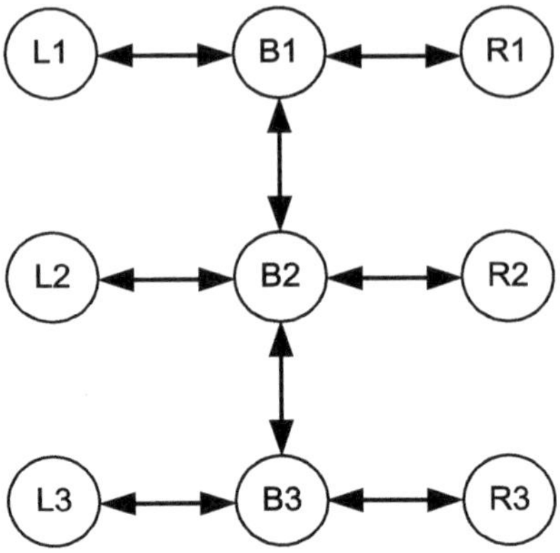


**Fig. 4.2** Network topology: the undirected diffusive tree graph maps the spatial arrangement of the hexapod legs. This is composed of a central backbone ($B_1$, $B_2$, $B_3$) and two side chains, each neuron here controlling a single leg ($L_1$: anterior left leg motor neuron; $\cdots$; $R_3$: posterior right leg motor neuron)

**Table 4.1** Phase displacements of each leg with respect to leg $L_1$ for specific gaits

| Gait | $\Delta\phi_{L_1,B_1}$ | $\Delta\phi_{R_2,B_2}$ | $\Delta\phi_{L_3,B_3}$ | $\Delta\phi_{R_1,B_1}$ | $\Delta\phi_{L_2,B_2}$ | $\Delta\phi_{R_3,B_3}$ |
|---|---|---|---|---|---|---|
| Slow | 0° | 60° | 120° | 180° | 240° | 360° |
| Medium | 0° | 90° | 180° | 180° | 270° | 360° |
| Fast | 0° | 0° | 0° | 180° | 180° | 180° |

Refer to Fig. 4.2 for details

Summarizing the relation above introduced, our reference dynamical system has a reaction diffusion nonlinear dynamics (Eq. 4.5), is made of $n_c$ identical cells of the type Eq. (4.3), locally connected through the laplacian matrix L and is organised as an undirected tree graph. This graph represents the set of motor neurons controlling the locomotion pattern in our walking machine: each leg will be associated to a given cell of the graph, as in Fig. 4.2. Given that each leg motion will be controlled by the oscillation generated by one cell, the locomotion gait is defined through the relative phase displacement among the different cells associated to the legs. So, according to a given gait, we'll impose specific phase shifts among the oscillators (Table 4.1), as in Eq. (4.4).

The results from Partial Contraction will be applied to such a network to obtain rigorous results on the stability of the phase shifts that will have to be imposed to obtain specific gaits. According to this theory [18], a nonlinear dynamical system is said to be contractive if perturbations in initial conditions or over trajectories are exponentially forgotten. Here we'll focus on Partial Contraction: this starts from the existence of a flow-invariant subspace $\mathcal{M}$, defined as a linear subspace such that any trajectory of the system which starts in $\mathcal{M}$, remains there forever: this is the trapping region that we define for the system steady state dynamics. The details on the methodology can be found in [20, 22]: the core of the partial contraction theory is to apply contraction tools to virtual auxiliary systems.

Once defined the flow invariant subspace $\mathcal{M}$, the methodology is based on the concept that if the diffusive side of the network system, projected on its orthogonal complement $\mathscr{V}$ is sufficiently higher than the system autonomous incremental dynamics: then the system behavior will be trapped on $\mathscr{M}$.

In our case, referring to Fig:4.2, $\mathscr{M}$ becomes:

$$\mathscr{M} = [R(\phi_{L1,B1})R(\phi_{L1,R1})R(\phi_{L1,L2)})R(\phi_{L1,R2})R(\phi_{L1,L3)})R(\phi_{L1,R3)})]^T$$

$\mathscr{M} \in \mathbb{R}^{n_t \times n_c}$ Here each block describes the desired phase shift with respect to the reference node $L_1$ ; also the traditional composition of rotational matrices applies; for instance we have, looking at Fig. 4.2 that:

$$R(\phi_{L1,L2}) = R(\phi_{L1,B1}) \cdot R(\phi_{B1,B2}) \cdot R(\phi_{B2,L2})$$

For generic network topologies the construction of such subspace is in general more complicated than in our special case of an undirected tree-like structure.

According to the Partial Contraction Theory, once defined $\mathscr{M}$ for each particular gait, we have to build its orthogonal complement $\mathscr{V}$ (in our case $\mathscr{V} \in \mathbb{R}^{n_t \times (n_t - n_c)}$. One fundamental result from Partial contraction (in our case of balanced, undirected network and positive semidefinite Laplacian matrix), can be summarized in the following relation:

$$k \cdot \lambda_{min}(\mathscr{V}^T \cdot L \cdot \mathscr{V}) > \sup_{x_i,t} \lambda_{max}\left(\frac{\partial f}{\partial x}(x_i, t)\right) \quad (4.7)$$

This formula provides a minimum value for the control gain $k$ in Eq. (4.5) such that the coupling among cells (represented by the projection of the laplacian on to $\mathscr{V}$) is enough to dominate the incremental dynamics of the uncoupled system (represented by the second member in Eq. 4.7).

This means that the dynamics of the system cannot expand in other directions than $\mathscr{M}$, so performing the desired phase-locked synchronization [19].

The choice of $k$ depends on the dynamics of the uncoupled system, through its $\lambda_{max}$, on the laplacian matrix L and on the orthogonal subspace $\mathscr{V}$, which in turn depends on $\mathscr{M}$. So, as the gait to be imposed changes, supposing that the network is unchanged, in principle a new $k$ value should be calculated through the following steps:

1. define $\mathscr{M}$ and consequently the laplacian L;
2. calculate $\mathscr{V}$;
3. calculate $\lambda_{min}(\mathscr{V}^T \cdot L \cdot \mathscr{V})$;
4. choose $k_{min}$ as in Eq. (4.7).

In [22] a useful step ahead was performed. This is summarized in the following:

**Theorem 2.1** *For the nonlinear dynamical system in Eq. (4.5), defined on a given undirected diffusive tree graph, for any imposed flow-invariant subspace $\mathscr{M}$, (i.e. for any phase shift among the oscillators), solutions of system 4.5 converge exponentially to $\mathscr{M}$ if it holds:*

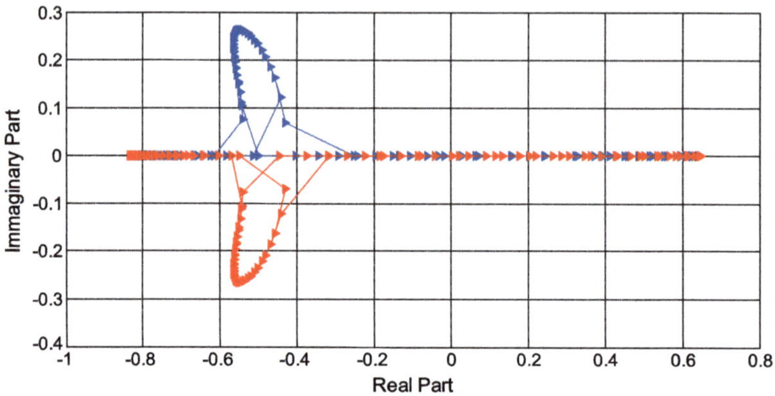


**Fig. 4.3** Trajectory of the Jacobian eigenvalues $\lambda_1, \lambda_2$ for the uncoupled system

$$k \cdot \lambda_1 > \sup_{x_i,t} \lambda_{max}\left(\frac{\partial f}{\partial x}(x_i, t)\right) \tag{4.8}$$

*where $\mathscr{V}$ is the orthonormal complement associated to $\mathscr{M}$ and*

$$\lambda_1 = \lambda_{min}(\mathscr{V}^T \cdot L \cdot \mathscr{V}) = \lambda_{min}(L)$$

*$\lambda_1$ is defined as the algebraic connectivity of the graph according to Fiedler's theory [24].*

The proof is given in [22].

**Remark** This simple result is really useful: we have a sufficient condition on a unique minimum k value that guarantees global exponential convergence to any desired locomotion gait. This value can be a priori determined by simply knowing the algebraic connectivity of the tree-graph (which is an invariant for that graph), and the maximum value, along the uncoupled trajectory of the dynamical system, of its largest Jacobian eigenvalue (see [22]). Once defined this value for k, we can start with any desired locomotion gait, then switch the connection matrix from one gait to any other desired one, having assured exponential convergence of the system trajectories to the new imposed phase-locked configuration. The result is particularly useful for freely choosing also other locomotion gaits which are in between the three more classical ones, i.e. slow, medium and fast. In fact a continuum transition among locomotion patterns is observed in various insects species, like stick insects.

This result drawn from simple considerations after [20] leads to important consequences on the global control of any CPG gait within the same neural topology. In the case of Fig. 4.2, the tree topology adopted reflects the schematic structure of a hexapod: the legs on the right and left side are connected to a central chain that we call backbone. In this case, numerical evaluation of the relevant parameters in Eq. 4.8 gives: $\lambda_{max} = 0.6464$, $\lambda_1 = 0.2679$ and then $k_{min} = 2.4110$. In Fig. 4.3 the Jacobian eigenvalues for the uncoupled cell are depicted.

**Fig. 4.4** The Minihex structure and a model of the Minihex implemented by using ODE dynamic simulator

The simulation results were performed using a dynamic simulation environment based on the Open Dynamic Engine platform [25]. The structure realised consists of a 12 degrees of freedom hexapod robot, whose physical parameters strictly resemble the Minihex, an hexapod robot used for the experimental campaign (see Fig. 4.4). In particular, the signals generated by the outer neurons of the network in Fig. 4.2 are used to set the position values of the twelve servomotors used as the robot leg actuators.

We have simulated the dynamics of the coupled system choosing $k = 2.5$ in Eq. (4.5), whereas the phase shifts are selected in accordance to Table 4.1. The nodes in the backbone are synchronized in phase. The results obtained are shown in Fig. 4.5. The waveforms representing the control signals coming from the six outer neurons of Fig. 4.2 are depicted at the top of the figure, whereas the corresponding stepping diagrams are depicted at the bottom: the hexapod migrates through different gaits. Starting from the slow gait, the pattern is changed toward first the medium and then the fast gait. After the switching between two gaits, via the imposition of new L parameters, the short settling time to the new gait can be appreciated.

The same strategy has been used to control the Minihex, using a position control for the twelve actuators, as mentioned above when dealing with the dynamic simulator. A video depicting the robot motion is reported in the web page [26].

## 4.3 Steering Control

The CPG control is essentially an open loop approach to the generation of locomotion in animals. It essentially imposes a particular set of dynamics to the animal limbs without needing, in principle of any kind of return signals.

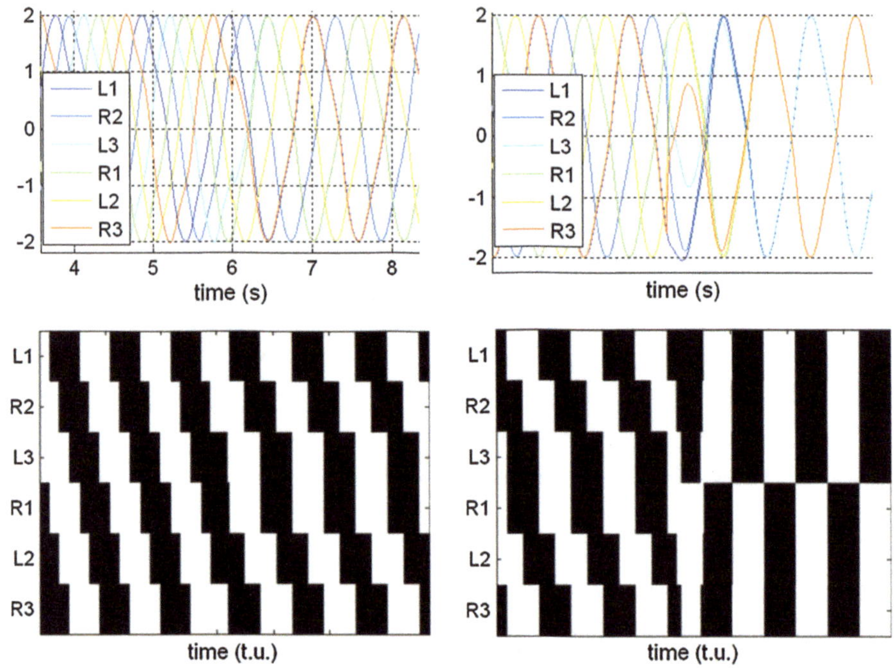


**Fig. 4.5** State variables of the leg actuators (*on the top*) and the stepping diagrams (*on the bottom*) for the six limbs of the hexapod migrating toward different gaits. *Black* bars indicate the swing phase

A classical example is represented by fast escape reactions e.g. in insects, where a stereotyped cyclic motion of the limbs can be easily noticed.

However locomotion can be controlled by high level signals coming from, i.e. visual stimuli, which change the locomotion according to environment stimuli. In this case, high level feedback is used to control the motion direction. This strategy will be discussed below. In walking flies, the steering control evoked by visual stimuli has been deeply investigated. Interesting aspects emerged from the biological experiments and are here summarized: as in straight walking, variation is achieved by concurrent changes in the step length and stepping frequency. Turning is achieved by decreasing the step lengths on the inner side. Specific parameters within the CPG network reported above can be identified at this aim.

## *4.3.1 The Role of Frequency in Steering Control*

A suitable control parameter for experimenting a steering control in the hexapod structure is the intrinsic frequency of particular cells in the graph. This experimental approach is derived from [12]: if we consider two identical neurons with different

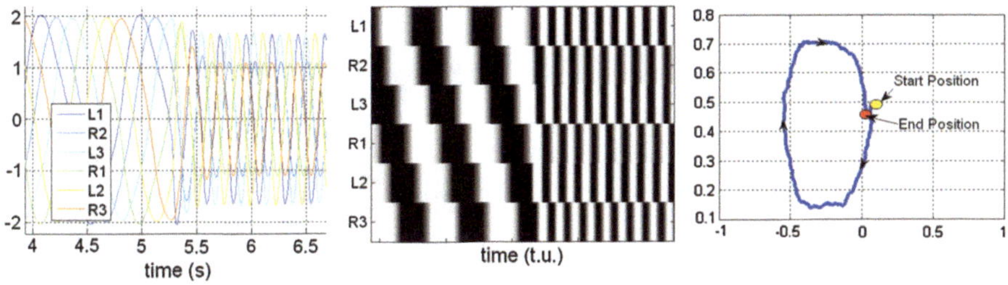


**Fig. 4.6** The waveforms (*left*), the stepping diagram (*middle*) and the body centre of mass position (*right*) when $\gamma = 6$ is applied on the L1, L2 and L3 legs at 5.4s starting from the slow gait, as recorded from the dynamic simulation environment

frequencies $f_1$ and $f_2$, i.e. such that $f_1 = \gamma f_2$, they, once connected will tend to oscillate with the same frequency $\bar{f}$; this in turn is a function of $\gamma$. For $\gamma$ values within the synchronization range, the two neurons will oscillate at a frequency depending on $f_1$ and $f_2$, and they will be synchronized with $\Delta\Phi = \Delta\Phi_L + \Delta\Phi_\gamma$, where $\Delta\Phi_L$ is the phase displacement set in the Laplacian matrix L and $\Delta\Phi_\gamma$ is a phase-shift due to $\gamma \neq 1$. In [12] an analytical relation among two harmonic oscillators was reported, which, the case of our oscillator in Eq. 4.3, becomes [21]:

$$\gamma = \left(\frac{\bar{\omega} + sin(\Delta\Phi_\gamma)}{\bar{\omega} - sin(\Delta\Phi_\gamma)}\right)^{-1} \tag{4.9}$$

This effect is really interesting: phase displacement among the neurons can be studied as an effect of the frequency disturbance of a neuron in a chain which propagates as an effect of diffusion.

At this aim it is enough to modify the cell dynamics as follows:

$$\dot{x}_{i,1} = \gamma_i(-x_{i,1} + (1+\mu)y_{i,1}) - sy_{i,2} + i_1), \tag{4.10a}$$

$$\dot{x}_{i,2} = \gamma_i(-x_{i,2} + (1+\mu)y_{i,2}) + sy_{i,1} + i_2) \tag{4.10b}$$

where the frequency modulation is directly applied to the reactive dynamics of the network i.e to the second term of the equation above. The diffusion coupling makes the whole network to find a new equilibrium condition, which includes an amplitude difference among the oscillators.

Figure 4.6 shows the dynamics of the neurons controlling all the hexapod legs. During the straight trajectory the locomotion pattern matches the slow gait and no frequency control is applied. Around $t = 5.4$s a gain $\gamma = 6$ is applied to the left side legs. This causes, as reported in the figure, a reduction in the step length on the contralateral side. In fact, as the frequency of a side is increased, the signal amplitude of the contralateral legs decreases and, concurrently, a new frequency and phase equilibrium is achieved. This effect is an induced deviation on the pre imposed phase which fades off with the resetting of $\gamma$ to its pre-defined values.

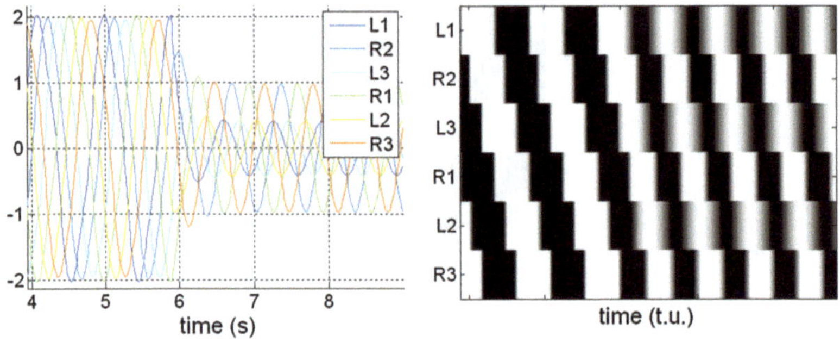


**Fig. 4.7** Steering control based on coupling weights starting from the slow gait. The lower amplitudes for the signals having $w_{ij} < 1$ can be appreciated; the invariance of the stepping diagram during the steering time can be also drawn. The bottom shows the turning behavior of the simulated robot

Experimentally $\gamma > 1$ causes a phase lead on the stimulated legs. Figure 4.6 reports also the trajectory performed by the simulated robot implementing a slow locomotion pattern with steering. The simulation results obtained with this approach fairly match biological observations in walking flies. Further details can be found in [22].

### *4.3.2 The Role of Coupling Weights in Steering Control*

Another strategy for the efficient steering control lies in adding coupling weights within the laplacian operator, i.e. assuming a dynamics of the type:

$$\dot{x}_i = f(x_i, t) + k \sum_{i \neq j \; j \in N_r} (R_{ij} w_{ij} x_j - x_i) \tag{4.11}$$

where $w_{ij} < 1$ are used in the connections of the neurons $j$ belonging to the same side as the turning direction. It was experimentally found [22] that the weight modulation has no effect on the pre-existing phase equilibrium: so the imposed gait is preserved. On the other side the frequency changes, although the effect is lighter than applying the frequency gain-based strategy.

Figure 4.7 depicts the waveforms for the labeled neurons in Fig. (4.2) and the locomotion pattern during the steering period for the fast gait. The value $w_{ij} = 0.5$ was selected for the connection linking the $L1$, $L2$, $L3$ limbs to the backbone, and $w_{ij} = 1.5$ for connection on the opposite side. We can note that the effect of the $w_{ij}$ parameters is the same in terms of frequency and amplitude variations for the slow gait. The turning behaviour obtained adopting this strategy on the dynamic simulation environment is depicted in Fig. 4.8 by recording the robot center of mass.

If a summarizing comparison can be made on the two approaches to steering control, the frequency based approach employs a slightly more complex structure;

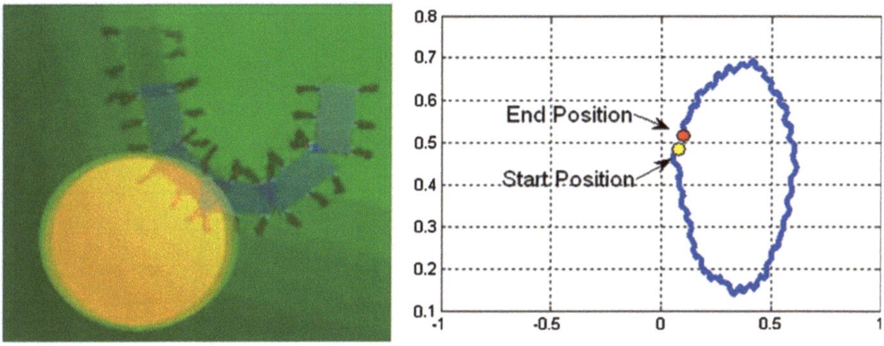


**Fig. 4.8** Steering control based on coupling weights: trajectory of the robot center of mass

it also involves a modulation of the speed of the different leg actuators; the effect is considered as a perturbation on the former conditions. On the other hand, the weight based steering approach involves a simpler structure, the ongoing gait retains its stability conditions, so the locomotion pattern is not affected by any additive phase error. Moreover, the new steady state frequency is very near to the one of the uncoupled oscillators.

### *4.3.3 Towards More Complex Topologies*

The phase induced modulation deriving by a frequency control opened the way to think at sophisticated control laws that could include a minimal number of parameters for locomotion control. In particular, once the network parameters are chosen to show a steady state locomotion pattern (for instance the fast gait), and once the desired phase-shift synchronization is reached, following the guidelines described earlier, our aim is to generate the transition toward other gaits, such as the medium and the slow one, without modifying the laplacian terms, but exploiting the phenomenon of a frequency induced phase shift modulation in the whole network. This is expected to take place by modulating the frequency gain of only a few neurons (called *drivers* in the following). The network needs to be slightly enlarged, with the addition of a more efficient control. The inclusion of a suitable symmetric number of interneurons (i.e. neurons not subject to frequency control) will act so as to lead to a kind of diffusion effect in the phase shift, useful to obtain the imposed phase displacements to reach the desired gaits in the drivers. The network so assumes the form depicted in Fig. 4.9. The connections among the cells are bidirectional and the arrows on them indicate the direction of the phase-shift propagation. The drivers, outlined by the bold arrows in Fig. 4.9, will be the controllers for each leg of the hexapod. Using the same strategy as in the previous network the laplacian matrix, in the case of the slow gait, takes on the following form, where the subscripts express the amount of phase rotation in

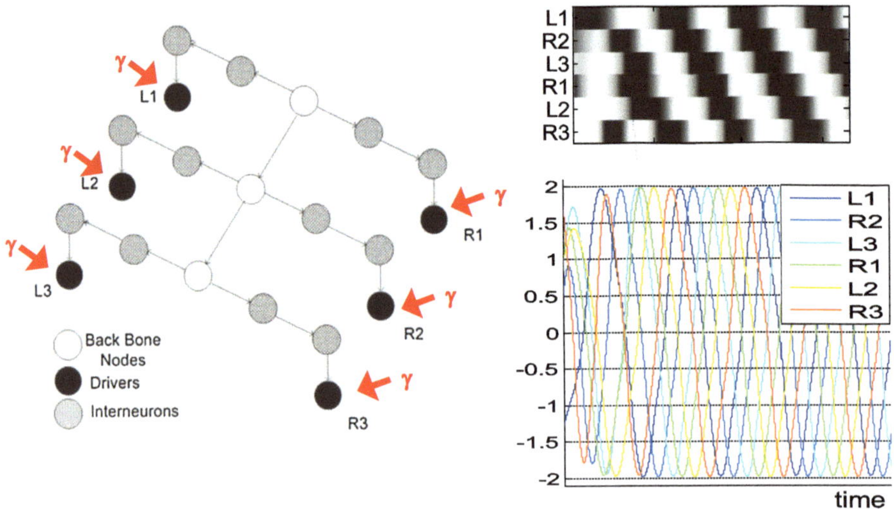


**Fig. 4.9** Left side: the increased topology of the hexapod where the driver neurons are outlined. Right side: the stepping diagram and the state variables generated by the motor neurons

degrees. The $L_1$ neuron is the reference for the calculation of the laplacian phase shifts. The order of construction for $L$ rows is: $[L_1\ B_1\ R_1\ L_2\ B_2\ R_2\ L_3\ B_3\ R_3]$.

$$L = \begin{bmatrix} I & -R_0 & 0 & 0 & 0 & 0 & 0 & 0 & 0 \\ -R_0 & 3I & -R_{180} & 0 & -R_0 & 0 & 0 & 0 & 0 \\ 0 & -R_{-180} & I & 0 & 0 & 0 & 0 & 0 & 0 \\ 0 & 0 & 0 & I & -R_{-240} & 0 & 0 & 0 & 0 \\ 0 & -R_0 & 0 & -R_{240} & 4I & -R_{60} & 0 & -R_0 & 0 \\ 0 & 0 & 0 & 0 & -R_{-60} & I & 0 & 0 & 0 \\ 0 & 0 & 0 & 0 & 0 & 0 & I & R_{-120} & 0 \\ 0 & 0 & 0 & 0 & -R_0 & 0 & -R_{120} & 3I & -R_{300} \\ 0 & 0 & 0 & 0 & 0 & 0 & 0 & -R_{-300} & I \end{bmatrix} \tag{4.12}$$

For this configuration the other parameters have the following values: $\lambda_1 = 0.087$; $\sup_{x_i,t} \lambda_{max}\left(\frac{\partial f}{\partial x}(x_i,t)\right) = 0.646$; $k_{min} = 7.36$.

The frequency control, by using the $\gamma$ parameter as in Eq. (4.10), is applied to the network in Fig. 4.9, to realize migration among different gaits: $\gamma$ can be in fact used as a parameter useful for controlling the connected system towards a desired phase configuration using the smallest number of modulation parameters. We show through simulations, that in the case of the hexapod, we'll succeed in migrating from the slow to the medium and to the fast gait acting only on the $\gamma$ value of four driver neurons. However, the network topology will have to be modified,

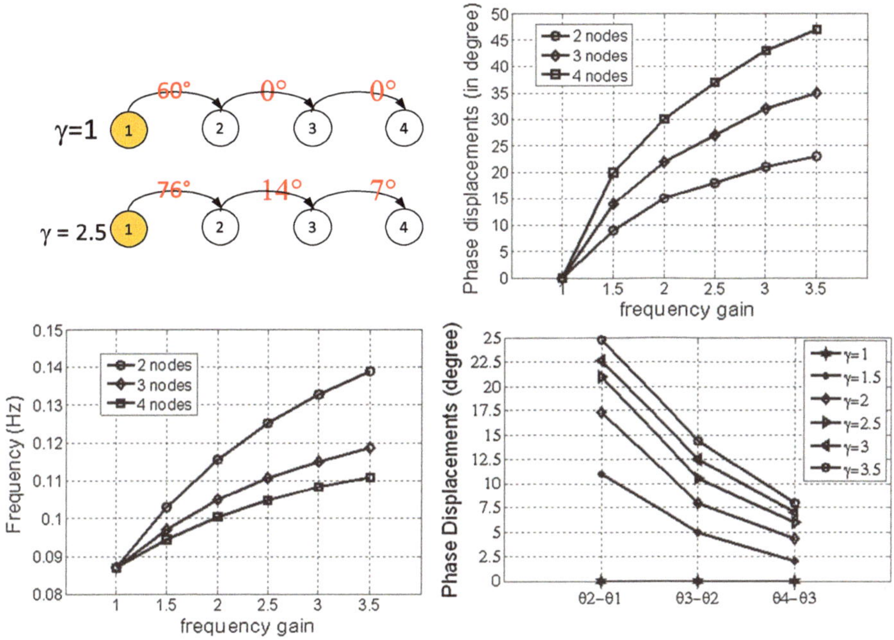


**Fig. 4.10** Top left: effect of the induced phase displacement from a basic phase configuration, for $\gamma = 1$ to a new phase displacement when $\gamma = 2.5$, for a four neuron chain, where the first neuron is the driver. The other pictures report: the absolute phase displacement obtained between the first and the last neuron in chains with 2,3 and 4 nodes (*top right*); the steady state frequency for the whole chain (*bottom right*); the phase displacements between each couple of neurons of a chain with six nodes

as shown below. Figure 4.10 depicts the result of the simulation analysis referring to a simple example, which shows how the control method works. Let us consider a chain of neurons, where only the first one is the driver (see Fig. 4.10). With respect to the imposed phase shift when $\gamma = 1$, if $\gamma = 2.5$ we have the diffusion of a phase displacement from the driver to the outer neurons (see Fig. 4.10) whose effect depends on the number of neurons in the chain. Figure 4.10 top right reports the cumulative phase displacement, as a function of the neuron number in the chain, whereas in the bottom left of the same figure the frequency gain for the steady state frequency attained in the whole chain is reported. For a four neuron chain, the bottom right panel of Fig. 4.10 reports the relative phase shift, parameterized with $\gamma$. The phase displacements monotonically decrease moving away from the driver. Moreover, increasing the interneuron number causes a decrease in the steady state frequency for the connected network and concurrently an increase in the cumulative phase displacement. If this chain takes the role of one side-chain in a wider network representing a CPG, the results above lead to a compromise between the number of oscillators in each side chain (which should be as small as possible), and the residual phase diffusing to the other side chains of the network. The goal is to reach a phase

**Table 4.2** Frequency gain on the drivers to reach specific gait: $\gamma_0 = 3.69$

| Gait | $\gamma_{L1,R1}$ | $\gamma_{L2,R2}$ | $\gamma_{L3,R3}$ |
|---|---|---|---|
| Slow | 1 | 1 | 1 |
| Fast | 1 | $\gamma_1 = 0.65\gamma_0$ | $\gamma_0$ |
| Medium | 1 | $1/0.65\gamma_1$ | $1/\gamma_1$ |

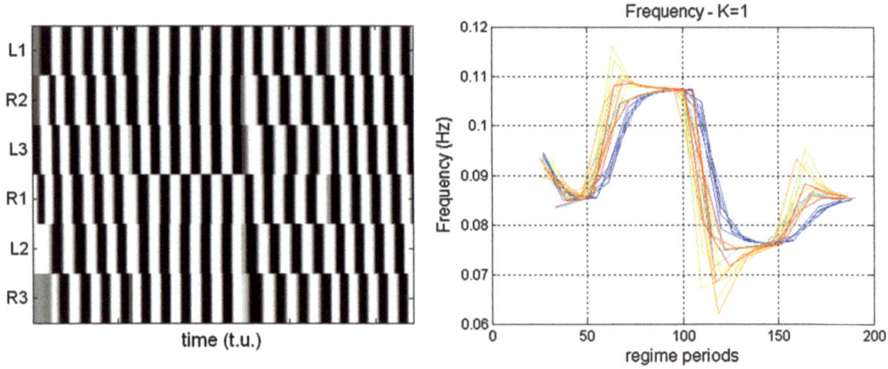


**Fig. 4.11** Left: stepping diagrams as a function of $\gamma$, which is changed step-size as in Table 4.2. The migration from the slow to the fast, through intermediate speed gaits can be appreciated through the progressive phase lead the L2-R2 and L3-R3 modules with respect to L1-R1

shift effect limited within the side chain where the driver acts. In this way, indeed, every leg can be driven autonomously, minimizing the action of the drivers outside the corresponding side chain. The network in Fig. 4.9 is the final result. For our experiment on the Minihex we took into account the drivers outlined in Fig. 4.9. The $\gamma$ values for each selected driver in order to generate the desired gaits are reported in Table 4.2.

Taking advantage of the symmetry of the CPG network we can fully control the locomotion pattern by selecting only one control parameter ($\gamma_0$ in Table 4.2) and acting on the frequency of a minimal number of drivers. The results are summarized in Fig. 4.11 for the three different gaits: these are obtained controlling the frequency gain altogether in only four drivers in the network (Fig. 4.9).

The right side in Fig. 4.11 depicts the trend of the frequencies of all the 21 neurons, where the frequency is changed step-size at given times. The frequency is evaluated at each cycle of the L1 oscillator. It can be seen that, as soon as the drivers are frequency-modulated, a diffusion driven transient takes place, which leads to a steady state condition where all the oscillators reach a common frequency. This transient can be also seen in the stepping diagram (left side of the figure), where the corresponding phase migration is appreciated. In particular, given that L1 and R1 are the reference oscillators, a phase lead in the L2-R2 and L3-R3 modules with respect to L1-R1 is observed in the first migration from the slow gait to the fast, and then a lag, from the fast gait back to the medium through intermediate speed gaits.

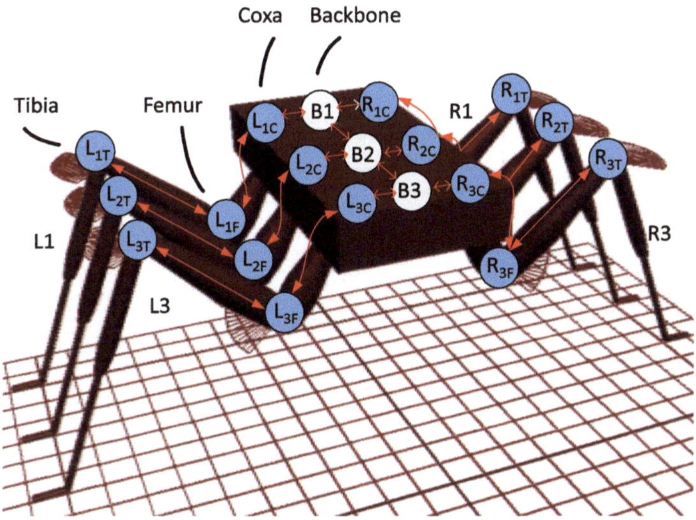


**Fig. 4.12** The improved neural network for steering

**Remark** The network topology has been experimentally designed so that the desired phase displacements are obtained among the driver neurons, which, in a topographical view, match with the feet of a hexapod robot structure. Since the driver neurons map the hexapod legs, the $\gamma$ value can depend on environmental feedback signals. The original CPG concept, conceived as an open loop control structure, can be easily extended in order to host a direct dependence of the selected gait on the low level environmental feedback as in [27].

## 4.4 Improving Robot Dexterity

According to the results derived in the previous Section, any desired locomotion pattern can be imposed, starting from any one and then switching to the other. Moreover also efficient steering strategies can be obtained via frequency control. This Section aims at facing the steering strategy by modulating the relative phase displacements among the joints within each single leg. However, we do not increase further the network topology: we adopt the same network structure as in Fig. 4.9, but in this case the three neurons in each of the different side chains control the three different joints which constitute the single robot leg, which in this case, possesses three degrees of freedom. The scheme is reported in Fig. 4.12.

Arbitrary turning trajectories in the walking robot can be obtained with a suitable phase control retaining the stability conditions on the overall neural structure: this allows the suitable degree of dexterity requested by the legs. The topology retains the backbone and the six side chains (corresponding to the different legs) as the previous

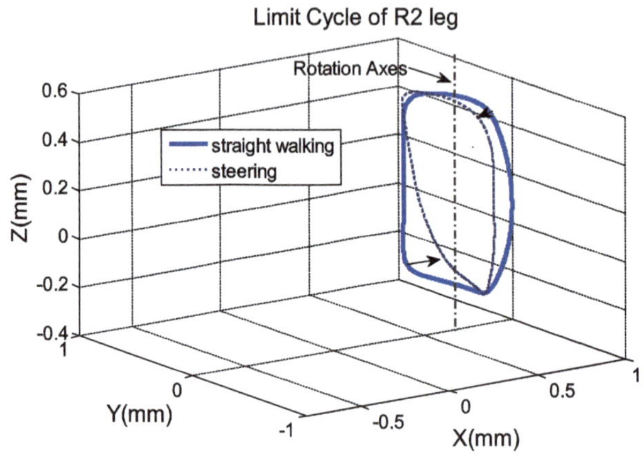


**Fig. 4.13** The regular limit cycle shown by leg $L_2$ and its phase shift induced modulation for implementing left turning $\phi_{FT} = 60°$

scheme. Each of these is made-up of three neurons, each one actuating the coxa, femur and tibia joint, respectively. The proposed control method uses the central part of the network (including the backbone, the coxa neurons and their synapses in Fig. 4.12) to fix the gait (slow, medium or fast) in terms of phase shifts in the stepping diagram, whereas the outmost neurons (including the femur and tibia neurons for each leg) are used to control steering. We experimentally show here that by modulating the femur-tibia synaptic connections, an efficient steering can be implemented. Being the single leg kinematics more complex, the design was performed via the Matlab Robotic Toolbox, whereas a dynamic simulation environment was employed to assess the steering control performance. The steady state behavior shown by the concurrent actuation of the three degrees of freedom (i.e. the concurrent dynamics of the three neurons) in each leg resulted in a regular limit cycle in the working space (Fig. 4.13). Details on the design can be found in [28].

This particular topology allows us to split the problem of gait selection and that one of steering implementation. In details, the three neurons of the backbone and the six of the coxa joints are used to impose, through appropriate phase shifts identical to those ones reported in Table 4.1, the selected gait (first row of Table 4.3). The phase shifts among the other neurons are modulated to generate, within the same selected gait, the desired steering control. In this case of steering control over a slow gait, we have to simply add particular phase rotations only on the connections Femur-Tibia ($\Delta\Phi_{F,T}$), as reported in the third row of Table 4.3. The approach uses only one control parameter: $\Delta\Phi_{F,T} = \Delta\Phi_{steering}$. Figure 4.15 shows the relation between the phase displacement imposed on $\Delta\Phi_{F,T}$ and the radius of the curved trajectory made by the robot in the XY-plane.

Figure 4.14 shows how large can be the effect in the orientation of the stance phase trajectory for a given leg, as a function of $\Delta\phi_{F,T}$. In the figure, circles (squares) represent the anterior (posterior) extreme position of the leg. Various orientations of

**Table 4.3** Phase Shift among joints during forward slow walking and left steering

| | $\Delta\phi_{B_1,L_{1C}}$ | $\Delta\phi_{B_2,R_{2C}}$ | $\Delta\phi_{B_3,L_{3C}}$ | $\Delta\phi_{B_1,R_{1C}}$ | $\Delta\phi_{B_2,L_{2C}}$ | $\Delta\phi_{B_3,R_{3C}}$ | $\Delta\phi_{C,F}$ | Left $\Delta\phi_{F,T}$ | Right $\Delta\phi_{F,T}$ |
|---|---|---|---|---|---|---|---|---|---|
| Forward slow gait | 0° | 60° | 120° | 180° | 240° | 360° | 90° | 0° | 0° |
| Steering left gait | 0° | 60° | 120° | 180° | 240° | 360° | 90° | $+\Phi_{steering}$ | $-\Phi_{steering}$ |

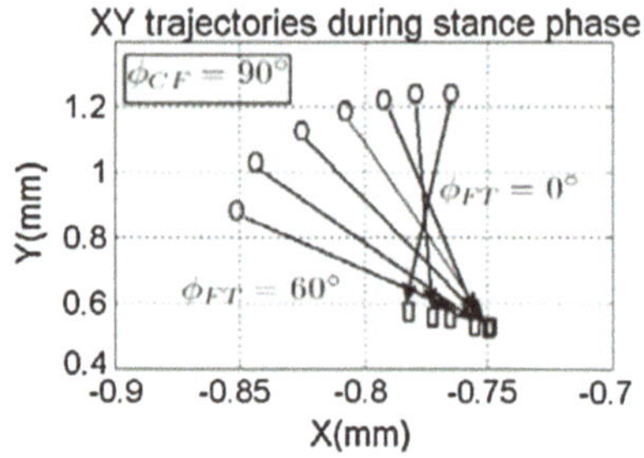


**Fig. 4.14** Step length and amplitude modulation for the leg $L1$ as a function of $\phi_{FT} \in [0 : 60°]$, maintaining $\phi_{CF} = 90°$. Variations are reported at steps of 10°

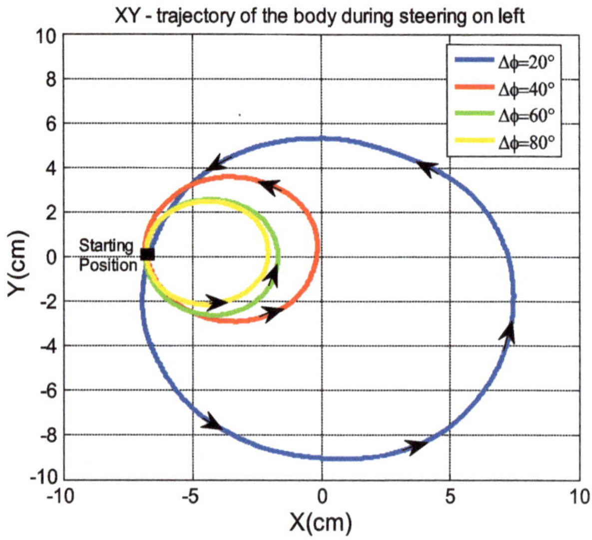


**Fig. 4.15** Trajectory of the body during left steering

the leg stride can be therefore controlled by the joint modulation of that parameter, making the overall robot motion similar to any desired trajectory experimentally found in insects. Therefore we have implemented an efficient steering control acting on only one parameter, retaining the issue of phase stability.

The experimental results have been tested in the Open Dynamic Engine (ODE) dynamic simulation environment to prove their validity. The simulated robot legs, in this case, were built strictly resembling the real anatomy of Drosophila leg, and Open Dynamic Engine libraries [25] were used to simulate motion in an environment whereas contact and friction forces act on the robot.

In Fig. 4.15, the temporal positions of the body for different values of $\Phi_{steering}$ are reported. Videos on these results are available in [29]. A series of snapshots drawn by this video is depicted in Fig. 4.16.

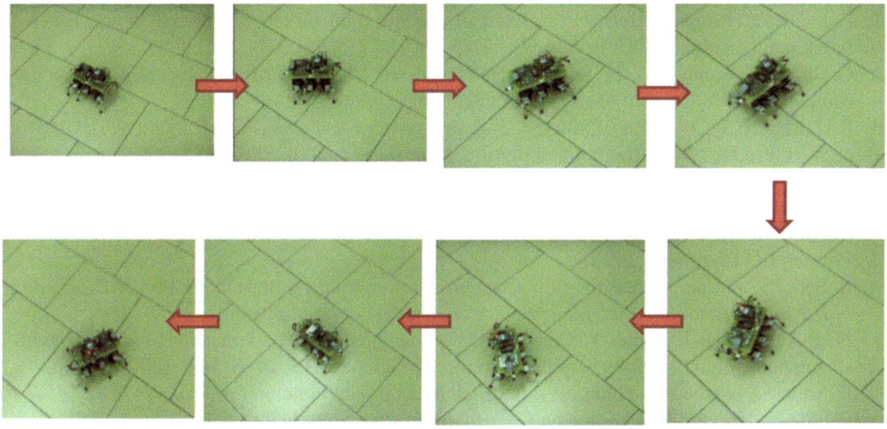

**Fig. 4.16** Snapshots showing the Minihex left steering

## 4.5 Conclusions

This Chapter discussed the fundamental aspects related to locomotion control in a multipodal robotic structure. The theory of Partial contraction has been exploited to guarantee global exponential phase convergence to any given locomotion gait imposed to the designed CPG structure. Moreover, different topologies were presented, and different approaches were discussed both to realize both migration through different locomotion gaits, and to implement efficient steering control. The theoretical analysis was performed referring to the particular neural structures adopted for the locomotion control in our legged prototypes; simulation and experimental results were also reported.

## References


1. G.N. Orlovsky, T.G. Deliagina, S. Grillner, *Neural Control of Locomotion* (Oxford Press, Oxford, 1999)
2. B. Klaassen, R. Linnemann, D. Spenneberg, F. Kirchner, Biomimetic walking robot scorpion: Control and modeling, in *9th International Symposium on Intelligent Robotic Systems*, pp. 101–108 (2001)
3. B. Klaassen, F. Kirchner, D. Spenneberg, A biologically inspired approach towards robust real world locomotion in an eight legged robot, in *Neurotechnology for Biomimetic Robots*, ed. by J. Ayers, J. Davis, A. Rudolph (Mit press edition, Cambridge, 2002)
4. T. Zielinska, J. Heng, Development of a walking machine: mechanical design and control problems. Mechatronics **12**, 737–754 (2002)
5. P. Arena, L. Fortuna, M. Frasca, L. Patané, A cnn-based chip for robot locomotion control. IEEE Trans. Circuits Syst. I 52:1162–1171 (2005)
6. R.D. Quinn, R.E. Ritzmann, Construction of a hexapod robot with cockroach kinematics benefits both robotics and biology. Connect. Sci. **10**, 239–254 (1998)

7. F. Fukuoka, Adaptive dynamic walking of a quadruped robot on irregular terrain based on biological concepts. Int. J. Robot. Res. **22**(3–4), 187–202 (2003)
8. G. Taga, Y. Yamaguchi, H. Shimizu, Selforganized control of bipedal locomotion by neural oscillators. Biol. Cybern. **65**, 147–159 (1991)
9. R.D. Beer, H.J. Chiel, R.D. Quinn, K.S. Espenschied, P. Larsson, A distributed neural network architecture for hexapod robot locomotion. Neural Comput. **4**, 356–365 (1992)
10. J.J. Collins, I.N. Stewart, Coupled nonlinear oscillators and the symmetries of animal gaits. J. Nonlinear Sci. **3**, 349–392 (1993)
11. M. Golubtisky, I. Stewart, P. Buono, J.J. Collins, A modular network for legged locomotion. Physica D **115**, 56–72 (1998)
12. K. Seo, J.J. Slotine, Models for global synchronization in cpg-based locomotion, in *Proceedings of IEEE International Conference on Robotics and Automation*, Rome, 2007
13. D. Ryczko, A.J. Ijspeert, A. Crespi, J.-M. Cabelguen, From swimming to walking with a salamander robot driven by a spinal cord model. Science **315**, 1416–1420 (2007)
14. P. Arena, A mechatronic lamprey controlled by analog circuits, in *Proceedings IEEE MED'01 9th Mediterranean Conference on Control and Automation*, Dubrovnik, Croatia, 2001
15. K.G. Pearson, Central programming and reflex control of walking in the cockroach. J. Exp. Biol. **56**, 173–193 (1972)
16. D.L. Glanzman, Ion pumps get more glamorous. Nat. Neurosci. **13**, 4–5 (2010)
17. M. Frasca, P. Arena, L. Fortuna, Bio-Inspired Emergent Control of Locomotion Systems. World Scientific Series on Nonlinear Science, Series A, vol. 48, ISBN 981-238-919-9, 2004
18. W. Lohmiller, J.J Slotine, On metric observers for nonlinear systems, in *Proceedings of IEEE International Conference on Control Applications*, Dearborn, MI
19. W. Wang, J.J. Slotine, On partial contraction analysis for coupled nonlinear oscillators. Biol. Cybern. **92**, 38–53 (2005)
20. Q.C. Pham, J.J. Slotine, Stable concurrent synchronization in dynamic system networks. Neural Netw. **20**, 62–77 (2007)
21. E. Arena, P. Arena, L. Patané, Frequency-driven gait control in a central pattern generator, in *Proceedings of International Conference on Applied Bionics and Biomechanics (ICABB)*, Venice, 2010
22. L. Patané E. Arena, P. Arena, Efficient hexapodal locomotion control based on flow-invariant subspaces, in *18th World Congress of the International Federation of Automatic Control (IFAC)*, Milan, Italy, 2011
23. L.O. Chua, T. Roska, The CNN paradigm. IEEE Trans. Circuits Syst. **40**, 147–156 (1993)
24. M. fiedler, Algebraic connectivity of graphs. Czech. Math. J. **23**, 298–305 (1973)
25. R. Smith, Open dynamics engine, 2010. www.ode.org
26. P. Arena, L. Patané, SPARK II EU project multimedia. www.spark2.diees.unict.it/HexapodalControl.html
27. P. Arena, L. Fortuna, M. Frasca, L. Patané, Sensory feedback in CNN-based central pattern generators. Int. J. Neural Syst. **13**(6), 349–362 (December 2003)
28. E. Arena, P. Arena, L. Patané, Modelling stepping strategies for steering in insects, in *Frontiers in Artificial Intelligence and Applications, Proceedings of the 21st Italian Workshop on Neural Nets*, vol. 234, pp. 275–283
29. P. Arena, L. Patané, Steering videos web page. www.spark2.diees.unict.it/SteeringInsects.html

# Chapter 5
# A Prototype 2N-Legged (insect-like) Robot. A Non-Linear Dynamical System Approach

**E. del Rio and M. G. Velarde**

**Abstract** A nonlinear closed lattice or ring is proposed as a central pattern generator (CPG) for controlling hexapodal robots. We show that the ring composed of six anharmonically interacting units coupled to the limb actuators permits to reproduce typical hexapod gaits. We provide an electronic circuit implementation of the CPG providing the corresponding gaits. Then we propose a method to incorporate the actuator (motor) and leg dynamics in the units of the CPG. With this electromechanical device we close the loop CPG—environment—CPG, thus obtaining a decentralized approach for the leg control that does not require higher level CPG intervention during locomotion in a non-smooth hence non flat landscape. The gaits generated by our CPG are not rigid, but adapt to obstacles faced by the robot.

## 5.1 Introduction

Animal locomotion seems to be driven by a central pattern generator (CPG), which is an intra-spinal network of neurons capable of generating a rhythmic output required for the limb control [1]. The study of CPGs is crucial for understanding both the global animal behavior and specific functions of such neural networks. Besides it is also important for designing neuro-inspired robots capable to move in an efficient manner in almost arbitrary landscapes or environments.

Research on the leg coordination and movement has been shifted from descriptive studies [2] to investigations of the neurophysiological mechanisms and control of

E. del Rio (✉) · M. G. Velarde
Departamento de Física Aplicada, E.T.S.I. Aeronáuticos, Universidad Politecnica de Madrid, Madrid, Spain
e-mail: ezequiel.delrio@upm.es

M. G. Velarde
Instituto Pluridisciplinar, Universidad Complutense de Madrid, Madrid, Spain
e-mail: mgvelarde@pluri.ucm.es

P. Arena and L. Patanè (eds.), *Spatial Temporal Patterns for Action-Oriented Perception in Roving Robots II*, Cognitive Systems Monographs 21,
DOI: 10.1007/978-3-319-02362-5_5, © Springer International Publishing Switzerland 2014

walking [3]. Leg movement seems to be an intrinsic part of the step pattern generator rather than merely the reflection of activity in a higher level CPG [4]. The architecture of CPGs is seldom observable in vivo. However, important aspects of their structure can be inferred from observation of gait features such as the phase of the gait cycle at which a given limb hits the ground. Then phenomenological models reproducing these features can be introduced and used for robot design. Inspired by this idea the use of oscillatory neural networks with different architectures has been suggested [5, 6].

In fact, most animal gaits possess a degree of symmetry and universal features not far from the behavior of lattice rings of coupled oscillators [7]. For instance coupled nonlinear oscillators can be considered as possible models for locomotor CPGs in insects and other animals [8–12]. Then transitions between different gaits can be modeled as switching between different activity patterns in a lattice ring.

Let us now briefly describe the symmetries observable in hexapodal gaits. Figure 5.1 illustrates three gaits of an insect (for more details see e.g. [9]. We use the following convention: the limbs on the left and right sides are numbered starting from the frontal leg and marked by letters L and R, respectively (Fig. 5.1a).

When an insect moves slowly, it normally adopts the so called *metachronal* gait (Fig. 5.1b).This gait can be described as a "wave" propagating forward from the rear of the animal (first on the left side, and then on the rigth side) according to the sequence:

$$L_3,\ L_2,\ L_1,\ R_3,\ R_2,\ R_1.$$

For this gait the adjacent limbs of each half of the insect body ($R_3$ and $R_2$, $R_2$ and $R_1$) are 60° out of phase. The limbs of each segment (e.g. $R_3$ and $L_3$) are half a period (or 180°) out of phase.

*Caterpillar* is a medium speed gait at which the motion of the left and right limbs are in synchrony (Fig. 5.1c) according to the sequence:

$$(L_3R_3),\ (L_2R_2),\ (L_1R_1).$$

We note that in normal circumstances, insects alternate in movement legs sharing the same segment [13], hence they rarely employ this gate.However, its use is frequently observed in centipede-like animals.

For relatively fast displacements, the insect adopts the alternating *tripod* gait (Fig. 5.1d):

$$(L_3L_1R_2),\ (L_2R_3R_1).$$

In the tripod gait, the ipsilateral anterior and posterior legs, and the contralateral middle leg move together in phase. The limbs of each segment are half a period (180°) out of phase and the adjacent limb on the right and left sides are also half a period (180°) out of phase.

Names like tripod or tetrapod gaits may induce misinterpretation as fixed or rigid gaits. This is not the case as insects (and other animals e.g. crustaceans) play free gait, i.e. the gait characteristics are dynamically changing according to the environment.

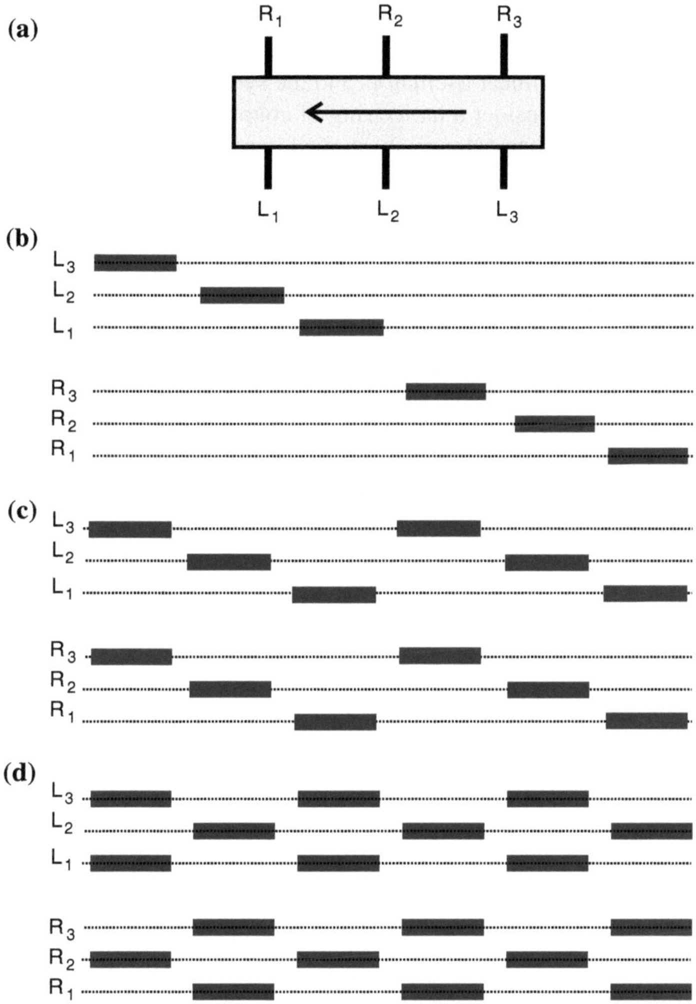


**Fig. 5.1** Typical hexapod gaits. **a** Leg numbering convention. Letters L and R indicate the left and right sides of the agent, respectively, while the subscript stands for the limb number. **b** Metachronal (low—speed) gait. Thick segments show the swing (transfer) phases (or the return stroke), dotted lines correspond the the stance (support) phases(or the power stroke). Only one leg at a time is lifted and moving forward. **c** Caterpillar (medium–speed) gait. **d** Tripod (fast–speed) gait

For example when a small obstacle is on the path the swing phase of a leg can be shifted or made shorter, thus ensuring smoothness in the animal's movements. This feature is a real challenge for robot design since a continuous real time adjustment of the CPG cycles is required, which in turn demands fast and sophisticated sensory-motor loops feedbacking information about the environment. The complexity of such an approach has led to shortcomings in most of present-day available walking robots.

In this report we approach the problem of locomotion control from the nonlinear dynamics viewpoint [14–16]. We take advantage of similarities between waves observed in coupled nonlinear oscillators and the symmetries found in animal gaits. In particular, here we consider a lattice ring of coupled oscillators whose nonlinear units or elements have been extensively studied in recent years. In our recent works [17–19], we have proposed combining a Toda inter-particle nonlinearity [20, 21] with a Rayleigh form of active friction [22] to help sustaining oscillations. We have shown theoretically and numerically the existence and stability of robust propagating waves in such a model problem using a hybrid electro-mechanical analog system.Several hardware implementations of a six-units model have been developed and tested [17]. Here we show that the excitation wave patterns in such a lattice ring of coupled Toda-Rayleigh oscillators have the same symmetries and waveforms as the above mentioned three forward-walking gaits adopted by insects (Fig. 5.1). In earlier works [5, 10, 12] gaits have been considered as fixed oscillatory rhythms. However, as note above insects (and other animals e.g. crustaceans) play free gaits, changing rhythms according to the environment. We propose a module composed by a mechanical leg, its actuator (an electrical motor) and an electronic circuit to form a CPG. In this way we incorporate the actuator and leg dynamics in the CPG hence closing the loop CPG—environment—CPG. Thus our approach departs from the concept of fixed gaits and naturally permits to efficiently perform during locomotion over a non-smooth, non-flat environment on a low CPG level with no direct participation of a possible robot upper "brain" or additional, higher level CPG.

## 5.2 The Leg Kinematics Model

Learning from Nature but not just copying or mimicking it, in this Section we propose a leg kinematics model inspired by the observation of a stick insect (see [23, 24] and references therein). The basic principles upon which we build the model are: (i) Minimization of the energy consumption during walking; (ii) Autonomous behavior of individual legs also in the presence of small obstacles; and (iii) Dynamical coupling of the legs permitting gait adaptability to the environment.

According to biological evidence,a positive feedback in the robot body—coxa ("hip") and femur—tibia ("knee") joints has been postulated [25, 26]. This positive feedback involving *elastic* components during walking leads to conservation of gravitational energy [27] observed in insects. At variance with such positive feedback, in our case conservation of the potential energy will not be due to the existence of elastic components in the hexapod robot, but it comes with the mechanical design of the legs.

In order to minimize the energy loss during (linear translatory) motions of the robot with a constant velocity we impose that the robot potential energy is approximately constant. It implies that the vertical position of the centre of mass of the robot remains at a given horizontal level. This conservation principle is, for example, satisfied by a wheel rotating over a flat horizontal surface with constant velocity. Such motion requires no energy supply. Below we apply the concept of passive dynamics to the

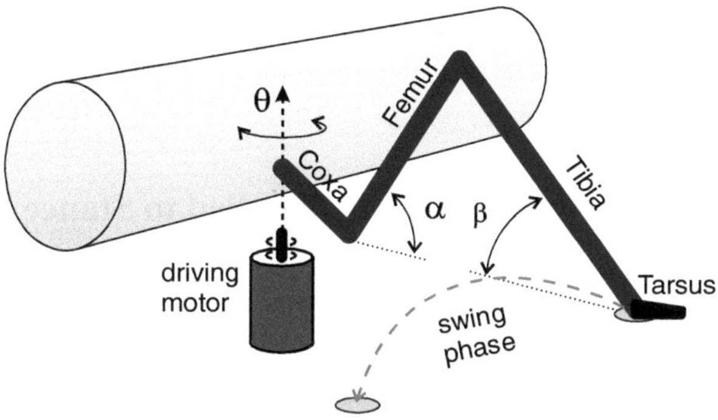


**Fig. 5.2** Schematic diagram showing the main modules of the robot leg inspired by observation of insects. Note the hinge character of the coxa-femur and the femur-tibia joints. A motor inside of the robot body (drawn here outside for easy visualization) drives angle $\theta$ controlling protraction-retraction of the coxa according to the leg moving phase. The other angles are functionally coupled to the evolution of $\theta$ (see main text)

robot design to drive automatically the robot legs in a way that the initially given potential energy of the robot remains unaltered without motor or controller dedicated to it.

## 5.3 The Leg Stance-Swing Cycle

Figure 5.2 sketches the main elements or modules of an insect leg: coxa, femur, tibia,and tarsus. The modules are coupled by joints in a way that the leg has nine degrees of freedom though for walking purposes only three are significant as shown by Cruse and collaborators [4, 28]. For the robot design we assume that the leg has only three degrees of freedom controlled by the angles: $\theta$—protraction-retraction of coxa,$\alpha$—depression-elevation of femur, and $\beta$—flexion-extension of tibia. An individual leg can only be in one of two possible functional states: (i) stance (or support) phase and (ii) swing (or transfer) phase.

In the stance phase the tarsus stays on the ground. The movement of the robot can be achieved by the counterclockwise rotation (in Fig. 5.2) of the angle $\theta$ by a driving motor placed inside the robot body. To keep the tarsus fixed on the ground (no sliding) the protraction of coxa (angle $\theta$) is accompanied by an appropriate continuous adjustment of the angles $\alpha$ and $\beta$. During the swing (or recovery) phase $d\theta/dt > 0$, the femur angle $\alpha$ increases, the leg is elevated over the ground, moves forward, and then the angle decreases again until the tarsus hits the ground, which generates the swing trajectory of the tarsus (Fig. 5.2). Thus the evolution of the angle $\theta$ functionally determines the behavior of $\alpha$ both in stance and in swing phases. This suggests that if a CPG can control just the angle $\theta$ of the legs producing different oscillatory modes (or gaits), then the other angles follow the dynamics of $\theta_n$ ($n = 1, 2, \ldots, 6$ for six legs). Then an appropriate synchronization of the stance-swing cycles of different legs can be considered in terms of synchronization of angles $\theta_n$ only.

## 5.4 Leg Design: Passive Dynamics Applied to Stance Phase

We are using the concept of passive dynamics in the sense of morphology of a mechanical system to reduce the number of actuators (and then the energy) required for motion.

Figure 5.3a illustrates schematically our model leg, where for simplicity we use only two segments: femur and tibia, and neglect the lengths of coxa and tarsus. The behavior observed in insects suggests that for walking over a flat horizontal surface, we can approximately assume that both femur and tibia lie in a vertical plane. At rest, i.e. when the robot stays on a horizontal plane, the elevation of the body over the plane is given by the constant $H_0$ (Fig. 5.3a) and the leg angles fixed by construction are $\theta = 0$, $\alpha = \alpha_0$, and $\beta = \beta_0$.

At each joint we introduce a circular winding reel (Fig. 5.3a). One reel is attached to the body during stance phase and another to the tibia.

According to the angle $\theta$ the former rotates always staying in the tibia-femur vertical plane. A wire connecting the two reels can be winded with no sliding on the reels when joints rotate. This mechanical coupling adds a constraint to the angles $\alpha$ and $\beta$ in the stance phase. We assume the reels to be circular cylindrical with radii $R_{\mathrm{B}}$ and $R_{\mathrm{T}}$ for the body and tibia reels, respectively.Then a change of value in one angle (e.g. $\Delta_\alpha$ in Fig. 5.4) leads to the corresponding change of the other angle (e.g. $\Delta_\beta$):

$$\gamma = \beta + \alpha \rightarrow \Delta\gamma = \Delta\beta + \Delta\alpha, \quad R_B \Delta\alpha = R_t \Delta\gamma \tag{5.1}$$

and finally,

$$\Delta_\beta = (k-1)\Delta_\alpha \tag{5.2}$$

where $k = R_{\mathrm{B}}/R_{\mathrm{T}}$ is the transduction coefficient of the reels. Note that for identical reels the tibia angle $\beta$ remains constant. Thus the mechanical coupling gives

$$\begin{aligned} \alpha_{\text{stance}}(t) &= \overline{\alpha} + \Delta_\alpha, \\ \beta_{\text{stance}}(t) &= \overline{\beta} + (k-1)\Delta_\alpha \end{aligned} \tag{5.3}$$

where $\overline{\alpha}$ and $\overline{\beta}$ are constants to be identified below.

As earlier noted, a leg movement in the stance phase imposes the condition of no tarsus sliding over the ground. Hence, during the stance phase, each the tarsus stays at a constant transversal distance $L$ (in general different for different legs) to the body (Fig. 5.3b) and the femur and tibia angles are coupled during the coxa protraction (changes of $\theta$) by the constraint

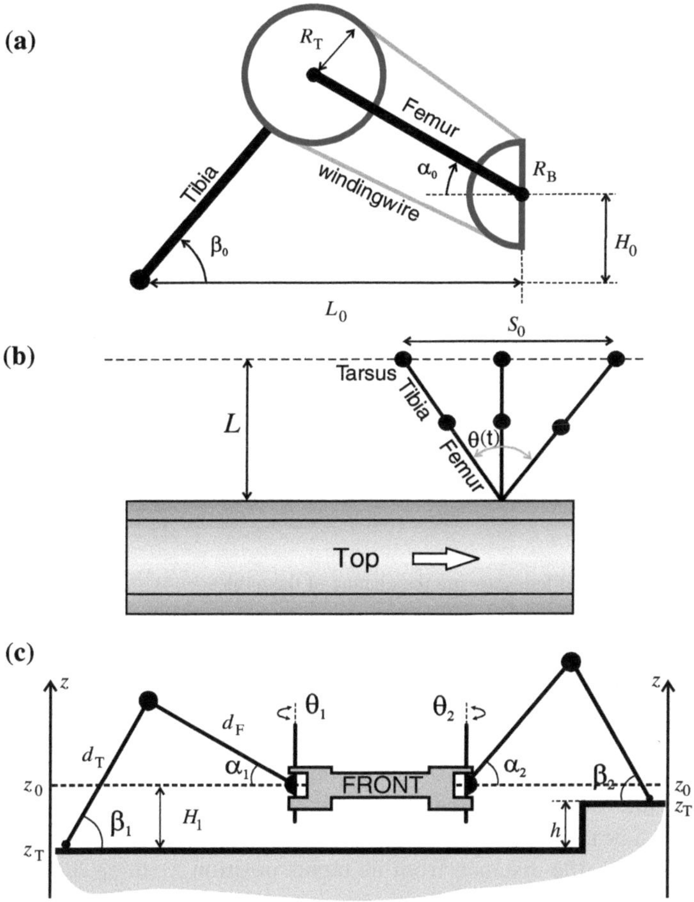


**Fig. 5.3** Schematic mechanics of the robot. **a** The leg configuration at rest. $H_0$ is the elevation of the robot body (body-femur joint) over the horizontal plane. Angles $\alpha_0$ and $\beta_0$ define initial inclination of femur and tibia, and $L_0$ is the distance from the body to the tarsus. The leg at each joint has a circular winding reel (not to scale) fixed to the body (during stance phase) and tibia (always). The reels are coupled through a winding wire. When one joint rotates, the wire winds in one reel while unwinding from the other with no sliding. **b** The robot top view. Relative successive positions of the $L_1$ leg for three different values of $\theta$ during the stance phase. $S_0$ is the robot step size. Note that the leg tarsus is fixed on the ground, while the robot body moves forward. **c** The robot frontal view. $H_n$ is the vertical distance from the $n$th tarsus $z_T$ to the body level $z_0$ (shown only for the anterior right leg, $n = 1$). All legs in the stance phase adopt a configuration of joints, i.e. values of $\alpha$ and $\beta$, such that the body level $z_0$ remains unaltered

$$d_F \cos \alpha(t) + d_T \cos \beta(t) = L / \cos \theta(t) \tag{5.4}$$

where $d_F$ and $d_T$ are the lengths of femur and tibia, respectively.

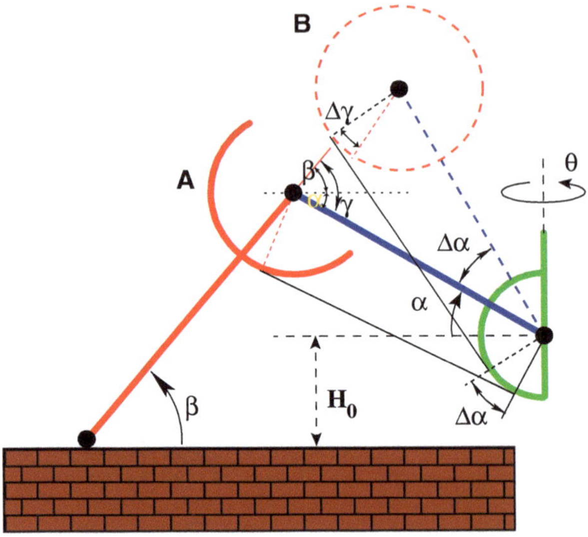


**Fig. 5.4** Two positions of leg, showing the change of the angles

Assuming that the body is the heavier part of the robot (recall that the motors driving the legs are inside the body), we aim at maintaining unaltered the vertical level of the robot centre of mass. This implies that during the robot movement with all legs being in stance phase (i.e on the ground) we should adjust the angles $\alpha$ and $\beta$ in a way that the vertical coordinate of the coxa of the corresponding leg $z_0$ remains constant (Fig. 5.3c).

For a given leg, the distance from its tarsus position $z_T$ to $z_0$ during the stance phase is given by:

$$H(t) = d_T \sin\beta(t) - d_F \sin\alpha(t). \tag{5.5}$$

Note that $H$ can vary from leg to leg and depends on the ground profile, i.e. on the presence of obstacles (Fig. 5.3c, left and right legs). At rest (5.5) is reduced to $H_0 = d_T \sin\beta_0 - d_F \sin\alpha_0$.

Expanding (5.5) in a Taylor series and using (5.3) we obtain

$$H(t) = H^{(0)} + H^{(1)}\Delta_\alpha + O(\Delta_\alpha^2) \tag{5.6}$$

with

$$\begin{aligned} H^{(0)} &= d_T \sin\overline{\beta} - d_F \sin\overline{\alpha}, \\ H^{(1)} &= (k-1)d_T \cos\overline{\beta} - d_F \cos\overline{\alpha}. \end{aligned} \tag{5.7}$$

By setting $H^{(1)} = 0$, we have

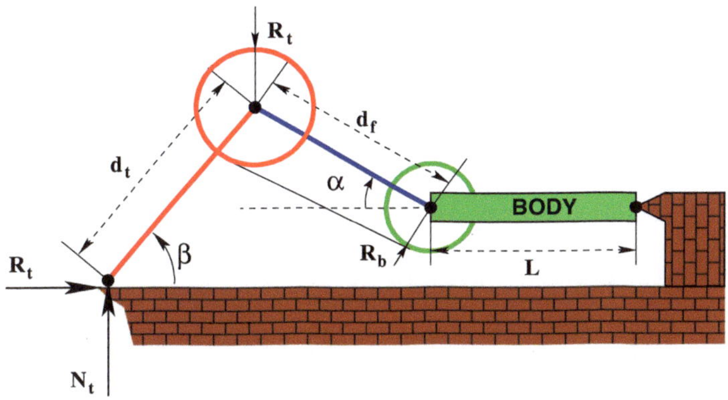


**Fig. 5.5** Rear vision of the body and a leg. The left part of the body are sustaining by the leg in a static equilibrium. $R_t$ and $N_t$ are the forces applied from the ground to the foot

$$\cos\overline{\beta} = \frac{d_{\mathrm{F}}}{(k-1)d_{\mathrm{T}}}\cos\overline{\alpha}. \quad (5.8)$$

With (5.8) satisfied, the variation of $H$ during the stance phase is of second order in $\Delta_\alpha$, which, as we shall see below, gives an error less than 1 % for an acceptable range of $\Delta_\alpha$. It is with such "tolerance" or error bar that we can say that energy is maintained "constant".

Given the height of an obstacle $h$ ($h > 0$ for a bump, and $h < 0$ for a trough), we can find the value of $\overline{\alpha}$

$$H^{(0)}(\overline{\alpha}) + h = H_0 \quad (5.9)$$

where

$$H^{(0)}(\overline{\alpha}) = d_{\mathrm{T}}\sqrt{1 - \left(\frac{d_{\mathrm{F}}}{(k-1)d_{\mathrm{T}}}\cos\overline{\alpha}\right)^2} - d_{\mathrm{F}}\sin\overline{\alpha}. \quad (5.10)$$

## 5.4.1 Statical stability

Note that when a leg is on the ground and the robot moves forward, there are no forces from the leg transversal to the body, because the system is in a flat potential region corresponding to a minimum. This means that a small friction between the tarsus and the ground would be enough to prevent sliding of the tarsus perpendicular to the body.

To study the static forces acting on a leg, let us consider the case shown on Fig. (5.5). In this figure $R_t$ is the friction force needed to maintain the leg-body system of the Fig. (5.5) in the static position of that figure. After the mechanical analysis of the system and lengthy algebraic computation we finally get the following expression for $R_t$

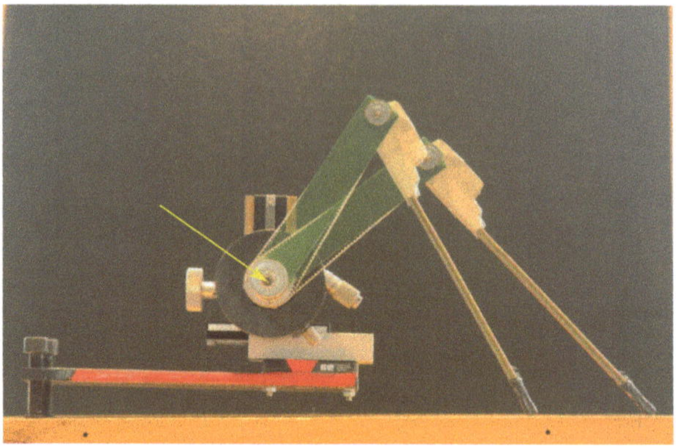

**Fig. 5.6** For different positions of the foot (tarsus), the height of the centre of mass (pointed by *yellow arrow*) doesn't change

$$R_t = \frac{g}{2}\frac{Ld_t(M+2M_f+M_t)(R_b-R_t)\cos(\beta) - d_f\cos(\alpha)\{L(M+M_f)R_t - d_t(M_f+M_t)R_b\cos(\beta)\}}{Ld_fR_t\sin(\alpha)+d_t\{L(R_b-R_t)\sin(\beta)+d_fR_b\sin(\alpha+\beta)\}} \tag{5.11}$$

where

$$C = \frac{d_f}{d_t} \tag{5.12}$$

and $M$, $M_f$ and $M_t$ are masses for body, femur and tibia respectively. Introducing the relation (5.8) in Eq. (5.11) we get

$$R_t = g(M_f+M_t)\frac{C(k-1)}{2}\frac{\cos(\beta)}{\sin(\alpha)+C(k-1)\sin(\beta)} \tag{5.13}$$

Note that the force $R_t$ does not depend on the body mass. This is a consequence of Eq. (5.8) and it means that the mass centre (and the gravity energy) of the body remain constant. Note also that the force in Eq. (5.13) goes to zero if the mass of tibia and femur $(M_f + M_t)$ goes to zero, so in this limit, it there is no friction force with the ground nor energy consumption to be standing.

A prototype with $k = 5/3$ and $C = 7/4$ has been made to check the properties just described (see Fig. (5.6)). There is good agreement with the theory.

For the bellware stretch of Fig. (5.5) we get

$$T_c = \frac{gLd_fd_t}{2}\frac{(M+2M_f+M_t)\cos(\beta)\sin(\alpha)+(M+M_f)\cos(\alpha)\sin(\beta)}{Ld_fR_t\sin(\alpha)+d_t\{L(R_b-R_t)\sin(\beta)+d_fR_b\sin(\alpha+\beta)\}} \tag{5.14}$$

and after using (5.8) we finally obtain

$$T_c = \frac{CgLd_f}{2R_t} \frac{(M + 2M_f + M_t)\sin(\alpha) + C(k-1)(M + M_f)\sin(\beta)}{(L + Ckd_f\cos(\beta))\{\sin(\alpha) + C(k-1)\sin(\beta)\}} \cos(\beta) \tag{5.15}$$

Note that if $R_b \to 0$ y $R_t \to 0$ then $T_c \to \infty$, so it is important to choose good bearings to eliminate friction in the joints of the legs.

## 5.5 Leg Design: Swing Phase

As earlier mentioned we are seeking a leg design permitting autonomously handling small obstacles solely by the lowest leg (intelligence) level. The robot legs should be able to react to the presence of obstacles on the path in a way that the robot body (i.e. position of the leg coxas) keeps constant its vertical coordinate $z_0$ (Fig. 5.3c). This can be achieved only if (5.8) is satisfied for the stance phases of all legs during all steps.

From (5.9) follows that the value of $\overline{\alpha}$ depends on the height of the obstacle under the leg tarsus. During walking over uneven ground, at each step the tarsus may be placed at different levels, unknown in advance. Consequently, the value of $\overline{\alpha}$ differs from step to step. On the other hand, (5.8) must be satisfied and hence the value $\overline{\beta}$ should be adjusted according to (5.8) at each step. This is only possible if the relationship (5.8) is satisfied during the whole swing phase, i.e.

$$\cos\beta_{\text{swing}}(t) = \frac{d_{\text{F}}}{(k-1)d_{\text{T}}} \cos\alpha_{\text{swing}}(t). \tag{5.16}$$

The latter means that the swing phase is not arbitrary, but instead the leg swinging in the air always stays "prepared" for the next stance phase.

A solution is to introduce during the swing phase a mechanical constraint on the leg mechanics to perform a suitable, *ad hoc* swing phase.

Figure 5.7 shows a possible mechanism allowing lifting up the leg over the ground satisfying (5.16) during the whole swing cycle. It consists of the stabilizing bar DC coupled by freely rotating joints to tibia and the lever CA at points D and C, respectively (Fig. 5.7A). The bar has the same length as the femur and stays always parallel to it.

During the stance phase, the sliding rail has two degrees of freedom (Fig. 5.7A) and hence has no effect on the leg dynamics described by (5.3), (5.4), and (5.8).At the beginning of the swing phase, the reel $R_B$ (Fig. 5.3a) becomes free, so (5.3) is no more valid.At the same time the rail is maintained in vertical position by the hammer (Fig. 5.7b), suddenly pushed to the left during the swing phase thus increasing the angle $\alpha_{\text{swing}}$. To complete the swing phase, which is detected by using a ground sensor, the sliding rail is shifted backwards until the tarsus hits the ground, perhaps, at a different vertical level. Then the reel is attached again to make the next stance phase of the leg movement cycle.

During swing the leg has no degree of freedom, and its motion follows the sliding rail displaced by the hammer. From geometrical considerations (Fig. 5.7b) we have

$$\overline{\text{OA}}\cos\beta_{\text{swing}}(t) = \overline{\text{OB}}\cos\alpha_{\text{swing}}(t). \tag{5.17}$$

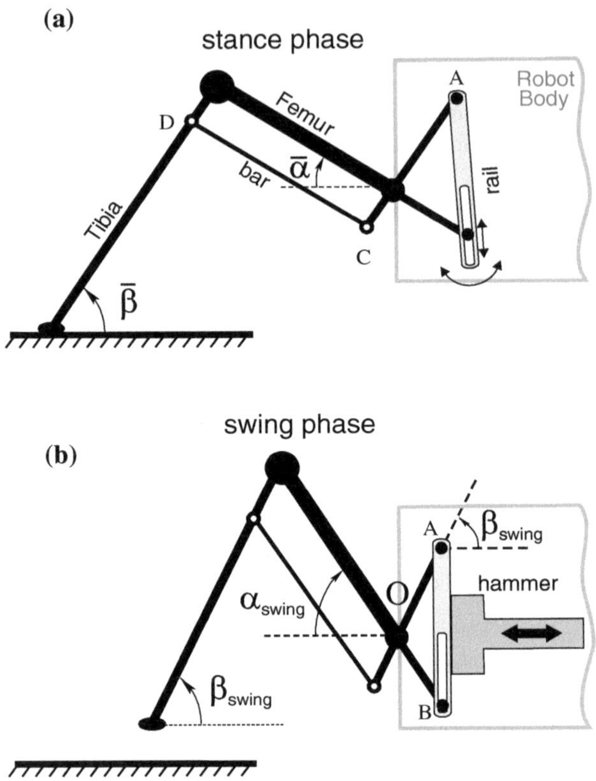


**Fig. 5.7** Leg model design (front view). Implementation of an "intelligent-like" swing phase. **a** During stance phase the mechanism produces no constraint on the dynamics of the $\alpha$ and $\beta$ angles. **b** In swing phase the device lifts up the leg

Choosing

$$\frac{\overline{\mathrm{OB}}}{\overline{\mathrm{OA}}} = \frac{d_F}{(k-1)d_T} \tag{5.18}$$

we satisfy (5.16) during the swing phase.

With the above described leg design,all $H_n$ remain approximately constant (albeit in general unequal, Fig. 5.3c) during the robot walk even over a rough ground. This ensures conservation of the vertical coordinate of the robot's centre of mass. Hence, energy is only necessary to lift and swing the robot legs that, with our design, demands relatively light energy consumption.

**Table 5.1** Typical parameter values for a stick insect [24, 29]

| Body elevation $H_0$ (mm) | Length of tibia $d_T$ (mm) | Length of femur $d_F$ (mm) | Step size $S_0$ (mm) | Transversal distance $L_0$ (mm) | Step frequency (step/min) |
|---|---|---|---|---|---|
| 7 | 14 | 8 | 15 | 12 | 18 |

## 5.6 Simulations of Leg Dynamics and Gaits Control

In the previous section we have shown that our kinematic model of the leg allows the gait control using only six degrees of freedom (one per leg) given by angles $\theta_1, \ldots, \theta_6$. Moreover, it is based on a conservation energy principle thus minimizing energy losses during the robot movement. Let us now illustrate how the model works by performing simulations of walking over different ground profiles. For simulations we use parameter values found for stick insects summarized in Table 5.1 [24, 29].

### *5.6.1 The Leg Dynamics*

Using parameter values from Table 5.1 we get the leg angles $\alpha_0 = 43°$ and $\beta_0 = 62°$ for $\theta = 0$ that we can take as $\overline{\alpha}$ and $\overline{\beta}$. Note that the parameters $\overline{\alpha}$, $\overline{\beta}$, and $L$ are related through (5.4), so we have only as free parameter $k$ that can be fixed by using (5.8). Finally, we have $\beta$ as a function of $\Delta_\alpha$

$$\beta = \arccos\left(\frac{d_F \cos\overline{\alpha}}{k d_T}\right) + k\Delta_\alpha. \tag{5.19}$$

Hence, from (5.4) we have $\alpha$ as a function of $\theta$. This means that according to (5.10) the body elevation $H(\theta)$ is determined only by $\theta$. Note also that the horizontal position $s$ of the coxa relative to the tarsus, during the stance phase, is given by

$$s(t) = L \tan\theta(t). \tag{5.20}$$

Thus we have a parametric relation between the body elevation and $s$ as shown in Fig. 5.8a for a flat surface with $\overline{\alpha} = 30°$. Note that during motion the leg preserves approximately the robot's potential energy. As earlier mentioned our proposed swing maintains (5.8). Accordingly, the behavior of the leg during the stance phase does not depend so much on the flatness of the ground, as seen in Figs. 5.8b and c, thus fullfilling item (ii) Sect. 5.2, introduction.

Figure 5.8d shows an enlarged view of the body trajectories shown in Figs. 5.8a,b,c. Note that the error in the horizontal position of the body increases with increasing step length.In our case, a small increment of potential energy $\Delta U$ must by related to

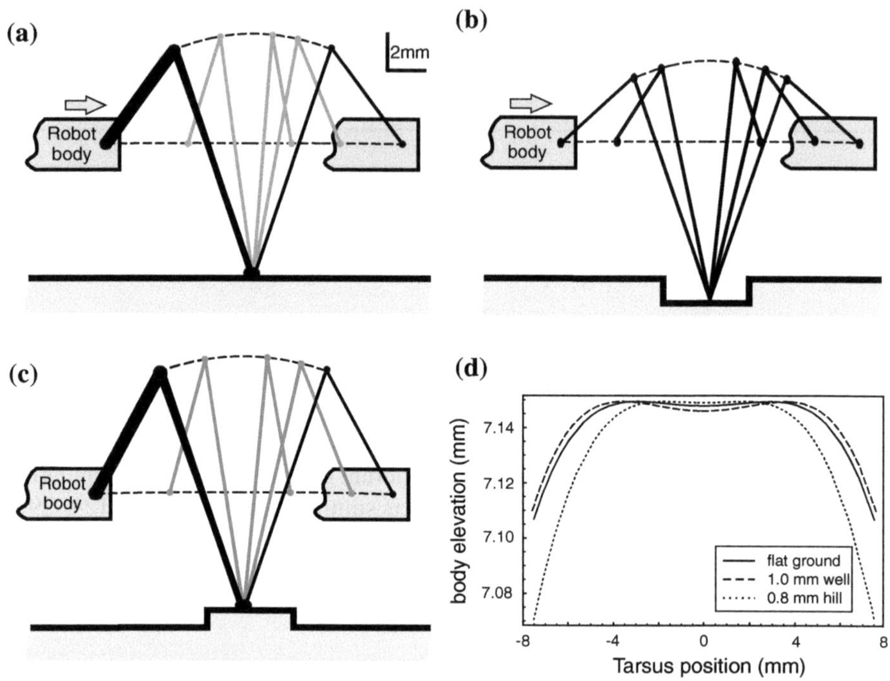


**Fig. 5.8** Leg behavior preserving approximately the potential energy during the stance phase of a step over different ground profiles. Translatory movement of the leg from initial (*thick black line*) to intermediate (*gray*) and to final (*light black*) position over a stance phase. The body moves forward with constant speed. **a** A step over a flat ground ($\overline{\alpha} = 45°$); **b** A step in a small trough ($\overline{\alpha} = 30°$); **c** A step over a small bump ($\overline{\alpha} = 65°$). **d** Body elevation $H(t)$ versus tarsus position $s(t)$ during the leg movement in the three cases. The deviation of the body elevation from its initial vertical position does not exceed unity

a sliding of length $\Delta L$ of the tarsus on the ground as

$$F_0 \Delta L = \Delta U \tag{5.21}$$

where $R_t$ is the total external force on the tarsus in the direction of its displacement. Note that in the limit of an ideal connecting wire and perfect solid rigid bodies, no elastic energy is stored in the robot and the internal forces do not work. As shown in Fig. 5.8d, the body elevation remains approximately constant so we have $\Delta U \approx 0$ and then $R_t \approx 0$. This means that a small friction force of the tarsus $F_{fric} > R_t$ with the ground will be enough to prevent sliding of the tarsus.

### 5.6.2 Gaits control

Our leg design allows handling small obstacles at the level of each individual leg (Fig. 5.8). This permits a simple control of the robot movement just by an appropriate driving of the angles $\theta_n(t)$ $(n = 1, 2, \dots, 6)$. In order to determine the evolution of the angles we assume that the robot employs one of the gaits shown in Fig. 5.1. The ratio between the durations of the stance ($T_{\text{stance}}$) and swing ($T_{\text{swing}}$) phases is given by:

$$\frac{T_{\text{stance}}}{T_{\text{swing}}} = \frac{N - m}{m} \tag{5.22}$$

where $N$ is the number of legs of the robot,and $m$ is the so called mode or gait number. For an hexapod $N = 6$ and we have $m = 1, 2$, and 3 that correspond to the metachronal,caterpillar, and tripod gaits, respectively.

Let us assume that the robot moves with a constant velocity $v_0$. Then independently on the gait number duration of the stance phase is

$$T_{\text{stance}} = \frac{S_0}{v_0} \tag{5.23}$$

where $S_0$ is the step length. For typical dimensions of a stick insect (Table 5.1) we have $v_0 \approx 4.5$ mm/s and $T_{\text{stance}} \approx 3.3$ s. Using (5.20) we obtain the dynamics of the protraction-retraction angle in the stance phase (when the tarsus is on the ground):

$$\theta_{\text{stance}}(t) = \arctan\left[\frac{S_0 - 2v_0 t}{2L_0}\right] \tag{5.24}$$

where $0 \le t \le T_{\text{stance}}$. According to (5.22) and (5.23) the duration of swing phase depends on the mode number

$$T_{\text{swing}}(m) = \frac{m}{N - m}\frac{S_0}{v_0}. \tag{5.25}$$

The swing phase of the metachronal gait is 2.5 and 5 times shorter than those for caterpillar and tripod gaits, respectively. The angle is then given by

$$\theta_{\text{swing}}(t) = -\arctan\left[\frac{S_0 - 2v_0 t(N - m)/m}{2L_0}\right] \tag{5.26}$$

where $0 \le t \le T_{\text{swing}}$. The coordination of the legs is achieved by an appropriate shift of stance-swing cycles (Fig. 5.1):

$$\theta_n(t) = \theta\left(t + \frac{(n - 1)mS_0}{(N - m)v_0}\right). \tag{5.27}$$

Figure 5.9 shows the angle $\theta_n$ ($n = 1, 2, \ldots, 6$) as a function of time for three different gaits and the corresponding stance-swing cycles very much like the qualitative relationships shown in Fig. 5.1. Note that usually insects adopt an appropriate gait for walking with a certain velocity, but they also can move with the same velocity employing different gaits.In this case (Fig. 5.9) the leg dynamics in the stance phase is the same for all gaits (5.23), (5.24), but the swing phase differs among them being faster for the slower gait (5.25), (5.26).

## 5.7 Electro-Mechanical CPG Integrating the Actuator and Body Dynamics in the Gait Control using a Toda Lattice

Figure 5.9 shows the evolution of the angles controlling the legs. A minimal CPG should be able to produce the corresponding signal driving the actuator motors. Let us first start with a description of a dynamical systems approach [16] to the generation of rhythms.

### *5.7.1 Toda-Rayleigh Lattice Ring and its Analog Circuit Implementation*

Toda [20, 21] provided exact solutions for a mechanical lattice system (and also for its electric analog system) (Fig. 5.10a). Six units are coupled by special springs whose force exponentially increases with the decrease of the inter-unit distance (Fig. 5.10b). Under appropriate limits the exponential Toda interaction reduces to the harmonic oscillator or to the hard sphere interaction.

The Toda system is a conservative one, whereas its circuit implementation unavoidably has energy losses [30]. Thus any excitation of the circuit decays in time and finally vanishes. Accordingly, to maintain oscillations we need to supply energy to the system. One energy-dissipation balance, earlier mentioned, was long ago proposed by Lord Rayleigh [22]. It includes a cubic nonlinearity (and active friction) in the original Toda system regulating the pumping-dissipation balance [17, 19]. Another possibility was proposed by Van der Pol who proposed using and anharmonic force [31, 32].

In view of the above we take a composite conservative-dissipative system. Such a Toda-Raleigh model [19] in its canonical form is given by:

$$\ddot{x}_n + \omega_0^2 (e^{x_n - x_{n+1}} - e^{x_{n-1} - x_n}) - \gamma(\mu - \dot{x}_n^2)\dot{x}_n = 0 \tag{5.28}$$

where $\omega_0$ is the frequency of linear oscillations, $\mu$ is the Rayleigh parameter, and $\gamma$ accounts for the strength of the Rayleigh cubic nonlinearity in the dynamics of the ring. In the limit $\gamma = 0$ we have the original Toda equation whose exact solution is a cnoidal periodic wave or a solitary wave [20, 21].

By allowing $\gamma$ to be positive or negative we introduce an input-output energy balance hence offering the possibility of sustaining oscillations. In the truly damped

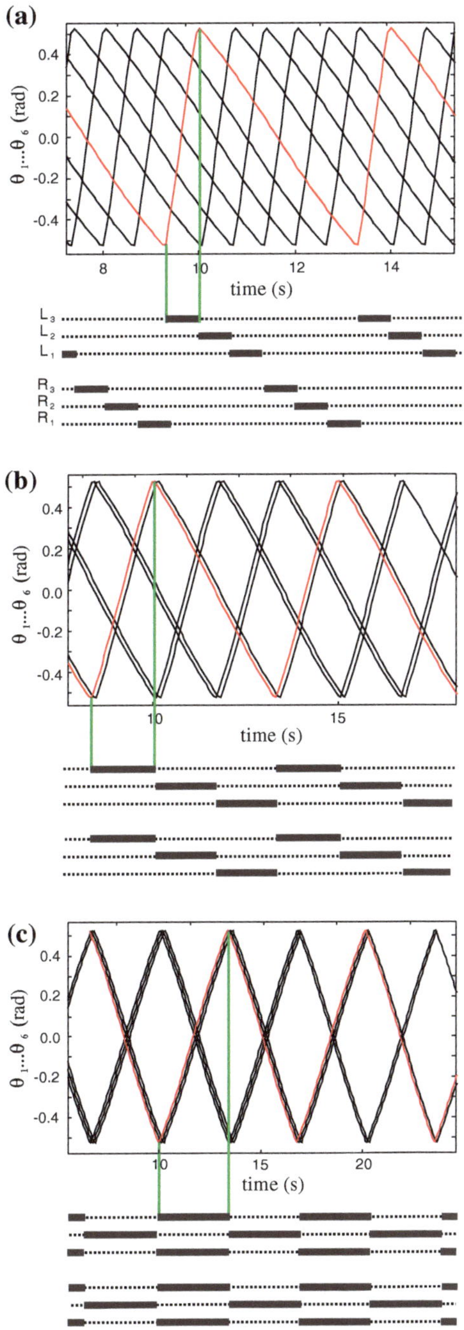


**Fig. 5.9** Hexapod Gaits Simulations (compare to Fig. 5.1) and time evolution of the angles controlling the protraction-retraction of the legs during walking for: **a** Metachronal, **b** Caterpillar, and **c** Tripod gaits. Parameter values taken from Table 5.1

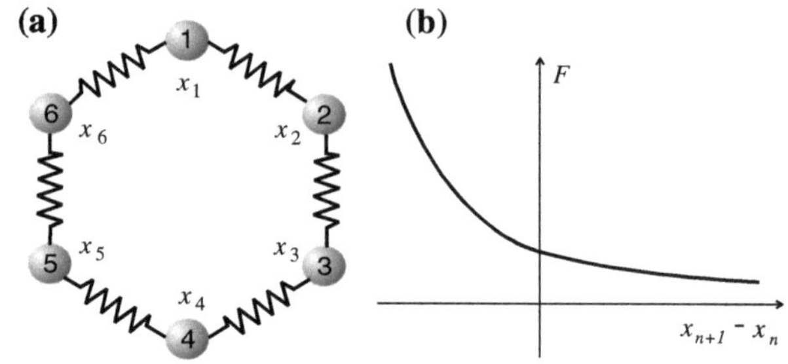


**Fig. 5.10** The Toda ring. **a** Six units are coupled in a lattice ring by "exponential springs". **b** Exponential coupling force acting between pairs of neighboring units

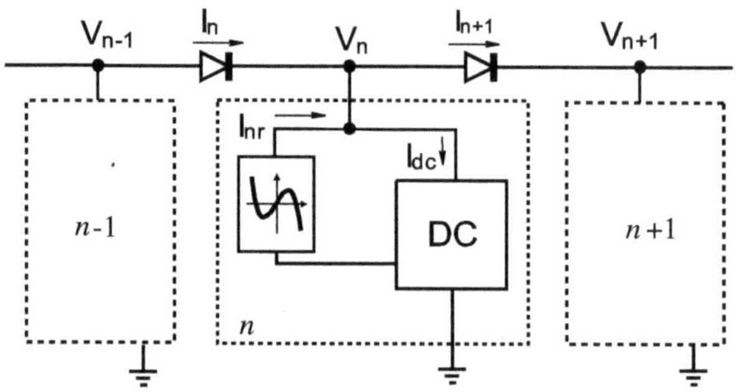


**Fig. 5.11** Block scheme of the Toda-Rayleigh ring. Each unit, $n$, (encircled by a dashed box) includes two main blocks: a double capacitor (DC) and a nonlinear resistor

case ($\mu < 0$) the system has only one motionless asymptotically stable solution $\{x_{n+1} - x_n = 1\}$. At $\mu = 0$ the system undergoes a symmetric Hopf bifurcation [19, 33], and for $\mu > 0$ the energy balance admits only a discrete set of solutions.($N-1$) different oscillatory modes (five for six-units lattice ring) appear in the system [17, 34]. These modes correspond to stable limit cycles coexisting in the $2N$ dimensional phase space of the system.They represent nonlinear waves, similar to waves shown for $\theta_n(t)$ in Fig. 5.9, propagating along the ring and can be labelled by their wave number $m$. For the six-units lattice ring (Fig. 5.10a) the mode $m = 1$ corresponds to a single-peak wave; $m = 2$ to a two-peak waves and $m = 3$ is a mode such that the nearest neighbours move in antiphase (also denoted "optical" mode at variance with the others denoted "acoustic" modes). The sign in the mode number defines the (clockwise or counterclockwise) direction of the wave propagation.Note that since each mode corresponds to a stable limit cycle, only one mode can be realized in the lattice ring at a time with no superposition admitted.

Figure 5.11 shows a circuit block-scheme for the Toda-Rayleigh lattice ring consisting of six units (5.28).The complete scheme for a one unit is showed in Fig. 5.12 and a detailed description of all components can be found in [19].

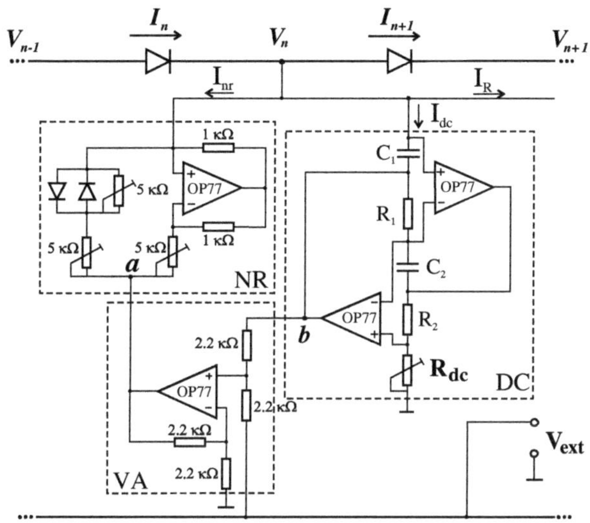


**Fig. 5.12** Detailed electronics of the Toda-Rayleigh ring

According to the current-voltage characteristics of a double capacitor (DC) we have [35]

$$\frac{d^2V_n}{dt^2} = aI_{dc} \tag{5.29}$$

where $a$ is a parameter depending on the inner components of the double capacitor. Using Kirchhoff's laws we get the equations governing the circuit [19, 36]:

$$\frac{d^2V_n}{dt^2} = a\left(I_n - I_{n+1} + I_{\text{nr}}\right) \tag{5.30}$$

where $I_{\text{nr}}$ is the current through the nonlinear resistor (Fig. 5.11). $I_n$ represents the current through the junction diode, that can be accurately modeled with

$$I_n = I_s \exp\left(\frac{V_{n-1} - V_n}{V_t}\right) \tag{5.31}$$

where the constants $I_s$ and $V_t$ depend on the inner diode structure. Thus using diodes we obtain the Toda exponential coupling (Fig. 5.10b) between neighboring units.The current through the non-linear resistor $I_{\text{nr}}$ is a cubic function of the voltage applied to its terminals $\Delta V$ that accounts for the Rayleigh energy pumping mechanism

$$I_{\text{nr}}(\Delta V) = b\left(\widetilde{V}^2 - \Delta V^2\right)\Delta V \tag{5.32}$$

where $b$ and $\widetilde{V}$ are constants. Finally we have the equation for the voltage $V_n$ of the unit $n$

$$\frac{d^2V_n}{dt^2} = a\left\{I_s\left[e^{\frac{V_{n-1}-V_n}{V_t}} - e^{\frac{V_n-V_{n+1}}{V_t}}\right] + I_{\rm nr}\right\}. \tag{5.33}$$

Comparing (5.28) and (5.32) with (5.33) one can see that dynamically the circuit described by the voltages $V_n$ is equivalent to the mechanical lattice ring $V_n \propto x_n$.

Figure 5.13 shows experimental traces of the three oscillatory modes ($m = 1, 2,$ and 3) generated by the Toda-Rayleigh six-units lattice ring.

To accommodate these oscillatory modes to the gate symmetries we change the initial ring geometry. Fig. 5.14a illustrates how a new twisted topology can be obtained. For the new topology we have the following sequence of units:

$$2 \to 3 \to 4 \to 1 \to 6 \to 5.$$

Note that this sequence is used only for mapping of the units to the limbs, whereas the actual Toda-Rayleigh circuitry remains unchanged. This procedure permits a direct mapping of the wave modes observed in the circuit to the gait symmetries described above. The three modes will lead to the limb movements with symmetries shown in the lower parts of Fig. 5.13. Comparing the gaits obtained with the Toda-Rayleigh CPG with the actual insect gaits shown in Figs. 5.1 and 5.9 we indeed see that the metachronal, caterpillar and tripod gaits are successfully generated by the CPG (Figs. 5.13a, b and c).

We can associate each limb with a single oscillator whose dynamics drives through an adaptor the corresponding actuator. The actuator motor rotates according to the voltage dynamics of the corresponding Toda-Rayleigh unit. Thus we can transform the voltage on the unit into angle value,

$$V_n \to \theta_n. \tag{5.34}$$

When the voltage derivative is positive we have the swing phase, whereas the negative derivative corresponds to the stance phase.Then the interlimb coordination will naturally follow from the coupling and dynamical interaction of the lattice units. Although having a practical potential such a CPG has a drawback.It produces the same rhythms (fixed gaits) thus having neither account for the dynamics of the robot legs nor for the body.

### *5.7.2 A Central Pattern Generator (CPG) Based on the Toda-Rayleigh Electro-Mechanical Circuit*

Let us now introduce a new circuit implementation for the anharmonic Toda lattice ring that includes the robot current state as a variable.Thus the new CPG will have a

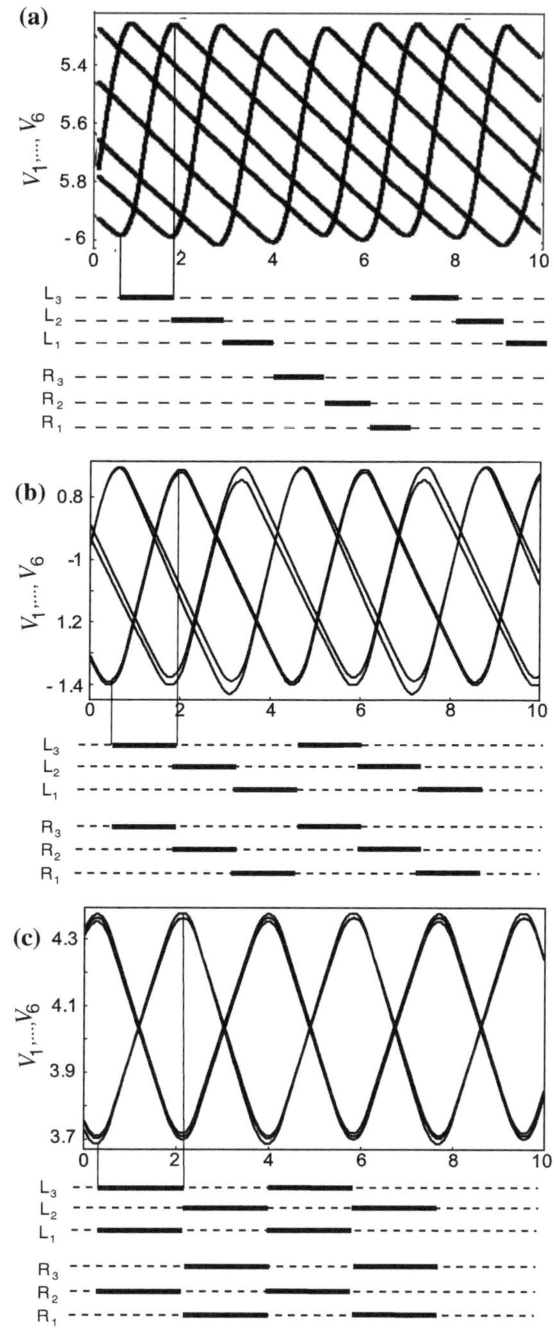


**Fig. 5.13** Oscillatory modes generated by a six-unit Toda-Rayleigh-circuit and their corresponding relations to hexapod gaits (compare to Figs. 5.1 and 5.9). Upper parts show oscilloscope traces of the voltages from all six units. Bottom parts show the corresponding phase relations. **a** The wave mode with $m = 1$ corresponds to the metachronal gait (Fig. 5.1). **b** The mode $m = 2$ corresponds to the caterpillar gait. **c** The optical-like mode ($m = 3$) models the tripod gait

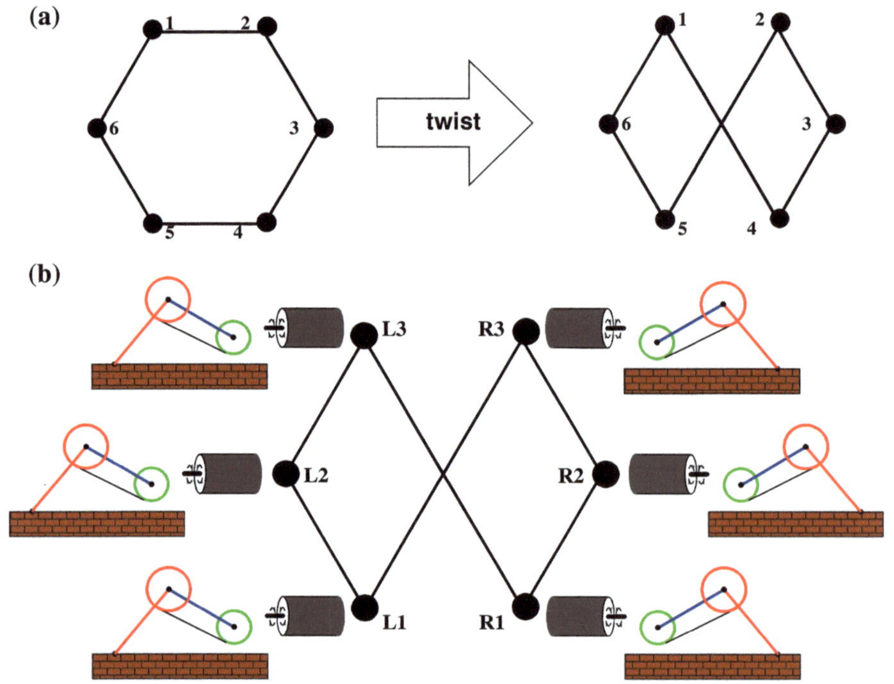


**Fig. 5.14** Electro-mechanical hexapod CPG based on the Toda-Rayleigh six-units lattice ring (see Fig. 5.10). **a** Twisted ring topology. **b** Connection scheme of the electro-mechanical Toda-Rayleigh CPG with no intermediate devices between the leg mechanics and electrical circuit

context-dependent dynamics so it will be able to change some characteristics of the gait "on the fly".

Let us start with the equation of the motor driving a leg.First we note that, in a quite general case, the dynamics of the motor obeys the following equation

$$J\frac{d^2\theta}{dt^2} + f\frac{d\theta}{dt} = I\Phi - T \tag{5.35}$$

where $\theta$ is the angle of the rotor of the motor defining the protraction/retraction of the coxa of the leg, (see Fig. 5.2), $I\Phi$ is the electromagnetic torque, $T$ is a constant load torque including the internal Coulomb friction, the parameter $f$ is due to the internal viscous friction in the motor that can be taken as a constant, and $J$ is the momentum of inertia. $J$ depends on the inner motor structure and geometrical distribution of the masses. Note that (5.35) includes the mechanical characteristics of the robot and $J$ is a parameter domain-dependent because it is affected by the global coupling between leg standing on the ground and the body of the robot. Thus we fullfill item (iii) Sect. 5.2, introduction. The current flowing through the motor windings $I$ is given by

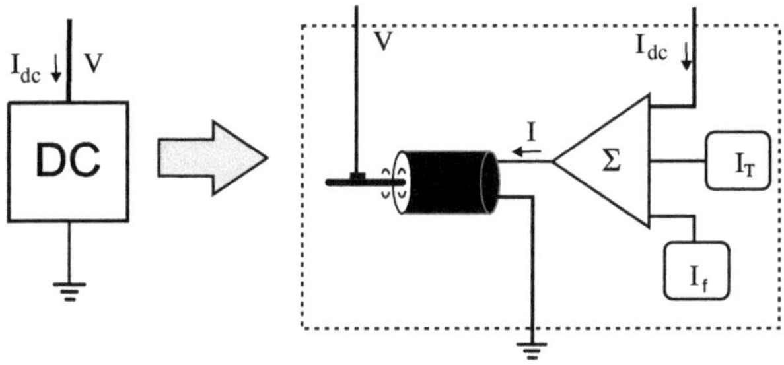


**Fig. 5.15** Block-scheme illustrating the DC of the CPG (Fig. 5.11) and the electro-mechanical circuit integrating the motor (leg) dynamics. Both blocks follow the similar Eqs. 5.29 and 5.41

$$\ell \frac{dI}{dt} + rI + \Phi \frac{d\theta}{dt} = V_{in}(t) \tag{5.36}$$

where $V_{in}(t)$ is the input voltage, and $\Phi$ is the magnetic flux across the air gap; $\ell$ and $r$ account for the self (inductance) and the resistance of the motor, respectively.

Usually, each motor leg is driven by an input voltage generated by the CPG (typically a Heaviside step function [37]),while the current through the motor is determined by (5.36). In our case, the motors are not driven by an input voltage but by the input current I according to (5.35). In the case of small motors that we use the magnetic flux is generated by a permanent magnet. Then, to a first approximation $\Phi = \Phi_0$ is a constant.

In order to integrate the dynamics of (5.35) in our Toda-Rayleigh lattice ring we use an angle-voltage converter coupled to the rotor of the motor (Fig. 5.15),

$$V_n(t) = V_0 \theta_n(t) \tag{5.37}$$

where $V_0$ is the converter constant. Usually this type of converters are optical devices and do not affect the motor behavior.Then the new variable $V_n(t)$ evolves according to (5.35):

$$\frac{J}{V_0} \frac{d^2V}{dt^2} + \frac{f}{V_0} \frac{dV}{dt} + T = \Phi_0 I. \tag{5.38}$$

Now we can split the current

$$I = I_f + I_T + I_{dc} \tag{5.39}$$

and adjust the values $I_f$ and $I_T$ in such a way that:

$$I_f = \frac{f}{\Phi_0 V_0} \frac{dV}{dt}, \quad I_T = \frac{T}{\Phi_0}. \tag{5.40}$$

Then from (5.38) follows:

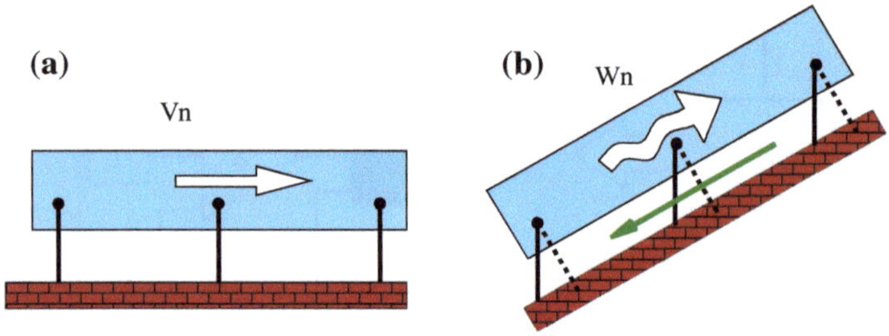


**Fig. 5.16** Scheme of the middle position of the leg (solid lines) for: **a** Horizontal walking and **b** Sloped walking

$$\frac{d^2V}{dt^2} = \frac{V_0}{J}\Phi_0 I_{dc}, \tag{5.41}$$

which is mathematically equivalent to the equation describing the behavior of the double capacitor (5.29) and Fig. 5.15. Thus we have been able to replace the DC circuit by its electro-mechanical analog. This demands replacing in (5.33) the factor $a$ by $V_0/J\Phi_0$, which now describes the inner mechanical properties of the leg-motor.

In this way, we obtain a network where instead of the electronic circuit with double capacitors we use the dynamics of the motors given by (5.41). Accordingly, the new CPG incorporates a body-leg-motor feedback loop. It produces swing-stance cycles for each leg, similar to those earlier described (Fig. 5.13), but now directly referring to the angles $\theta_n$ (Fig. 5.9).

For illustration of the electro-mechanical distributed CPG with feed-back through the legs, let us consider a sloped ground (Fig. 5.16.b). In this case the robot shout be affected by a tangential force due to its weight, that by means of the legs, produces an additional torque applied to the motor rotors.

This torque modifies the dynamical system (5.33) by the addition of the new term $-\alpha g$, where $g$ is the gravity acceleration and $\alpha$ is a suitable constant.

$$\frac{d^2W_n}{dt^2} = C\left\{I_s\left[e^{\frac{W_{n-1}-W_n}{V_t}} - e^{\frac{W_n-W_{n+1}}{V_t}}\right] + I_{\text{nr}}\left(\frac{dW_n}{dt}, V_{\text{ext}}\right) - \frac{1}{R}W_n - \alpha g\right\} \tag{5.42}$$

The sloped solution, $W_n$, is related by the solutions of (5.33) by $W_n = V_n - \alpha g$, so the oscillation of the legs are shifted related to the horizontal ground (Fig. 5.16), so the locomotion system produces adaptation to the context.

It is important to note that adaptation is a response of the dynamical locomotor system, by it self, to a external force. This response does not need sensor, because the system includes the forces on the legs to produce the CPG.

If the robot includes sensors like inclinometers, the brain of the robot can compare both signals, one from the inclinometer and other from the locomotor system. If both signals are equivalent the brain of the system must infer that the external force affecting to the robot is due to the sloped floor. Otherwise, the brain must look for

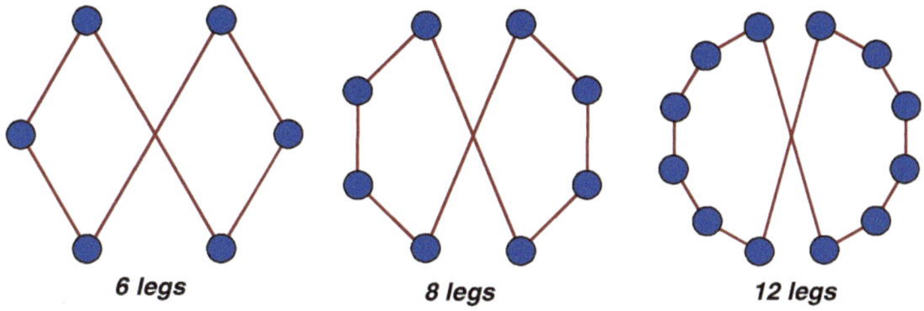


**Fig. 5.17** A prototype 2N-legged (insect-like) robot

an additional external force. In this sense, this version of the CPG will produce additional information to help to the brain of the system.

## 5.8 Conclusions

We have proposed how to use a nonlinear Toda-Rayleigh lattice ring (Fig. 5.10) as a central pattern generator (CPG) for controlling hexapodal robots. We have shown that in order to model three different insect gaits, we can use a lattice ring composed of six-units.

First, we have provided a mechanical leg design with three degrees of freedom per leg, two of which are functionally constrained to the leg protraction-retraction angle (Figs. 5.3 and 5.7). Then we have shown how a suitable swing phase can be used to prepare the following stance phase in order to keep smooth robot movements even in the presence of obstacles.

We have shown how a simple leg works maintaining the robot's potential energy practically constant. In this case our CPG is able to control the phases between the legs and also to control the flexion/extension angle between the femur and the tibia (see Fig. 5.8).

To illustrate our results we have built a "circuit" of nonlinear oscillatory units and have shown that the dynamical behavior of the circuit reproduces the phase relationships found in gaits of a six-legged animal (compare Figs. 5.1, 5.9 and 5.13).

We have also proposed how to incorporate the actuator (motor) dynamics in the CPG. With this approach we close the loop CPG – environment – CPG, thus obtaining a autonomous system for the leg control.Our model system does not require further higher level intervention to such CPG for locomotion in a non-flat ground.

In this case, our CPG includes the motor device.and then, motor device is involved with a *goal-directed behavior* to produce a specific gait or control the stability in a sloped landscape.

We have verified experimentally, that the behavior of the proposed electro-mechanical block is similar to the DC block used in previous oscillatory units.

Finally, we note that our method based on the nonlinear Toda lattice ring is not limited to the case of six legs. It is possible to implement a 2N leg robot by coupling N blocks each with two legs as it is illustrated in Fig. 5.17.

## References


1. E.R. Kandel, J.H. Schwartz, T.M. Jessell, *Principles of Neural Science* (McGraw-Hill, New York, 2000)
2. C.A. Wiersma, *Invertebrate Nervous Systems* (University Chicago Press, 1968)
3. H. Cruse, What mechanisms coordinate leg movement in working arthropods? Trends. Neurosci. **13**, 15–21 (1990)
4. J. Dean, T. Kindermann, J. Schmitz, M. Schumm, H. Cruse, Control of walking in the stick insect: from behavior and physiology to modeling. Newblock Auton. Robots **7**, 271–288 (1999)
5. P. Arena, L. Fortuna, M. Branciforte, Reaction-diffusion CNN algorithms to generate and control artificial locomotion. IEEE Trans. Circuits Systems I, **46**, 253–260 (1999)
6. H. Cruse, T. Kindermann, M. Schumm, J. Dean, J. Schmitz, Walknet - a biologically inspired network to control six-legged walking. Neural Net. **11**, 1435–1447 (1998)
7. G. Yiang, W.Y. Schooner, J.A.S. Kelso, A synergetic theory of quadrupedal gaits and gait transition. J. Theor. Biol. **142**, 359–391 (1990)
8. J.J Collins, I. Stewart, Coupled nonlinear oscillators and the symmetries of animal gaits. Nonlinear Sci. **3**, 349–392 (1993)
9. J.J. Collins, I. Stewart, Hexapodal gaits and coupled nonlinear oscillator models. Biol. Cyb. **68**, 287–298 (1993)
10. J.J. Collins, I. Stewart, A group-theoretic approach to rings of coupled biological oscillators. Biol. Cyb. **71**, 95–103 (1994)
11. M. Golubitsky, P.L. Buono, I. Stewart, J.J. Collins, A modular network for legged locomotion. Physica D **115**, 56–72 (1998)
12. M. Golubitsky, P.L. Buono, I. Stewart, J.J. Collins, The role of symmetry in animal locomotion. Nature **401**, 693–695 (1999)
13. J. Gray, *Animal Locomotion* (Weidenfeld and Nicolson, London, 1968)
14. L.O. Chua, *CNN: A Paradigm for Complexity* (World Scientific, New Jersey, 1998)
15. G. Manganaro, P. Arena, L. Fortuna, *Cellular Neural Networks. Chaos, Complexity and VLSI Processing* (Springer, Berlin, 1999)
16. V.I. Nekorkin, M.G Velarde, *Synergetic Phenomena in Active Lattices. Patterns, Waves, Solitons, Chaos* (Springer, Berlin, 2002)
17. E. Del Rio, V.A. Makarov, M.G. Velarde, W. Ebeling, Mode transitions and wave propagation in a driven-dissipative toda-rayleigh ring. Phys. Rev. E **67**, 056208–056217 (2003)
18. W. Ebeling, Makarov, V.A and M.G. Velarde, Soliton-like waves on dissipative toda lattices. Int. J. Bifurcation Chaos **10**, 1075–1089 (2000)
19. V.A. Makarov, E. del Rio, W. Ebeling, M.G. Velarde, Dissipative toda-rayleigh lattice and its oscillatory modes. Phys. Rev. E **64**, 036601 (2001)
20. M. Toda. *Theory of Nonlinear Lattices* (Springer, New York, 1981)
21. M. Toda. *Nonlinear Waves and Solitons* (Kluwer, Dordrecht, 1983)
22. J.W. Rayleigh, *The Theory of Sound* (Dover reprint, New York, 1945)
23. G. M. Nelson, R.J. Bachmann, D.A Kingsley, J.T. Offi,T.J Allen, R.D. Quinn, R.E. Ritzmann, Parallel complementary strategies for implementing biological principles into mobile robots. Int. J. Robotics Res. **22**, 164–186 (2003)
24. M. Schumm, H. Cruse, Control of swing movement: influences of differenly shaped substrate. J. Comparative Physiol. A **192**, 1147–1164 (2006)
25. H. Cruse, C. Bartling, T. Kindermann, *High-pass filtered positive feedback for decentralized control of cooperation*, (Springer, Berlin, 1995), pp. 668–678

26. T. Kindermann, Behavior and adaptability of a six-legged walking system with highly distributed control. Adapt. Behav. **9**, 16–41 (2002)
27. A. Schneider, H. Cruse, J. Schmitz, A biologically inspired active compliant joint using local positive velocity feedback (lpvf). IEEE Trans. Syst. Man Cyb. Part B: Cyb. **35**, 1120–1130 (2005)
28. J. Schmitz, J. Dean, T. Kindermann, M. Schumm, H. Cruse, A biologically inspired controller for hexapod walking: simple solutions by exploiting physical properties. Biol. Bull. **200**, 195–200 (2001)
29. V. Durr, J. Schmitz, H. Cruse, Behaviour-based modelling of hexapod locomotion: linking biology and technical application. Arthropod Struct. Devel. **33**, 237–250 (2004)
30. A.C. Singer, A.V. Oppenheim, Circuit implementations of soliton systems. Int. J. Bifurcation Chaos **9**, 571–590 (1999)
31. B. Van der Pol, On relaxation-oscillations. Phil. Mag. **2**, 978–983 (1926)
32. B. Van der Pol, Forced oscillations in a circuit with non-linear resistance. Phil. Mag. **3**, 65–70 (1927)
33. Yu. A. Kuznetsov, *Elements of Applied Bifurcation Theory*. (Springer, New York, 1995)
34. V.A. Makarov, M.G. Velarde, A. Chetverikov and W. Ebeling, Anharmonicity and its significance to non-ohmic electric conduction. Phys. Rev. E **73**, 066626–066612 (2006)
35. P. Horowitz, W. Hill, *The Art of Electronics* ( Cambridge University Press, Cambridge, 1987)
36. N. Islam, J.P. Singh, K. Steiglitz, Soliton phase shifts in a dissipative lattice. J. Appl. Phys. **62**, 689–693 (1987)
37. S. Still, K. Hepp, R.J. Douglas, Neuromorphic walking gait control. IEEE Trans. Neural Netw. **37**, 496–508 (2006)

# Part III
# Software/Hardware Cognitive Architectures

# Chapter 6
# A Robotic Simulation Framework for Cognitive Systems

**P. Arena, L. Patanè and A. Vitanza**

**Abstract** The insect brain computational model introduced in Part I of the book was there demonstrated through simple simulation results which showed the performance of the main blocks involved. In this chapter a dedicated software/hardware framework for cooperative and bio-inspired cognitive architectures, were the brain computational model was embedded, is presented. Here a complete description of the system, named Robotic Simulations for Cognitive Systems (RS4CS), will be introduced in order to show potentialities and capabilities. Moreover, the design choices and implementation issues related to the proposed robotic programming environment will be here addressed. The framework can be interfaced with robot prototypes mediating the sensory motor loop or with 2D and 3D kinematic and dynamic simulation environments.

## 6.1 Introduction

Over the last years, the advances in intelligent agents and robots have been incredible. Promising improvements in future scientific applications, such as the creation of cognitive agents that can make their own mental representations, improving their capabilities are expected. The main goal of this research line is to create agents that are, as much as possible, aware of what they are doing, and can adapt robustly to changing environments and requirements dealing with new and unexpected sit-

P. Arena (✉) · L. Patanè · A. Vitanza
Department of Electrical, Electronic and Computer Science Engineering, University of Catania, 95125 Catania, Italy
e-mail: parena@dieei.unict.it

L. Patanè
e-mail: lpatane@dieei.unict.it

A. Vitanza
e-mail: avitanza@dieei.unict.it

P. Arena and L. Patanè (eds.), *Spatial Temporal Patterns for Action-Oriented Perception in Roving Robots II*, Cognitive Systems Monographs 21,
DOI: 10.1007/978-3-319-02362-5_6, 

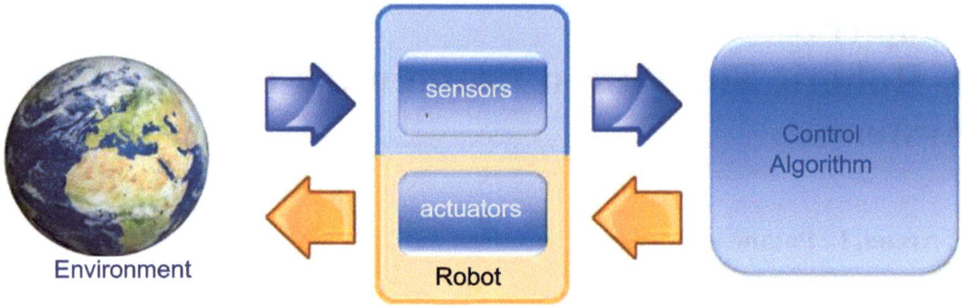


**Fig. 6.1** Generic functional robotic schema

uations. Robot perception, world modeling, prediction, selective attention, control and learning, planning and acting are the main capabilities required in a cognitive architecture.

Besides the necessary functions for sensing, moving and acting, a cognitive robot will exhibit the specific capabilities enabling it to focus its attention, to understand the spatial and dynamic structure of its environment and to interact with it, to exhibit a social behaviour and communicate with other agents at the appropriate level of abstraction depending on the context. According to these peculiarities the design of cognitive architectures needs to identify structures and components in a flexible and adaptable way.

A flexible Robotic System has to show suitable solutions for a multitude of problems in the simplest manner. The possibility to develop new behaviours in a rapid and transparent way, and to furnish a lot of ready-made libraries for different algorithms are the backbones for a good general-purpose *Robotic Framework*.

In general, in robotic programming the common functional schema can be summarized in Fig. 6.1, where the correlation between real environment, robot and control algorithms are highlighted. In particular, a robotic controller receives information from the environment using robot sensors and, after the selection of an appropriate strategy, the selected bevahiour is transduced using actuators to perform specific actions in the environment.

## 6.2 Mobile-Robot Software/Hardware Frameworks

An overview about already existing robotic simulation environments is important to identify and discuss different solutions in order to compare them all together and also in relation with the proposed framework. An interesting architecture has been proposed in the project called *Player project* [1]. It furnishes a Free Software to carry on research in robot and sensor systems. *Player* represents the robot device server providing a robot device interface and a network transparent robot control as a hardware abstraction layer. In this way, it is possible to create a controller that is language and device independent. Among the same project Stage and Gazebo are the 2D and 3D robotic simulators. Player provides a simple interface to the robot sensors and actuators over the IP network. The client program talks to Player over a TCP socket, reading data from sensors, writing commands to actuators, and configuring devices on the fly. This mechanism is present in the proposed framework regarding, for example, the communication between the algorithms and the simulator, but, in order to decouple and mask the differences, a unique robot-interface is given, making the algorithm free from hardware specifications. Player supports a variety of robot hardware, at first only the ActivMedia Pioneer 2 family, but now several other robots are supported because its modular architecture makes it easy to add support for new hardware.

Another example of a distributed object-oriented framework is Middleware for Robotics (MIRO) [2]. It was based on CORBA and allows a rapid development of software on heterogeneous environments and supports several programming languages. It can be divided into two inseparable main parts: a set of services to address the several sensors/actuators and a framework of classes that implement design patterns for mobile robot control. In this way, sensors and actuators can be modelled as objects with specific methods used to control and query data. Thereby, robots can be viewed as aggregations of these kinds of objects to give information in an agent-like manner. Although the view of abstraction is similar to our approach the main difference is that those objects provide services instead of abstract interfaces, whereas in the proposed framework robot interfaces give a useful abstract level to mask hardware or simulated implementations. Besides, the use of CORBA middleware represents the disadvantage of this solution in terms of memory overhead and processing power, even if it allows a rapid development of reliable and safe software on heterogeneous computer networks.

Furthermore, an interesting framework recently developed on robot control architectures is XPERSIF [3]. Like MIRO, it is a component-based and service-oriented architecture through the Component Based Software Engineering (CBSE) approach. It uses Ice middleware in the communication layer to provide a component model furnishing an efficient communication pattern [4]. The use of components allows to divide system into functional parts, and permits their access through logical interfaces; for these reasons we can classify them in three basic groups according to their functionality (i.e. basic, organizational and aggregate components). To allow soft real-time capabilities, two different types of communication are provided; in particular, *Operations* are used for real-time services, while *Commands* are used for more relaxed services and they are implemented as non-blocking Remote Procedure Calls (RPCs). From this prospective XPERSIF is more similar to the proposed framework, where there are blocking and non-blocking commands to satisfy strict requirements. Furthermore, a lot of libraries are supplied in order to collect all functionalities into organizational components.

Finally, another interesting framework to analyze is ORCA [5]. It is a robotic open-source architecture useful to develop component-based robotic systems and use them together in order to obtain more complex systems. These features are provided using common interfaces, libraries and a repository of existing components. Orca is similar to XPERSIF system and also our Architecture in the use of interfaces and libraries providing a simple but powerful component-based structure. Orca2, that is the Orca most recent version uses, as XPERSIF, the Ice middleware.

Reviewing the state of the art in mobile-robot control architectures, it is clearly found out that a suitable software implementation platform for developing mobile robot software is not available up to now. It is due to two distinct factors: on the one hand the different environments distributed on the market suffer from several limitations regarding flexibility, scalability, and portability, moreover there is no support for multi-robot applications; on the other hand the other available architectures are developed in research projects and they can not be easily used or extended for different applications.

## 6.3 Simulation Environments

During the last years the needs of a reliable simulation environment has became a very critical factor with the increase of complexity in different algorithms and the proliferation of complex cooperative strategies. Starting from these considerations, an ad hoc Dynamic robotic simulator has been provided for complicated 3D scenarios, to provide generic and flexible tools.

Several available simulators had been analyzed in order to choose the best solution to furnish a simple but powered instrument for application testing. A lot of interesting 3D robot simulators are available such as Gazebo [6] (that is the 3D version of Player/Stage simulators). It provides a suite of libraries for sensors and models of robots, in a typical Client/Server paradigm. To simulate the dynamic, Gazebo uses ODE library (such as many of these 3D robot simulators) to perform accurate simulation of rigid-body physics. It is a multi-robot simulator for outdoor environments, able to simulate a small population of robots, equipped with sensors and objects in a three-dimensional world. The flexibility of the simulator collides with the difficulty to simulate large groups of robots maintaining high performances.

USARSim (Unified System for Automation and Robot Simulation) is another interesting product to analyze [7]. It is a high fidelity simulator for urban search and rescue (USAR) for the investigation of multi-robot coordination and human-robot interaction (HRI). It uses unreal game engine for the dynamics and visualization and Karma engine for physics simulations. In order to concentrate forceson robotic relevant issues, the simulator leverages the strengths of a commercial game engine, delegating rendering aspects to it. An important strength of this simulator is the availability of interfaces compatible with controllers to allow the migration of code from simulation to real robots and vice versa without modifications.

Moreover, interesting aspect is the chosen policy about the realization of simulation environment: virtual arenas are created through AutoCAD model of real arenas, distinguishing several parts in simulated environments. For example, geometric models are seen as static objects and for this reason immutable and unmovable (for example floors, walls, etc.); instead obstacles simulation are objects that cannot be moved or manipulated. Light simulation is used to simulate the light environment in the arena, and the special effects are instruments to simulate particular objects such as mirrors or glasses; finally in victim simulations it is possible to simulate human victims. About robot models, in the simulator are already built five different Pioneer robots (P2AT and P2DX, the Personal Exploration Rover (PER), the Corky and a typical four-wheeled car), but it is possible to build new ones.

Another example of powered development environment is Webots, a commercial simulator, produced by Cyberbotics Ltd. [8]. It supports many kinds of robots, in fact there are a lot of models of commercially available robots (such us bipeds, wheeled robots and robotic arms), but it is possible to introduce new ones written in C, C++, Java or third party software, moreover an environment and robot editor are provided. Webots is not a robotics software platform, but rather a simulation engine with prototyping capabilities, and it is the main deficit of it.

Moreover, ARGoS [9] is another interesting 3D discrete-time simulator for multi-robot systems, designed for the Swarmanoid project simulations. It is entirely written in C++ and properly designed as a modular and pluggable simulator for solving swarm multi-robot applications and based on the use of free software libraries. It uses multiple physics engines and supports the three different types of robot: eye-bots, hand-bots and foot-bots. Being ARGoS a modular architecture new robots, sensors, actuators and physic engines can be added. It is possible thanks to the use of an XML configuration file, in this way new modules can be automatically referred using the XML file. Another important feature of this architecture is the possibility to transit between simulated and real robots in a total transparent way.

Finally, XPERSim simulator is the 3D simulator integrated into XPERSIF framework [10]; it uses Ogre engine (Object-Oriented Graphics Rendering Engine 3D) for the rendering of the simulation and Open Dynamics Engine (ODE) to calculate the dynamics. XPERSim provides an accurate and realistic physics simulation in a flexible and reasonable computational cost, a lot of sensors and actuators libraries are provided and it supports multiple simulated camera high frame rates. In order to improve the latency, given by the distributed nature of the system, while the client is rendering a scene it will receive the new information, or better the server does not wait the client request but sends images continuously. In same way it is possible to realize multiple client connection decoupling the physics and graphics, and for this reason, it was realized a XPERSim Server to calculate dynamics and a TeleSim view Client to render the new information.

The realization of a robotic simulation environment is convenient in robotic programming situations where simulated investigations can reduce development time and provide a rapid and useful platform for multi-cooperative strategies, such as in our scenarios.

The use of a powerful tool is demanded to finely reproduce the dynamic constraints among physical bodies interacting within a multi-body robotic structure in an unstructured environment. For this reason, ODE results the best trade-off between fidelity and computational performance and thereof it is the most used engine, as discussed before.

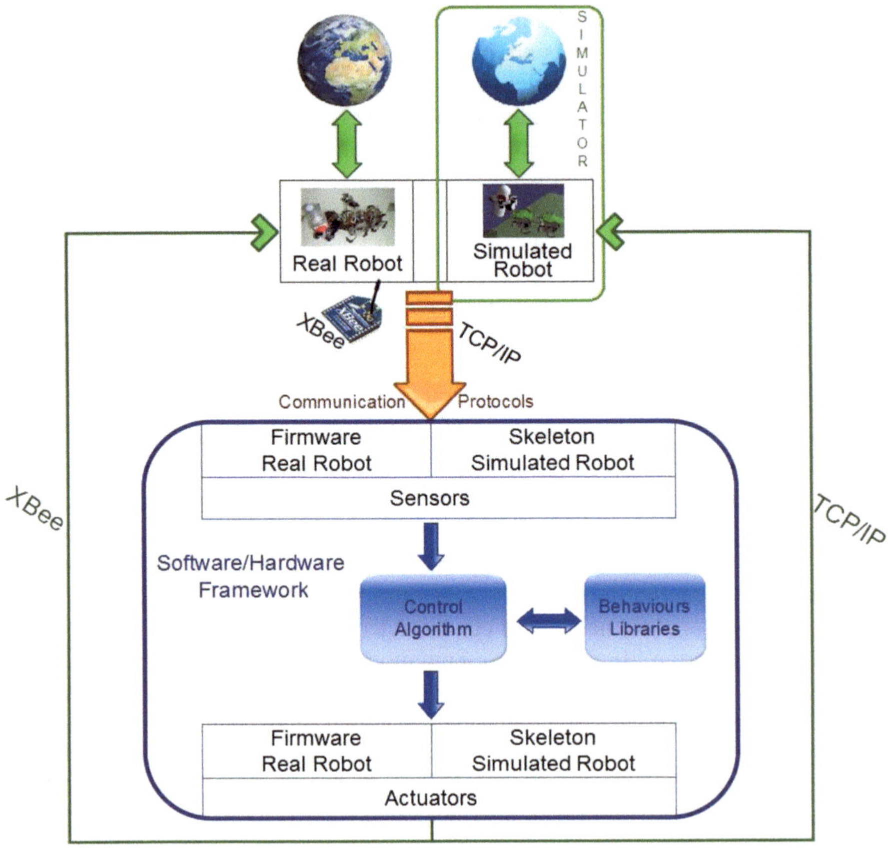


**Fig. 6.2** RS4CS–Block diagram of our proposed robotic system: the modules related to the framework and simulator and relative interactions are highlighted

## 6.4 Architecture Description

The main idea, proposed in this work, is to develop a software architecture based on C++, whose scheme is shown in Fig. 6.2. It is composed of *modules* that interact with other blocks, already developed, and *libraries*, with common functionalities useful for robotic control algorithms. It identifies an adaptive structure for rapid development of reusable components, and for creating cognitive and bio-inspired architectures that are able to investigate their behaviour, using classical approaches or proving neural structures based on spiking processing neural networks [11].

Although, intense efforts to define a common framework have been carried out in literature during the last years, due to the diversity of robotic applications, the development of a unique universal framework is up to now an open issue. On the other hand ***Modularity*** and ***re-usability*** have been identified as major features for robotic applications. The former arises from the need to divide an application in smaller *modules*, or better mutually decoupled software units with direct interfaces. The latter is related to the possibility to decrease the overall development time, by reassembling and using the components designed in other applications. Both concepts are directly interconnected, in fact splitting behaviors into modular units can improve re-usability and understandability, making easier the testing and validation phases [12]. Another important peculiarity of robotic architectures is the need of real-time interactions with dynamic and unpredictable environments; to satisfy all requirements the framework design must include real-time control systems to supervise sensors and actuators, to react to critical situations, also supporting concurrency. For this reason, usually these systems are decomposed in hierarchical-modular components to identify a layered structure in order to reduce complexity using abstraction concepts. A hierarchical structure allows to identify generic interfaces that are hardware independent, so that the same algorithms can run on different scenarios (several different robots with peculiar hardware architectures).

The separation between simulation environment and control algorithm is one of the main properties of our approach, as shown in Fig. 6.2. The need to maintain transparent linking with real hardware platforms induces to prefer a Client/Server communication paradigm, in order to decouple high-level controller from low-level actions.

These characteristics are fundamental in our scenarios where different kinds of robots are implemented. In particular, we are testing algorithms on two different classes of robots:

- ***Roving***, commonly used as test-bed to evaluate the performance of cognitive algorithms mainly devoted to navigation control. An example of rover used in our applications is the *Pioneer P3AT*, a high-performance and wheeled mobile robot for indoor and outdoor applications, produced by ActivMedia and software-compatible with all MobileRobots robots. It can be equipped with different kinds of sensors, such as sonar, bumpers, a laser rangefinder, a GPS receiver and a pan-tilt-zoom colour camera.
- ***Hybrid robots***, used to extend the cognitive capabilities of a system beyond navigation, by using a mechanical structure able to show different basic behaviours.

For the last class, a bio-inspired hybrid mini-robot, named *TriBot I* [13, 14], has been realized. It is composed by three modules, the first two are wheeled-modules, whegs with three-spoke appendages with a design that improve the stability of the structure [15]. The last module is composed by two standard legs with 3 degrees of freedom each connected to the main body through another actuated degree of freedom. Thanks to the interoperation of these modules, the TriBot is able to face with irregular terrains overcoming potential deadlock situations, to climb high obstacles compared to its size and to manipulate objects. Instead, *TriBot II* is the second prototype of the robot TriBot. The major differences between them lie in the manipulator design and in the motors used to actuate the whegs. In the two prototypes, two different configurations of the manipulator have been used: the first one is inspired by the

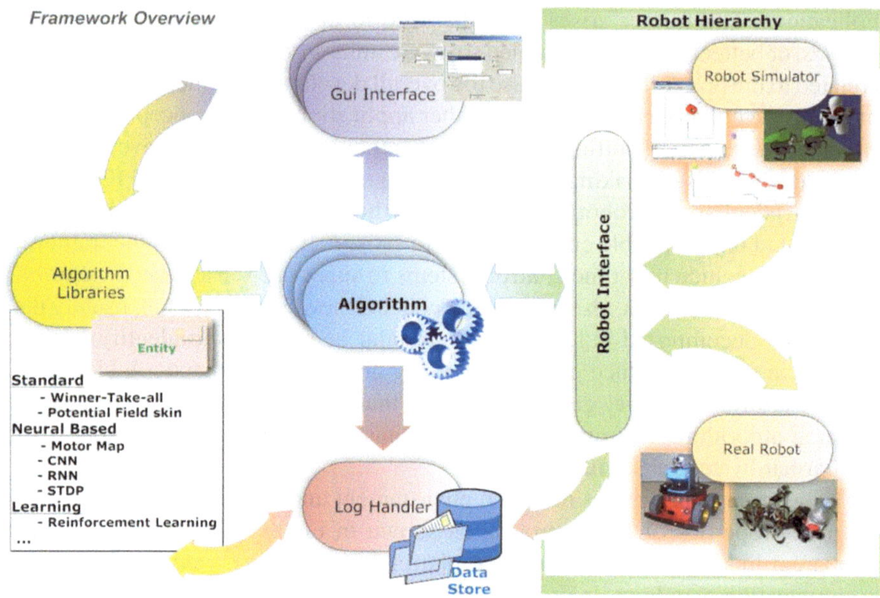


**Fig. 6.3** RS4CS–Overview of the interactions between components in the software architecture for cognitive systems

hexapod robot Gregor [16], whereas the second one uses the same leg configuration of the Tarry robot [17] to improve the working space of the manipulator.

In a modular architecture the mechanisms used by each module to access to the shared computational resources are extremely important together with the communication rules. Modularity and functional separation, together with re-usability and robustness, are the most basic software design principles that can be ensured in software applications. The aim of modularity, as shown before, is to encapsulate all the physical and logical characteristics of the main entities to decouple specific implementations, defining a set of access functions. For this reason, it is possible to divide our structure in specific parts, in order to decouple its functionalities.

The Architecture can be structured as reported in Fig. 6.3 where five main elements have been identified: The *Graphical User Interface (GUI)* provides tools to display real-time data while allowing the user to control robot and execute algorithms. It is directly connected to the *Algorithm module* in order to obtain data to show and to convey external commands during executions. Interconnected with the Algorithm module there are other two important parts of the architecture, the *Algorithm libraries*, useful to obtain specific and peculiar functionalities related to Algorithm implementations and the *Log Handler* dedicated to log fundamental information to create historical traces. Finally, *Robot Hierarchy* part gives an abstract view of robots, decoupled from specific implementations [18].

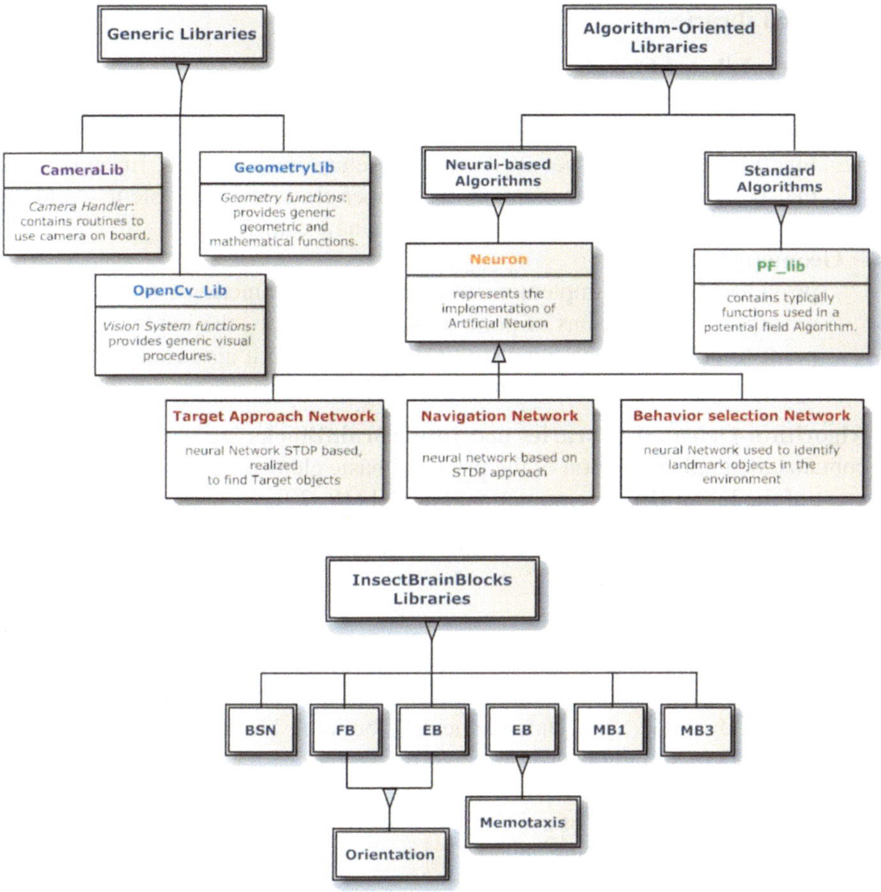


**Fig. 6.4** RS4CS–Algorithms library main components

As shown in Fig. 6.3, the framework can be interfaced with different kinds of robotic platforms, both robot prototypes and kinematic or dynamic simulated environments.

## *6.4.1 Algorithm Libraries*

This module is dedicated to collect all common and useful functions, in order to provide proper libraries including typical structures that are commonly present in our control algorithms.

It is very important for re-usability concept, the possibility to reuse implemented structures like Neural Networks in different ways and in different applications. Moreover to maintain a certain degree of flexibility a schema of functional blocks have been considered as shown in Fig. 6.4:

- **Generic Libraries**
  - **CameraLib** and **OpenCv_Lib**
    provide a complete set of specific routines for the vision system. A camera is a complex object, so the easiest way of handling it is via libraries, which contain functions to return or set attributes or toreturn image values. Encapsulating the specific device management in a proper class, it is possible to avoid the need of modifying code when installing a new camera.
  - **GeometryLib**
    provides a generic implementation of geometry functions. It contains, for example, the definitions of point and force concepts, the implementation of point-to-point distance, point-to-line distance, and it also contains transformation functions to convert measurement units.

- **Algorithm-Oriented Libraries** and **InsectBrainBlocks** are the libraries which contains the implementation of the principal basic elements used for the development of the Insect Brain computational model [19, 20].

(i) **Standard Algorithms** there are libraries which contain typical functions used in traditional algorithms.
(ii) **Neural-based Algorithms**

  - **Neuron**
    represents the implementation of the basic block used to model biological neuron dynamic; using this elementary building block is possible to create networks of spiking neurons.
  - **Navigation Network**
    is an implementation of a neural Network based on Spike Timing Dependent Plasticity (STDP) approach [21].
  - **Behaviors selection Network**
    implements a Neural Network used to choose the basic behavior that the robot can use when a particular landmark has been identified. Landmarks are distinguished on the basis of colour and shape information coming from camera.
  - **Target Approach Network**
    implements a neural network based on the STDP approach [21] and using neuron definitions present in the previous libraries. In particular, the network is divided in two parts: the first one is used to avoid obstacles, has major priority and uses distance and contact sensors as input, while the second one allows robot to approach target objects using vision information and target sensors as input.

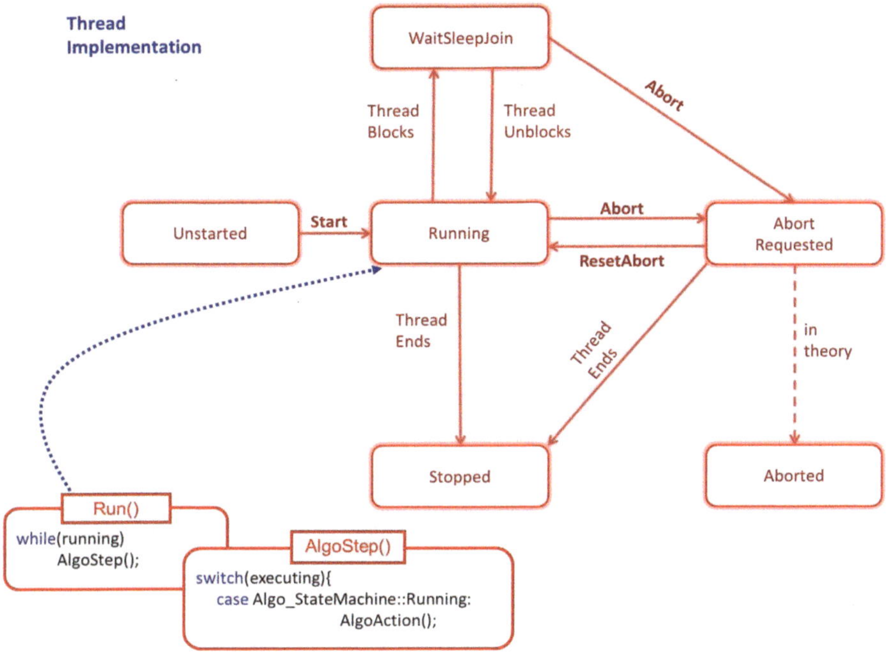


**Fig. 6.5** RS4CS–Algorithm core

## *6.4.2 Algorithms*

This module can be considered as the core of the Architecture, since it contains all instruments for implementing an algorithm. The *Algorithm* superclass[1] can be seen as a wrapper class to provide a simple and rapid implementation of specific algorithms. In particular, it is important to underline that an algorithm is implemented as a thread repeated periodically to perform peculiar actions in each step, as shown in Fig. 6.5. In fact, the difference between algorithms is only in various actions encapsulated in the *AlgoStep()* function (Fig. 6.6); for this reason, the superclass includes all of thread utilities functions such as thread management, with function to start, stop or resume threads. All implementations extend it in order to perform own specific actions.

In other words, this class is mainly a convenience wrapper around Thread concept, so that it can easily create an own algorithm. It provides a number of supports to make it easier to write object-oriented threaded codes. An overview of algorithms implemented in the framework up to now is shown in Fig. 6.7, even if this list is continuously evolving.

[1] A *superclass* is seen as a base class: a generic class from which other classes, called *subclasses*, are derived. Moreover, it establishes a common interface and allows the extending classes to inherit its attributes and methods.

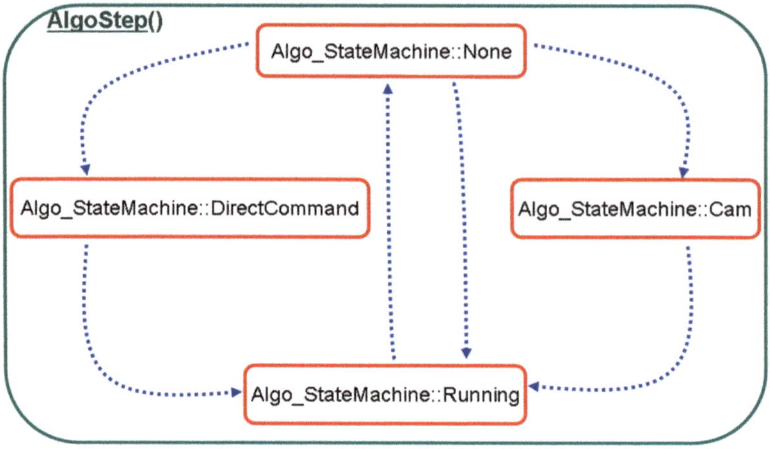


**Fig. 6.6** AlgoStep() state machine

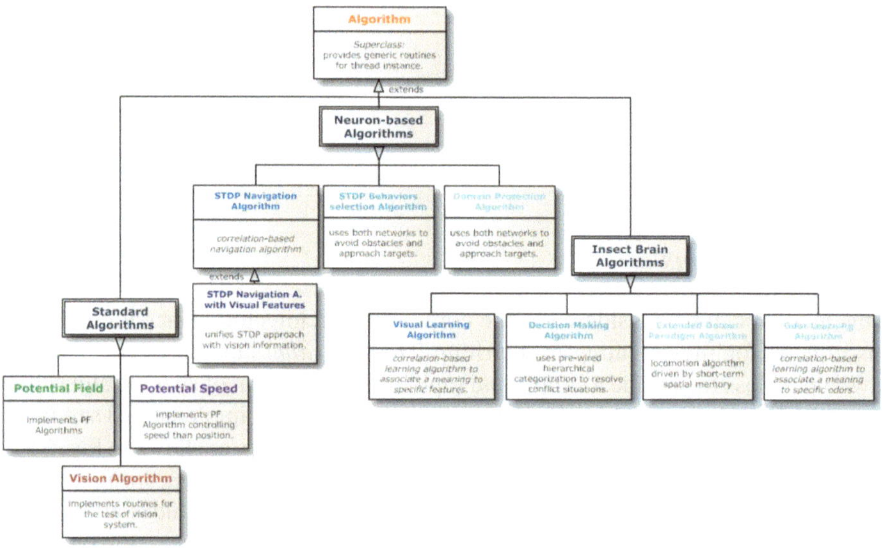


**Fig. 6.7** RS4CS–Algorithms overview

- **Standard Algorithms**
  - **Potential Field** and **Potential Speed**
    contains respectively the implementation of a classical Potential Field and Speed Algorithm; the latest differs from the previous one, since it evaluates the speed instead of the position of the robot, using the potential field algorithm.

- **Neural-based Algorithms**
  - **STDP Navigation Algoritm**
    implements a correlation-based navigation schema, based on Spike Timing Dependent Plasticity (STDP) paradigm. It is a simple algorithm which, using the neural network implemented in STDPNet library, allows a robot to approach or avoid objects.
  - **STDP Navigation Algorithm with Visual Features**
    contains the implementation of an algorithm to reach a specific target. To perform safe navigation the algorithm gives major priority to the network sub-part dedicated to the obstacle avoidance, using the camera information to explore the environment in order to identify a landmark.
  - **STDP Behaviors Selection Algorithm**
    implements an algorithm that uses sensors information to implement STDP step and a camera to obtain vision information about environment. In particular, it uses the previous algorithm to approach objects. When the robot is in front of a particular one it utilizes the camera to analyze the scene and identify the landmark. Then a specific behavior (i.e. take, climb, avoid), will be selected. For this reason this algorithm extends the previous Algorithms using robot basic behaviours and providing new ones.
  - **Domain Protection Algorithm**
    provides an algorithm to reach a target, considered as food to defend. For this reason, once the robot has taken possession of the place where is the food, it tries to defend it controlling continuously its domain by camera information, and if it intercepts an intruder it tries to push it out.

- **Insect Brain Algorithms**

contains all experiments predicted to test and validate Insect Brain computational model [19]. In particular, since flies are able to extract visual clues from objects like colour, center of gravity position and others that can be used to learn to associate a meaning to specific features (a reward or a punishment).

  - **Visual Learning**
    simulates how the fly treats visual inputs and learn through classical and operant conditioning the proper behaviour depending of the object visual clues [22, 23].
  - **Decision Making**
    wants to validate the process that has been trained to avoid objects with specific visual features; in presence of a conflict the fly have to decide which features are the most relevant to make a choice. The Decision Making strategies is guided by a pre-wired hierarchical categorization of the features that, for instance, leads the fly to give more importance to colour with respect to shape [24].
  - **Extended Detour Paradigm**
    is used to show that Drosophila possesses a short-term spatial memory; flies can remember the position of an object for several seconds after it has been

removed from their environment. The detour consists into temporarily attract the fly away from the direction towards the chosen target, putting a distractor in the arena for a few seconds while the target is switched off. When the distractor is eliminated, the fly is able to aim for the former target [25].

- **Odor Learning**
  implements an odor learning in a classical conditioning experiment [26].

### *6.4.3 Robot Hierarchy*

In this section the design and implementation of the Robot hierarchy will be described. It results as a collection of classes dedicated to decouple the specific robot implementations. Using the advantages of object technology, the Robot class is implemented as an abstract class, which is not determined for making instances of this class but for the next inheritance of members representing individual robot types.

These classes are organized in a three-level hierarchy in order to exploit C++ modularity and inheritance, as shown in Fig. 6.8. The superclass implements the generic abstract robot interface (level 1) to supply generic interface with software environment, in the second part we can see the specific device interfaces classes (level 2) to decouple algorithms from interfacing with hardware (real robot) or simulator (simulated robot) (level 3). The interfaces provide functions to modify attributes and to retrieve their values, but they also provide all functions useful to perform specific functionalities, such as acquiring information from sensors or executing actions.

- **Pioneer** It represents the implementations of a class useful to interact with Pioneer's specific library (i.e. ARIA development tools [27]). It supplies all functions to allocate a robot instance and runs indifferently on the real robot or in the simulation [28]. In this case, the library (ARIA) provides to emulate the behavior of the Pioneer robot in the simulation environment. It also includes interfaces to emulate the sensory system including sonar distance and laser sensors.
- **TriBot** It represents the robot class adapted for TriBot bio-inspired robot. It interacts with the lower level of robot hierarchy and provides independent routines to interlock TriBot Protocol, used to support communications with the real robot, and with Skeleton,[2] used to interact with the Dynamic Robotic Simulator.
- **Rover** It represents the robot class used to interact with a dual drive roving robot.
- **Hexapod** Robot class designed for interacting with a hexapod robot used for the testing of the Central Pattern Generator-based controller for legged locomotion.

The separation from the low level hardware is more complicated and for this reason it needs to develop a device driver for each external used device. In particular, it is obvious how a specific robot class, present in the second layer of the hierarchy, uses hardware or simulated interfaces to mask low level communications.

[2] A *Skeleton* is a server side interface analogue to the robot device interfaces, used to correctly invoke request to simulated robot.

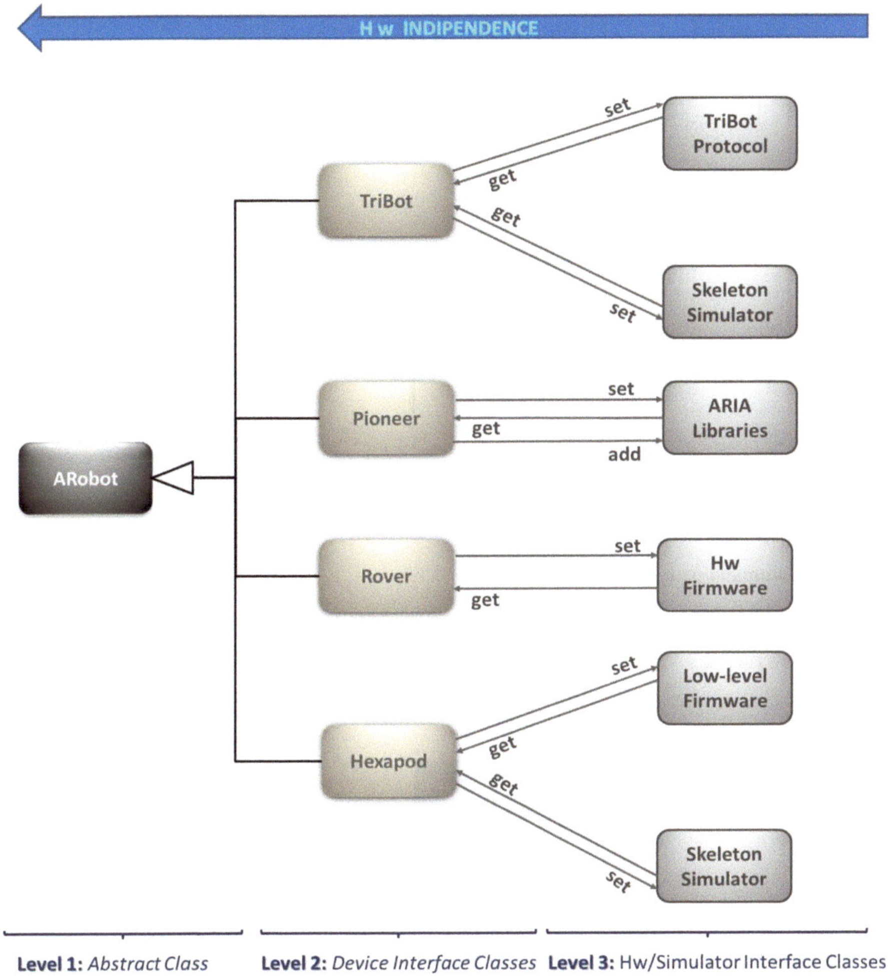


**Fig. 6.8** Hierarchy of classes involved in the implementation process. Pioneer and TriBot device interfaces are unfolded to their related specific classes for illustrative purposes. *Boxes* are classes and *arrows* imply inheritance relations. ARobot Class is the generic robot interface, it is the superclass for the specific robot classes (i.e. PioneeR, TriBot, Rover or Hexapod). Instead, these latter classes properly mask, if necessary, the differences among hardware robotic devices and simulated skeletons (e.g.TriBot Protocol vs. Skeleton Simulator)

## *6.4.4 Graphical User Interface (GUI)*

Graphical interfaces allow users to interact with the algorithms using specific input commands and showing information and results through the manipulation of graphical elements.

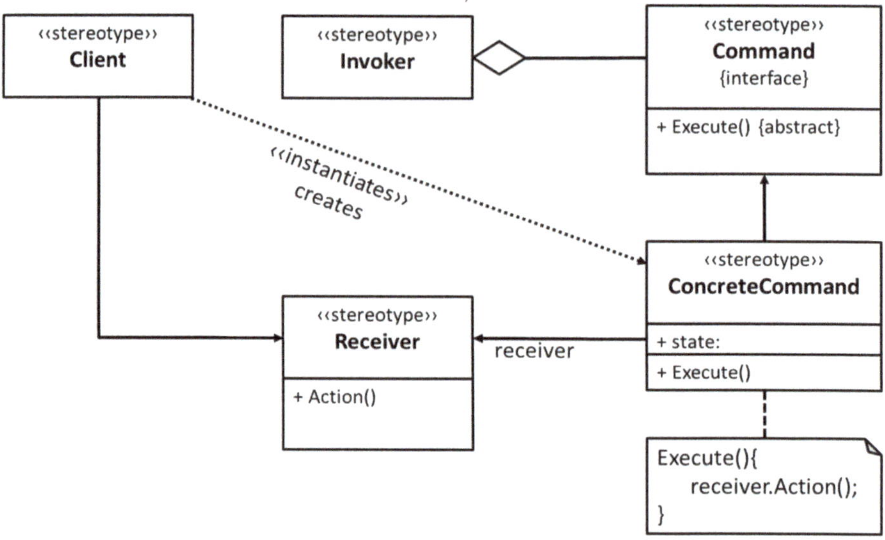


**Fig. 6.9** Command pattern structure

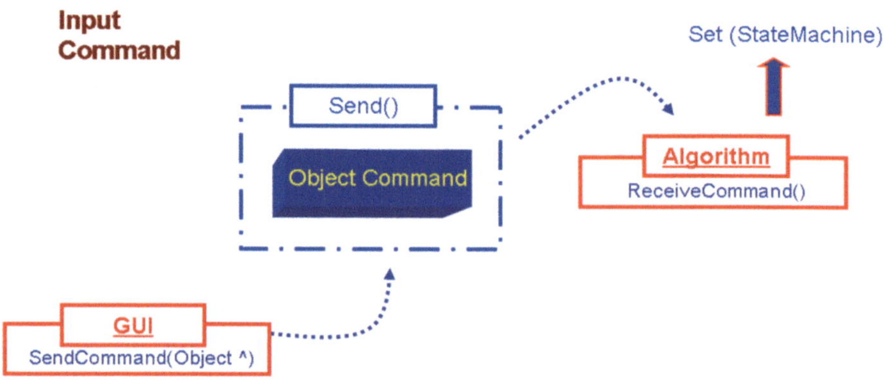


**Fig. 6.10** Communication between algorithms and graphical interfaces: GUI sends an object (*Send-Comman(Object)*ˆ) which contains all the parameters related to the specific command request. The algorithm receives and parses the command (*ReceiveCommand()*), setting the state machine in order to satisfy the request

In order to decouple the two mechanisms, the *Command pattern* was applied and both commands and graphical elements are seen as specific exchanging objects.

The classical structure of the pattern is shown in Fig. 6.9, whereas the message passing mechanism applied for our purposes, is shown in Fig. 6.10.

In particular, once received the command, the GUI creates the correspondent *object command* for Algorithm; which will consume it, setting the *internal state machine*.

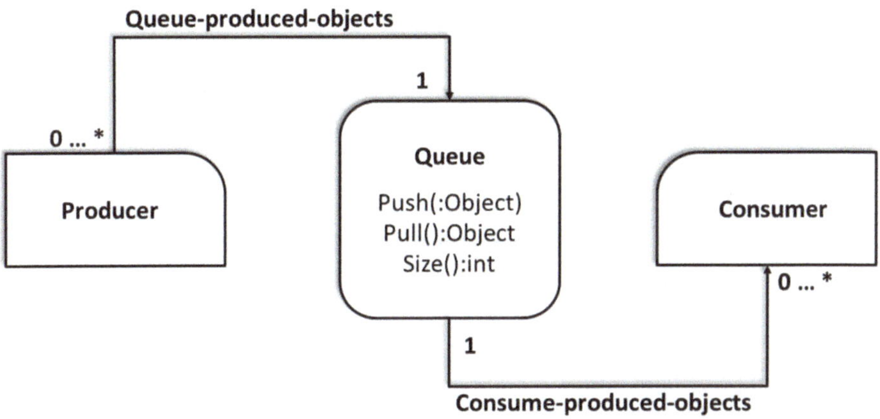


**Fig. 6.11** Producer/Consumer mechanism. In particular, the relationship between Producer–Consumer and queue is depicted in figure

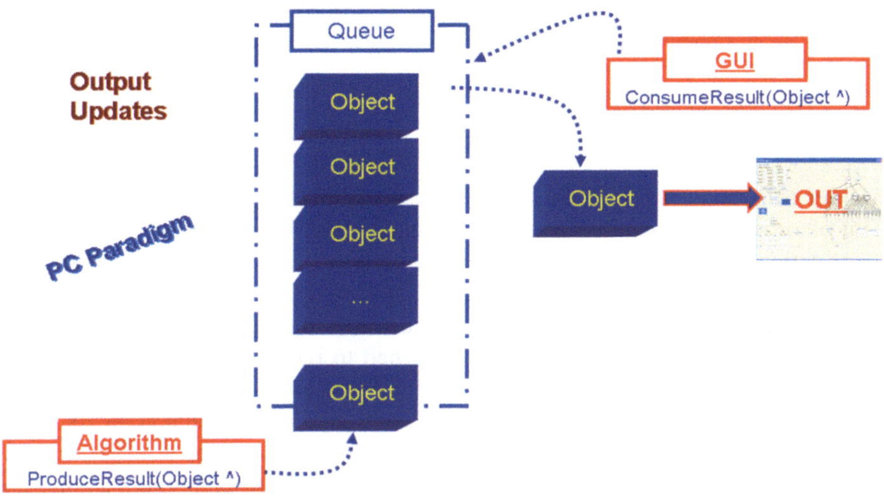


**Fig. 6.12** Objects queue for graphical Updates. During the running of the algorithm steps, all the output data are collected in specific formats (*ProduceResult(Object)*^); after that the output object is picked up by the GUI (*ConsumeResult(Object )*^) and the information used to update the visual output interface

As regards Graphical Updates, a mechanism to guarantee the decoupling between the execution of the algorithms and graphical refreshes is needed; for this reason a *Producer/Consumer mechanism* is used (see Fig. 6.11). During the algorithm execution all data are collected and sent to the common queue; where GUI periodically takes-off the upper object in order to update output graphical information. This communication mechanism is shown in Fig. 6.12.

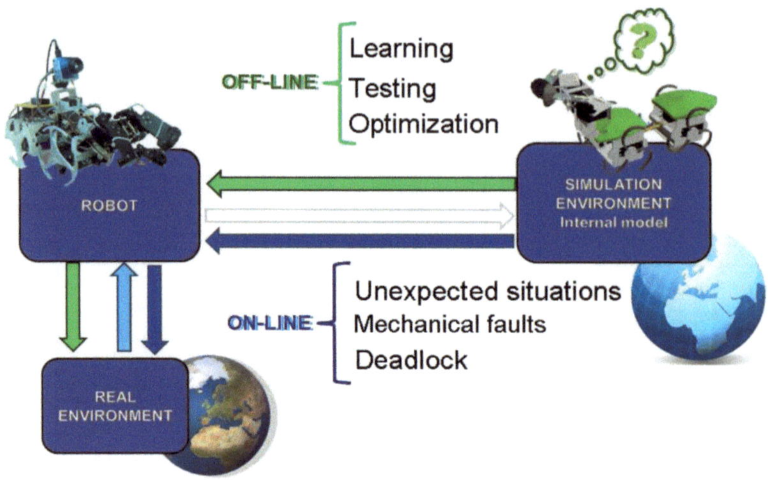


**Fig. 6.13** The use of a virtual dynamic environment for internal simulation of the robot introduce advantages both for off-line and on-line processing

## 6.5 Dynamic Robotic Simulator

In order to validate the functionality of our algorithms and to allow the development of versatile applications a simulation environment was needed. Starting from investigations about the simulation issues and robotic supports, an ad hoc Dynamic robotic simulator has been provided for complicated 3D scenarios, to simulate a generic and flexible tool for high performance 3D simulations of autonomous mobile robots. For this reason, the tool is designed to recreate a complete replica of the real environment and situations in which robots can be found, in order to learn how to achieve their goals. The simulator is developed in particular contexts where dynamism, performance and accuracy are necessary prerequisites.

In particular, the novelty of this approach lies in the representation of an *internal model*, or better the imitation of the interior state of the robot and the perception of the surrounding environment reconstructed through sensory feedback. This representation can be used off-line for parameter optimization, but also on-line to resolve deadlock situations in order to preserve the system from waste of resources and reduce energy consumption as depicted in Fig. 6.13.

The simulator becomes the instruments to learn new skills, to test capacity and performance of the robot to investigate preliminary study of new behaviors, in order to develop cognitive strategies. These characteristics, on the other hand, take inspiration from experience of everyday life, in a biologically plausible way. For example, we often *think* the best way to execute a task, before to perform it in a new environment or situation, according to our experiences. Starting from this key idea the simulator represents a platform where cognitive bio-inspired structures can be simulated.

The simulator is written in C++ using Visual Studio 2005 environment, and it uses Open Dynamics Engine (ODE) as physics engine to simulate dynamics and collision detection and Open Scene Graph (OSG) as high performance 3D rendering engine. Particular attention was given to provide versatile mechanisms to introduce and simulate robots, environments and objects. Like Gazebo, a CAD design is used to create simulated scenes. In fact, the main novelty of this approach consists in the extreme extensibility to introduce models. To import robot models in the simulator, a procedure was developed that starts from models realized in 3D studio MAX and provides, using NVIDIA Physics Plugin for 3D Studio MAX, a COLLADA (COLLAborative Design Activity) description of the model to permit the correct transfer in simulated environment. In this way, the possibility to simulate own environments and robots is guaranteed.

Moreover, the Client/Server paradigm and the possibility to establish the graphical model granularity permits the decoupling of simulation capabilities from robot controllers, so another advantage of this structure is the ability to simulate a large number of robots. Thanks to a Client/Server communication model this results very flexible and so it is perfectly interfaced with the architecture showed before. The simulator provides a simulated control connection accessible via a TCP port, that is similar to the real robot serial port connection, making transparent the interconnection to simulator or to real robot. Its flexibility is interconnected with the possibility to customize the simulation environment and simulated objects using a configuration file. The simulator includes a built-in parser based on TinyXML Library and a lot of tags are furnished to describe all setting parameters. In the first version of the simulator, only a limited set of sensor types are provided. At the moment only TriBot I and TriBot II CAD are imported in the simulation tool; the introduction of other robots and features has been planned in the next future.

### 6.5.1 How to Create Environment for ODE Simulator

Many suitable instruments are needed to model and create the mechanical model of the robot starting from the real physical robot. For this purpose a CAD (Computer Aided Design) design program ($Autodesk^{TM}$ Autocad 2008) is used to obtain the 3D design of the robot which can be divided in two parts: *visual* and *physic*, in order to decouple the visual representation of the simulator from the real physic model simulated by ODE engine. The policy chosen to create environment and robot model looks like USARSim approach where the environment is created using AutoCAD model of the real arena.

After that, using another modeling software ($Autodesk^{TM}$ 3DStudio MAX) and two *plugins* (PhysX Plug-In for Autodesk 3ds Max of Nvidia [29] and COLLADA MAX Nextgen by Fcollada [30]), it is possible to obtain the complete model of the robot. In particular, the former plugin provides the physical model of a body while the latter furnishes the visual representation in a standard format (COLLADA that is a XML schema). The decision to use a standard XML-like format to configure sim-

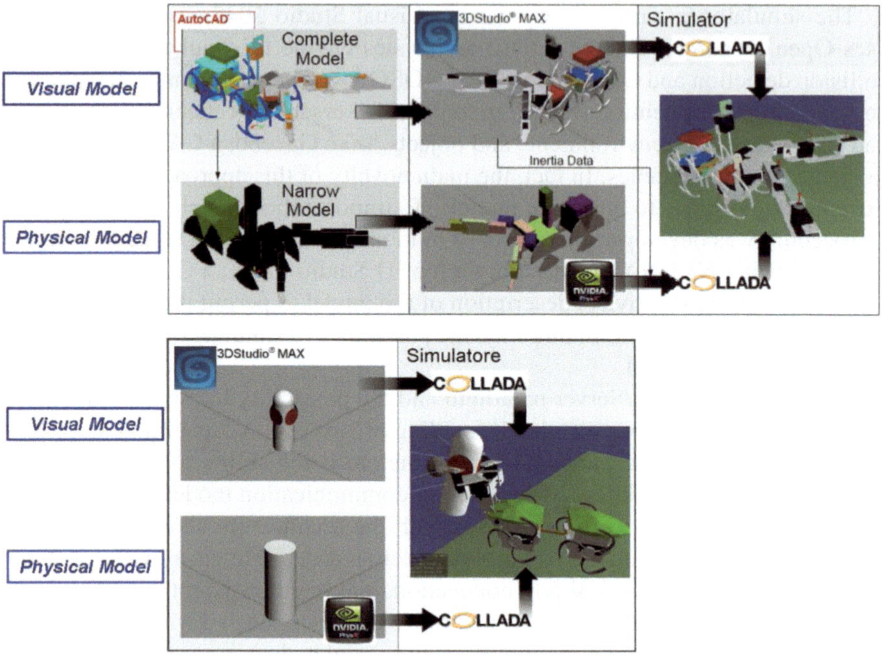


**Fig. 6.14** Pipelined process models design of the Tribot robot and a generic object using the Collada standard

ulator reminds the ARGoS approach, giving the possibility to include new modules in a transparent way.

In relation to the realization of environments or objects, this procedure will became extremely simple to be followed, so the models can be designed directly using 3DStudio MAX.

As shown in Fig. 6.14, the complete model is composed by two different parts: *physical* and *visual* for performance issues; for example the visual model can be more 'complex' than the physical one without big losses in performance. This mechanism is needed to simplify the model used by the ODE engine in order to avoid the increase of computational load as regards collision detection algorithms.

## *6.5.2 Logging System*

A *Tracing Mechanism* has been introduced in the Simulator, to help to collect data and traces paths during simulation execution, giving a mechanism to log critical information during the debugging phase.

The flexibility of the simulator's classes structure gives the opportunity to obtain TraceFiles about each of the simulation variables, setting the configuration XML-file

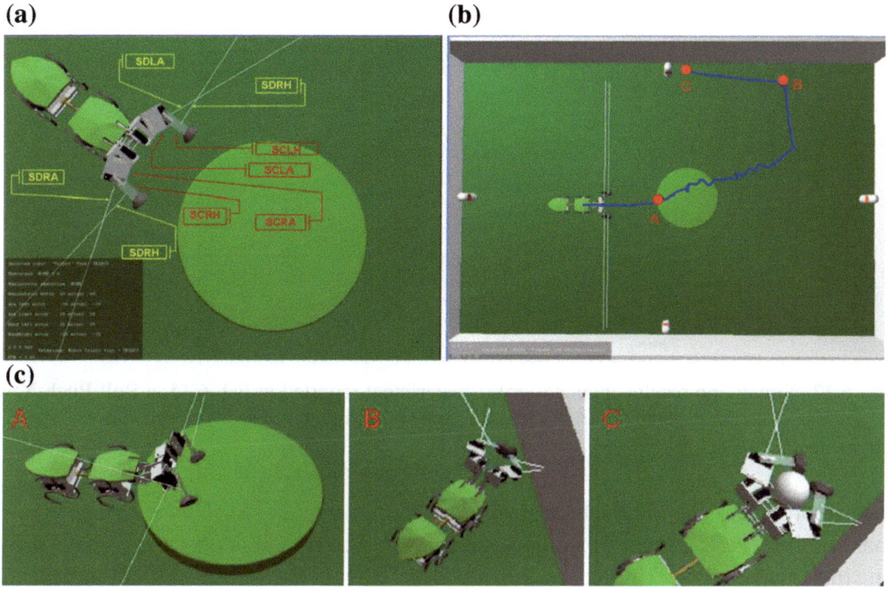


**Fig. 6.15** Example of robot navigation in the simulator. **a** the TriBot robot model with simulated sensors highlighted: distance sensors—**SDLA** (Left Arm Distance Sensor), **SDLH** (Left Hand Distance Sensor), **SDRA** (Right Arm Distance Sensor), **SDRH** (Right Hand Distance Sensor) and contact sensors—**SCLA** (Left Arm Contact Sensor), **SCLH** (Left Hand Contact Sensor), **SCRA** (Right Arm Contact Sensor), **SCRH** (Right Hand Contact Sensor). **b** Navigation path done by the robot in the simulation environment, where three interesting situations occurred during the path followed are underlined. **c** *A*—The robot tries to climb an obstacle in front of it, *B*—The robot, arrives near the wall and tries to avoid it, *C*—The robot takes an object, found in the environment

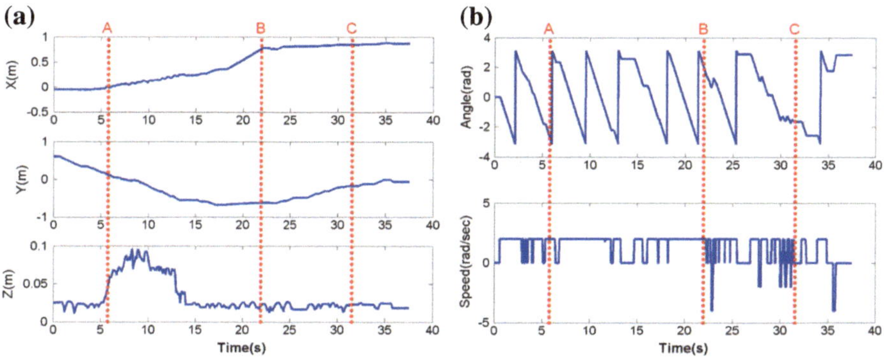


**Fig. 6.16** Simulation results obtained for the experiment reported in Fig. 6.15. **a** trend of the position of the anterior module of the robot in the 3D environment obtained during the simulation **b** Angle position and speed evolution of the anterior left wheel motor

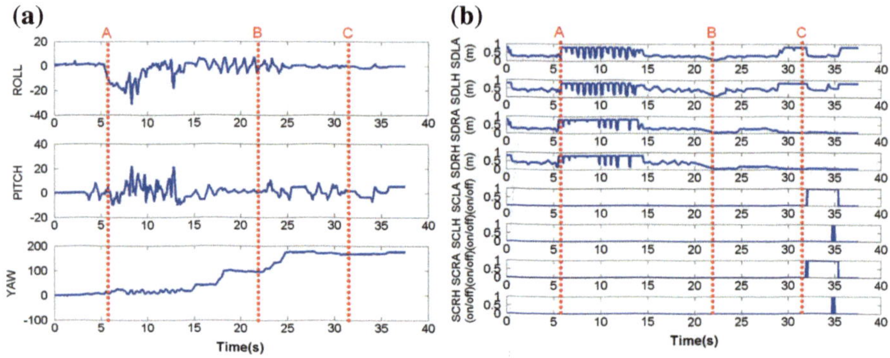


**Fig. 6.17** Simulation results obtained for the experiment reported in Fig. 6.15. **a** Roll-Pitch-Yaw angles related to the robot anterior module, **b** Distance and contact sensor time evolution

in an appropriate way. These files are created by a *Tracer* and a module (*TraceEngine*) that handles file tracing of specified variables. Up to now it is possible to have only one instance of a TraceEngine, and so, only one TraceFile during a simulation. During the Parsing phase, when the configuration file is analyzed, the different Traces are registered on TraceEngine module, and during the simulation, this module will collect desired variables step by step. The *Tracers* already implemented in this version of the simulator are:

- **RigidBodyPositionTracer** to trace the position of a physical body in the model, in terms of absolute position respect geometrical center.
- **RigidBodyRotationTracer** to trace the position of a rigid body, in terms of Eulero's angles in according to Yaw-Pitch-Roll convention.
- **MotorTracer** to log parameters of a motor used in the models of the robot. The variables traced are: position, velocity, supplied couple.
- **SensorTracer** to log sensors' parameters. On the basis of the sensor the variable traced assumes different meanings.

Using programs such as Matlab or GNUPlot it is possible to obtain interesting plots manipulating these files, in order to monitor and analyze relevant information about simulation results. The following figures show an example of robot navigation in the simulator, plotting some interesting information about the path.

In the Fig. 6.15, an example of the visual output of the simulator is provided, referring to the TriBot robot (Fig. 6.15a) and its navigation trail underlined in the virtual arena (Fig. 6.15b) are showed, highlighting three particular positions during the path (Fig. 6.15c), in order to analyze output traces files. In the Figs. 6.16e, 6.17, the first figure (Fig. 6.16a) shows position parameters (*x, y, z* coordinates) of the robot during the path, whereas remaining figures give an explanation of several tractable attributes respectively, wheel motor angle and speed (Fig. 6.16b), Roll-Pitch-Yaw angles relating to the robot (Fig. 6.17a) and contact information and distance value for low sensors (Fig. 6.17b).

## 6.6 Conclusions

In this Chapter the realization of a flexible software/hardware framework, useful to develop cognitive and bio-inspired architectures and the realization of a flexible robotic simulation environment have been described. Here the various blocks identified within the insect brain computational structure are embedded as modules together with other, more traditional algorithms, within the core of the architecture. Thanks to the use of a modular approach, the proposed framework results robust, expansible and general-purpose for a rapid development of reusable applications. The use of generic libraries and common interfaces assures complete independence from hardware or simulated resources. The functional separation helps to isolate the application itself from graphic interfaces and the underlying hardware. Moreover, an ad hoc Dynamic robotic simulator has been provided for complicated 3D scenarios as a powerful tool. It is convenient to underline that the flexibility of the *Client-server paradigm* used for communication permits to be easily interfaced by the RS4CS framework.

## References


1. B.P. Gerkey, R.T. Vaughan, A. Howard, The player/stage project: tools for multi-robot and distributed sensor systems, in*Proceedings of the 11th International Conference on Advanced Robotics*, pp. 317–323 (2003)
2. G.K. Kraetzschmar, H. Utz, S. Sablatnög, S. Enderle, G. Palm, Miro - middleware for cooperative robotics, in *RoboCup 2001: Robot Soccer World Cup V* (Springer-Verlag, London, 2002), pp. 411–416
3. I. Awaad, R. Hartanto, B. León, P. Plöger, A software system for robotic learning by experimentation, in *SIMPAR '08: Proceedings of the 1st International Conference on Simulation, Modeling, and Programming for Autonomous Robots*, (Springer-Verlag, Berlin, 2008), pp. 99–110
4. M. Henning, A new approach to object-oriented middleware. IEEE Internet Comput. **8**(1), 66–75 (2004)
5. A. Makarenko, A. Brooks, T. Kaupp, Orca: components for robotics, in *IEEE/RSJ International Conference on Intelligent Robots and Systems (IROS'06)*, (2006)
6. N. Koenig, A. Howard, Design and use paradigms for gazebo, an open-source multi-robot simulator, vol. 3 (2004) pp. 2149–2154
7. M. Lewis, J. Wang, S. Hughes, Usarsim: Simulation for the study of human-robot interaction. J. Cogn. Eng. Decis. Mak. **1**(1), 98–120 (2007)
8. Webots. http://www.cyberbotics.com. Commercial Mobile Robot Simulation Software
9. M. Dorigo, C. Pinciroli, M. Birattari, in *IRIDIA* Argos Technical report, Universit Libre de Bruxelles, Brussels
10. A. Iman, L. Beatriz, Xpersim: a simulator for robot learning by experimentation, ed. by S. Carpin et al., in *SIMPAR*, vol. 5325. Lecture Notes in Computer Science, (Springer, 2008), pp. 5–16
11. P. Arena, L. Patané, P.S. Termini, A. Vitanza, R. Strauss, Software/hardware issues in modelling insect brain architecture, in *4th International Conference on Intelligent Robotics and Applications (ICIRA)*, (Aachen, Germany, 2011)
12. B. Siciliano, O. Khatib (eds.), *Springer Handbook of Robotics* (Springer, Berlin, Heidelberg, 2008)

13. P. Arena, S. De Fiore, L. Patané, M. Pollino, C. Ventura, Insect inspired unsupervised learning for tactic and phobic behavior enhancement in a hybrid robot, in *WCCI 2010 IEEE World Congress on Computational Intelligence* (Barcelona, Spain, 2010), pp. 2417–2424
14. P. Arena, L. Patané, M. Pollino, and C. Ventura. Tribot: a hybrid robot for cognitive algorithm implementation, in *Proceedings of the 18th IEEE Workshop on Nonlinear Dynamics of Electronic Systems (NDES 2010)*, pp. 1–6
15. R.T. Schroer, M.J. Boggess, R.J. Bachmann, R.D. Quinn, R.E. Ritzmann, Comparing cockroach and whegs robot body motion, in *Proceedings of the IEEE International Conference on Robotics and Automation*, (New Orleans, 2004), pp. 3288–3293
16. P. Arena, L. Fortuna, M. Frasca, L. Patané, M. Pavone, Implementation and experimental validation of an autonomous hexapod robot, in *Proceedings of IEEE International Symposium on Circuits and Systems*, (Kos, Greece, 2006) pp. 401–406
17. Tarry project. Tarry robot home page. http://www.tarry.de/
18. P. Arena, M. Cosentino, L. Patané, A. Vitanza, SparkRS4CS: a software/hardware framework for cognitive architectures. Paper presented at the 5th SPIE's international symposium on microtechnologies. (Czech Republic, Prague, 2011), pp. 1–12
19. P. Arena, C. Berg, L. Patané, R. Strauss, P.S. Termini, An insect brain computational model inspired by *Drosophila melanogaster*: architecture description. Paper presented at the international symposium on neural networks (2010), pp. 1–7
20. P. Arena, L. Patané, P.S. Termini, An insect brain computational model inspired by *Drosophila melanogaster*: simulation results. Paper presented at the international symposium on neural networks (2010), pp. 1–8
21. P. Arena, S. De Fiore, L. Patané, M. Pollino, C. Ventura, in *SPIE* Stdp-based behavior learning on the TriBot robot, vol. 7365 (2009), p. 736506
22. R. Ernst, M. Heisenberg, The memory template in *Drosophila* pattern vision at the flight simulator. Vision Res. **39**(23), 3920–3933 (1999)
23. G. Liu, A. Seiler, H. Wen, T. Zars, K. Ito, R. Wolf, M. Heisenberg, L. Liu, Distinct memory traces for two visual features in the *Drosophila* brain. Nature, **439**(5), 551–6, (2006)
24. S. Tang, A. Guo, Choice behavior of *Drosophila* facing contradictory visual cues. Science **294**(5546), 1543–1547 (2001)
25. K. Neuser, T. Triphan, M. Mronz, B. Poeck, R. Strauss, Analysis of a spatial orientation memory in *Drosophila* (2008)
26. S. Scherer, R.F. Stocker, B. Gerber, Olfactory learning in individually assayed *Drosophila* larvae. Learn. Mem. **10**, 217–225 (2003)
27. ActivMedia Robotics. Aria - Advanced robotics interface for applications.
28. ActivMedia Robotics. Pioneer 3 operations manual (2006)
29. The Nvidia PhysX Technology. http://www.nvidia.it/object/physx
30. Collada MAX NextGen. https://collada.org/mediawiki/index.php/colladamax

# Chapter 7
# Robotic Platforms

**I. Aleo, P. Arena, S. De Fiore, L. Patanè, M. Pollino and C. Ventura**

**Abstract** To evaluate the capabilities of the insect brain model different robotic platforms have been considered. The different blocks of the cognitive architecture, inspired by MBs and CX, can be used as control systems both for legged and wheeled robots. This chapter reports the characteristics of the robotic platforms including information on the mechanical structure, sensory system, software and hardware low level control architecture. In particular the wheeled robot considered is the Pioneer P3AT, a commercial platform for indoor and outdoor applications, suitably modified to host new sensors and control boards. Concerning legged and hybrid robots, a series of robots have been developed to exploit the insect brain main functions on different platforms.

## 7.1 Introduction

The General software Hardware framework named RealSim4CS presented in the previous chapter and developed for the implementation of the proposed Insect-inspired control architecture, opened the way to the application of different elements of the complete control structure to several robotic platforms.

Figure 7.1 shows the different robots used for the evaluation of the insect brain model. In particular Roving robots have been used for testing orientation and navigation strategies adopted by the fly; hybrid robots where considered to better understand the interaction of different basic behaviours and for some motor learning purposes; legged systems where used for CPG and low-level motor learning whereas multilink

I. Aleo · P. Arena · S. De Fiore · L. Patanè (✉) · M. Pollino · C. Ventura
Department of Electrical, Electronic and Computer Science Engineering,
University of Catania, 95125Catania, Italy
e-mail: lpatane@dieei.unict.it

P. Arena
e-mail: parena@dieei.unict.it

P. Arena and L. Patanè (eds.), *Spatial Temporal Patterns for Action-Oriented Perception in Roving Robots II*, Cognitive Systems Monographs 21,
DOI: 10.1007/978-3-319-02362-5_7, 

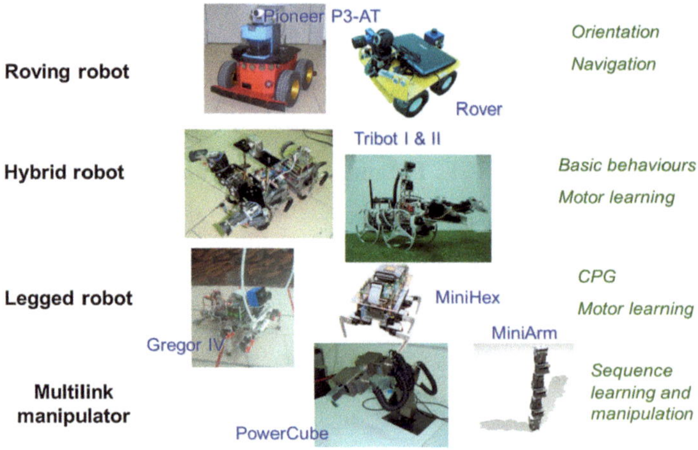


**Fig. 7.1** Gallery of the different robotic platforms used as testbed for the insect brain model

manipulators where also taken into account for sequence learning and manipulation tasks.

In the following sections details about the characteristics elements of each group of robots are provided.

## 7.2 Pioneer P3-AT

The Pioneer P3-AT is a commercial robot for autonomous navigation in unknown environment hosting a complete multi-sensory system. It is a classic four wheeled rover controlled through a differential drive system and it uses 100 tick encoders with inertial correction to compensate for skid steering.

The platform uses a client-server based control architecture. In the model, the robot server works to manage all the low-level details of the mobile robot systems. These include operating the motors, collecting and reporting sonar and wheel encoder data and others. With this client/server architecture, robotics applications developers do not need to know many details about a particular robot server, because the client isolates them from this lowest level of control.

The robot (see Fig. 7.2) is equipped with a microcontroller for a low level control of all activities regarding motion control, sensory system management and server implementation for client/server architecture.

The robot includes the following equipped systems:

- 8 sonars;
- motion control with odometry error compensation;
- three batteries for up to three hours autonomy;

**Fig. 7.2** Roving robot P3AT and embedded sensory systems

- 10 bumpers;
- gyroscope;
- digital compass;
- laser scanner;
- pan-tilt-zoom color camera;
- on board pc with wireless and wired Ethernet-based communications for autonomous applications;

We have customized the standard configuration including a hearing circuit, a CNN-based camera with panoramic lens, named Q-Eye (see Chap. 9 for details) [1], and a gray scale color sensory placed under the robot used to detect a black spot on the ground. These additional sensors are managed from the onboard computer using a microcontroller-based bridge.

The hearing sensor is inspired by the cricket hearing system and allows efficient localization of a specific sound source with a very simple analog circuit. It is inspired by phonotaxis in cricket: its ability to approach sound sources. Female crickets are able to recognise the species specific pattern of the male calling song, produced by opening and closing their wings, and move towards it. For *Gryllus bimaculatus* these songs consist of four 20 ms syllables of 4.7 kHz sound waves, separated by 20 ms intervals, which generate a "chirp", produced several times a second. Females appear to be particularly selective for the repetition rate of syllables within each chirp. The Ears circuit [2, 3] consists of two microphones and a circuit board fine-tuned to the carrier frequency of the cricket song. The output from each ear is an analog signal in the range from 0 to 5 V. The input to the circuit is given by two microphones separated by a distance equivalent to a quarter of the wavelength of the carrier frequency (4.7 kHz), i.e. 18 mm.

The block scheme of the complete Software/Hardware framework is reported in Fig. 7.3.

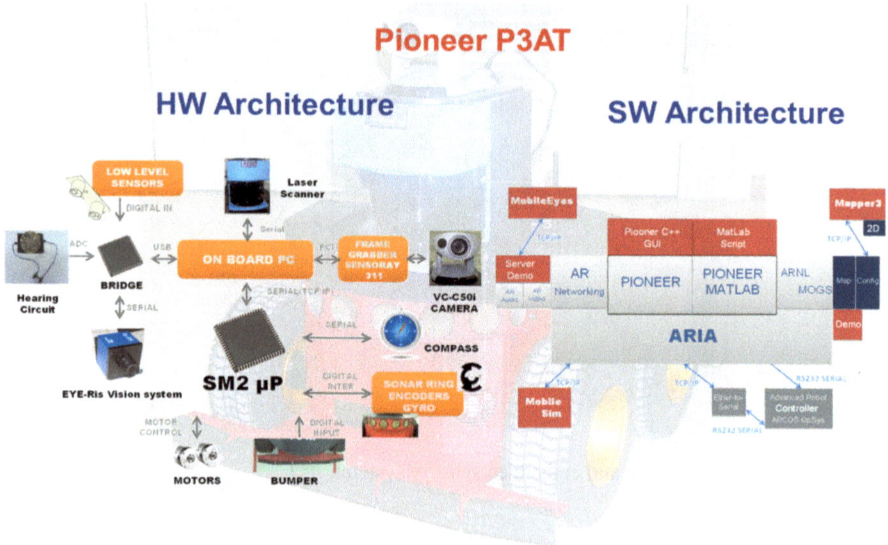


Fig. 7.3 Software Hardware architecture of the roving robot P3AT

In the following, some features are presented and discussed. For more details see [4].

## 7.2.1 The P3-AT Software Architecture

The architecture developed for the P3-AT robot follows a modular programming paradigm. The basic architecture has been expanded introducing a standard interface for VS2005 and Matlab environments. The core of the architecture is the ARIA (Advanced Robotics Interface for Applications) library that maps to the client software all the robot or simulator (named MobileSim) features. ARNL (Advanced Robotics Navigation and Localization) provides the tools to quickly and easily create, edit, and deploy the maps (using also the Mapper3 software) and floor plans required for advanced robotics applications. ARNL is a complete localization and autonomous navigation solution, as well as a comprehensive suite of development tools and example applications by which, with your LRF-enabled robot, you create and perform highly advanced mobile robotics tasks. Thanks the GUI server and MobileEyes tools, it is possible to drive the robot through a graphics user interface.

In order to guarantee an easy way to access at the robot facilities, a dynamic link library (dll), named Pioneer, has been developed. The library allows a standard interface with Visual Studio C++ Application and Matlab scripts. The main blocks included in the library together with a brief explanation are reported in the following:

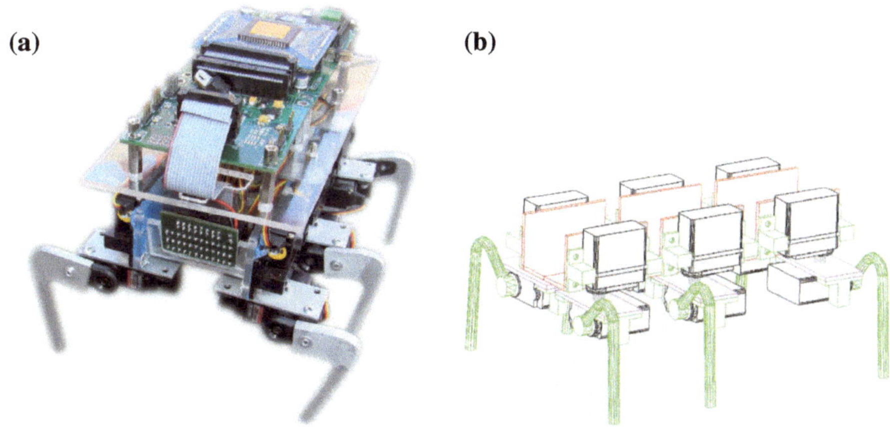

Fig. 7.4 The robot MiniHex II. **a** Prototype, **b** mechanical model

- **Sensors methods**: a collection of routines to manage sensors (sonar, compass, etc.);
- **Localization routines**: a collection of methods to solve a localization problem;
- **Robot's parameters analysis methods**: a collection of routines to analyze the robot state (sensors, battery level, etc.);
- **Utilities**: utility routine (trajectory calculation, etc.);
- **Action command**: the methods to drive the robot;
- **Camera functions**: the routines to manage the camera mounted on the robot;
- **Algorithms procedures**: a basic navigation algorithms implemented;
- **Connections handlers**: a collection of methods to manage the robot connection.

The platform is adapted for indoor and outdoor experiments as reported in [5, 6].

## 7.3 Hexapod MiniHex II

The hexapod MiniHex II (Fig. 7.4a) is the last prototype of the MiniHex series. Compared to the previous version of the robot (see for details [7]), the mechanical structure (Fig. 7.4b) is identical. The real difference is the improvement of the electronic circuitry. In particular the new robot uses a DC-DC converter to generate the Dual Voltage (instead of two more batteries as in the previous release), an Xbee module for the wireless communication and a communication bus that guarantees more flexibility and modularity to the structure.

The MiniHex II is equipped with three infrared sensors used to avoid bumps with walls or objects. The robot communicates with a PC via wireless, using a Zigbee protocol with two DigiMax Xbee modules, one on the PC and one on the robot. There are two ATMEL microcontrollers on the system: an ATMEGA162 and an

(a)

(b)

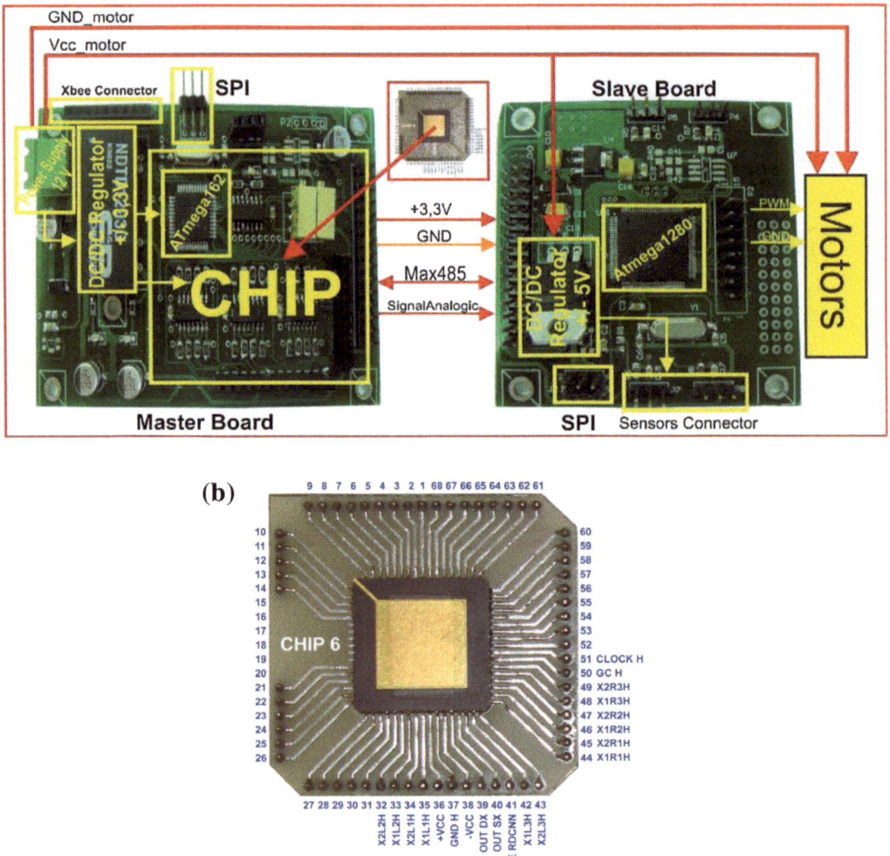


**Fig. 7.5** Hardware of MiniHex II. **a** Board with AT162 and AT1280, **b** VLSI Chip model

ATMEGA1280 (Fig. 7.5a) used to control the wireless communication, the PWM for the 12 motors Hitec HS-85MG, the ADC conversion of the infrared sensors and the communication with the analog Chip (Fig. 7.5b). The CNN-based CPG chip generates the locomotion patterns needed to control the robot movements [7]. Six different gaits are available: Fast, Medium and Slow gait for Forward and Backward motion. The robot is completely autonomous for the power supply. A 4.8 V, 0.3 A Ni-Mh battery pack is used for the motors and one 9 V 0.25 A Ni-Mh battery for the electronics.

The robot is a suitable platform and can be used both to better understand the locomotion patterns implemented by the CNN that try to reproduce the real patterns of locomotion of the cockroach and to test new high level cognitive algorithms on a legged platform. The wireless module is used to switch between the different patterns implemented on the VLSI Chip and for monitoring the status of the system.

**Table 7.1** Main characteristics of the MiniHex II

| | MiniHex II |
|---|---|
| Height | 10 cm |
| Length | 15 cm |
| Width | 10 cm |
| Weight | 0.8 kg |
| Motors | 12 × Hitec HS-85MG |
| | 4.8 V DC |
| Batteries | 1 × 0.25A@9 V |
| | 1 × 3A@4.8 V |
| $\varepsilon$ | 12.5 |

In Table 7.1, the main characteristics of the MiniHex II are shown. To measure energy efficiency an increasingly accepted parameter is the "Specific Resistance" [8] that for the MiniHex II, has been obtained during the movement at maximum speed on a flat terrain ($\varepsilon = 12.5$).

The MiniHex II robot was conceived since the beginning of its design as a test bed for demonstration of reliability of spatio-temporal dynamics to generate locomotion patterns. At this aim the CNN-based analog chip was used as locomotion controller. On the other side the VLSI implementation is quite fixed and does not allow any further modification of the neural dynamics there embedded. In order to study the effect of different neural controllers, a new controller design was performed witch used a low cost microcontroller. This strategy allows to simply modify the neuron number, the synaptic connectivity and all the other parameters that are responsible for the neural plasticity. The architecture was conceived as a master/slave configuration where the master controller is connected via wireless to a host PC able to change relevant neural parameters.

## *7.3.1 The MiniHex Software Architecture*

The MiniHex architecture, Fig. 7.6, is composed of three micro-controllers. The first, called ATMega162, is devoted to motion control and software interface using a wireless communication with a PC through XBee modules. The second one, called UC3B1256, generates the locomotion patterns and the last, called ATMega128, is used only for motors driving.

The UC3B1256 generates the locomotion patterns following the Central Pattern Generator paradigm (CPG) [9]. In particular, it is a 32 bit based microcontroller and the CPG equations are integrated using an Eulero integration method. At 66 MHZ up to 4,000 integration steps per second are available with a 0.01 integration step.

In order to permit the communication among different nodes, a master-slave based protocol has been developed. The robot is the slave and, in this case, a Personal Computer is used as master.

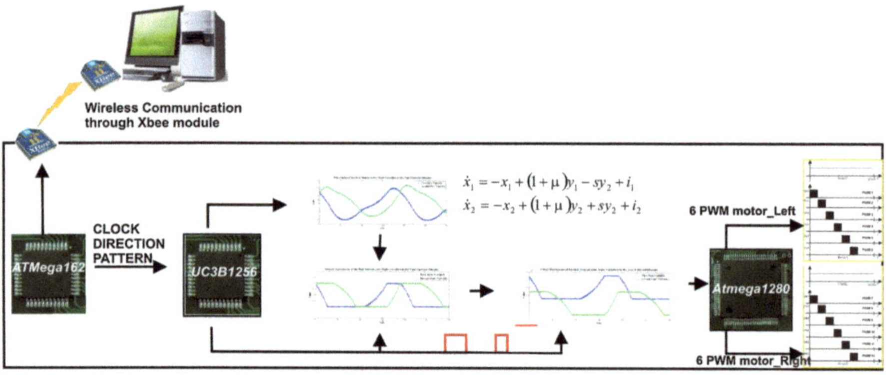


**Fig. 7.6** The Minihex software architecture

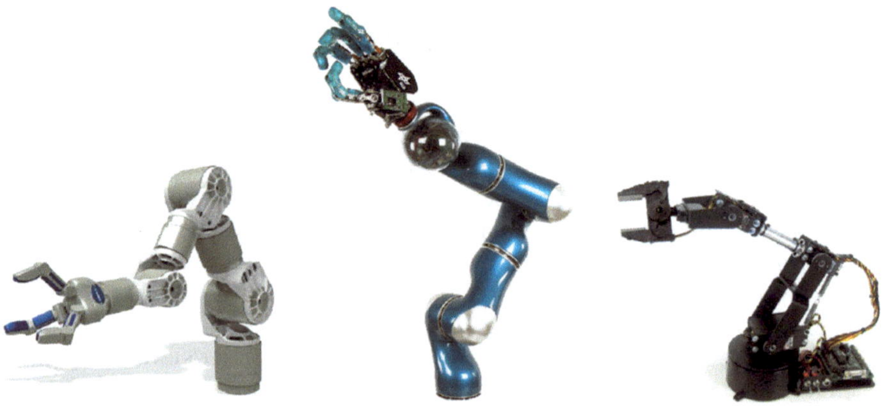

**Fig. 7.7** Serial manipulators for research purposes: on the *left* LWA3 from Schunk, LWRIII from DLR on the *center* and on the *right* ALB5B from Lynxmotion

## 7.4 MiniArm Robot

Serial manipulators are, nowadays, commonly used in industrial applications, for example in the automotive market. Their *manipulation* capabilities and kinematic properties are already known and well studied in the past decades. Nevertheless their use in service robotics is almost completely limited in big static structures with high costs and a really closed control environment. Due to these characteristics they are often not well suited for academic research purposes. Exceptions are Schunk [10] and DLR [11] modular robots, for *high-end* users, and Lynxmotion Robotic Arms [12], mostly for *low-end* users and hobbyists (see Fig. 7.7).

Inside the SPARK II project, a standard serial manipulator for research activity, called PowerCube, is used to apply the cognitive algorithms to different robotic platforms. The PowerCube is a commercial platform provide by Schunk [10]. Moreover,

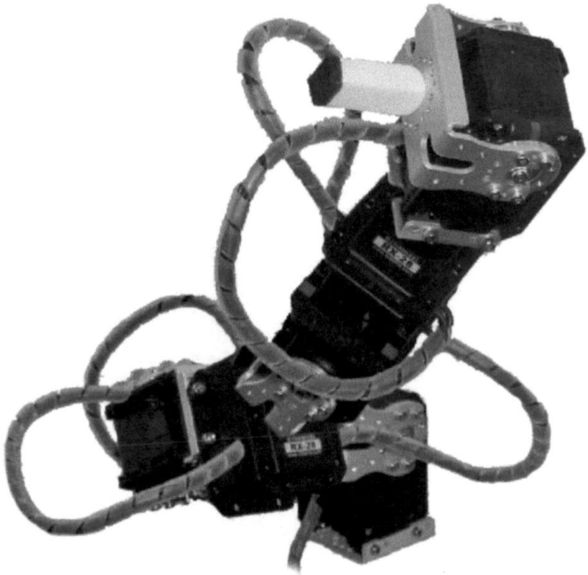

**Fig. 7.8** Picture of MiniARM first prototype realized with 6 DoF

in order to decrease the development time, an efficient small version of PowerCube, called MiniARM, has been developed. Moreover, for the same reason, a simulation environment has been developed for both platforms. In the next chapters, a brief and general description of the arm platforms used inside the project is provided in order to outline the capability of each robot.

### *7.4.1 MiniArm Architecture*

The MiniARM, shown in Fig.7.8, is a serial manipulator with rotational joints, completely custom built, developed in order to analyze and compare algorithm performances with low implementation effort: the underling idea for the design was the realization of a robust test-bed for high-level control algorithms.

The chosen actuators for the entire system are the Dynamixel smart servos from Robotis [13] and an embedded microcontroller board with an AVR32 UC3A 32bit microcontroller architecture, from Atmel, is used to control all the servos and retrieve sensory information through a RS-485 four-wires bus. A sensor unit with a three-axis accelerometer, a gyroscope and a digital compass is linkable to any of the present servos through the bus. All the high level control algorithms can be developed directly in the PC (Windows, Linux and MacOS operating systems are supported) using a *USB-to-Serial* communication and thanks to a Matlab Graphical User Interface and a developed serial library. An external USB camera directly connected to the PC is

**Fig. 7.9** High definition rendering of the realized CAD model for the MiniARM

supported by a GUI. The compatibility with all the most common simulation tools and control environments is given by a standard 3D model reconstruction, realized in a CAD drawing software (Solid Works [14]) using IGES model distributed by Robotis, together with a simplified low-count vertex VRML mesh and the serial control library.

This section describes the proceedings in the realization of the MiniARM with the hardware design considerations and the implemented software libraries. A general and modular approach is investigated and an external example of multi-purpose sensor node is presented.

### *7.4.2 Mechanical Design*

The MiniARM serial kinematic chain is designed to be redundant in space, within its dextrous operating space, and the mechanical structure has been realized for high payload/weight ratios (1 : 2, 1 : 3).

The geometry of the manipulator and the assemble of the entire structure have been studied through a CAD project, whose shaded rendered model is shown in Fig. 7.9. The final developed configuration shows the subsequence of orthogonal seven rotational axis servos. Almost all the structure interconnection frames are made with light commercial aluminium parts. The gripper module has a custom design.

**Table 7.2** Mechanical and electrical characteristics of Dynamixel

| Model | Stall torque (kg/cm) | Operating voltage (V) | Speed ($s$/60 degrees) |
|---|---|---|---|
| RX-28 | 37.2 [@16 V] | 12–16 | 0.126 [@16 V] |
| RX-64 | 64.2 [@16 V] | 16–18 | 0.169 [@16 V] |

Multiple hardware tests have been performed in order to test the goodness of the mechanical structure. A fully functional first prototype of the overall architecture has been realized. The estimated payload is of about 300 g when fully stretched. A simple pointer as been realized as end-effector for inverse kinematic problem solving using multiple approaches. The overall robot weight has been estimated to be about 0.7 Kg (slightly depends on the last module) and it is 45 cm long, together with the end-effector module. Preliminary power supply test shows a maximum power consumption of 3.5 A with full payload. Despite the complexity of the structure the wiring is extremely simple thanks to the communication bus presence. The mechanical structure has been assembled using aluminium brackets (2 mm thick) and hexagonal mounting screws ($M2$ and $M2.5$ and $M3$).

The robot mechanical structure involves two different kind of servos RX-28 and RX-64. The RX-28 is a very powerful compromise between high *stall-torque* and relatively *high-speed*. Looking at Table 7.2 and thinking the driving torque to be half of the holding one the overall structure should be able to drive at least 0.5 Kg at 20 cm of distance.

In order to implement a mixed Dynamixel bus with both RX-28 and RX-64 the chosen power supply voltage is 16 V.

### 7.4.3 Hardware Architecture

The manipulator low-level control has been realized with a microcontroller based (MCU) custom made board shown in Fig. 7.10. The embedded MCU is an AVR32 UC3A1512 from Atmel. The UC3A is a cost effective 32-bit architecture that reaches 66 MHz of clock and counts a lot of different peripherals (e.g. SPI, USB, DAC for HiFi, ADC mutiplexed, Ethernet, USART). The wireless outboard communication is managed with a 2.4 GHz transceiver (AT86RF230 from Atmel) fully compliant with the IEEE 802.15.4 physical layer. The firmware Integrated Development Environment (IDE) is the AVR32 Studio. It is based on the open-source project, Eclipse, and it is distributed from Atmel. The firmware upgrade is possible with a JTAG interface provided by Atmel JTAGICE MKII programmer. The RS-485 interface has been implemented with an external transceiver (from ST Microelectronics).

The sensory system in a robotic structure is a necessary interface to the external environment and it is able to give to the robot proprioceptive cognition capabilities. Under this point of view a sensor network structure has been studied and

**Fig. 7.10** Developed AVR32 MiniARM control board

a seven Degrees Of Freedom (*DoF*) Inertial Measurement Unit (*IMU*) has been developed. The MiniARM sensor board development involved a *tri-axis* accelerometer device (ST Microeletronics LIS3LV02DQ), a *single-axis* analog gyroscope (ST Microeletronics LISY300AL) and a *tri-axis* digital compass (Honeywell HMC5843). The MCU is a ATMEGA8 of the AVR 8-bit architecture, from Atmel. A final prototype of the realized board is shown in Fig.7.11, The realized board layout is 900 $mm^2$ (30 mm × 30 mm) and the internal circuitry requires 6.5 V which are then regulated at 5 V (the relatively high voltage is needed for the communication voltage) and 3.3 V. The board is fully compliant with the RS-485 Dynamixel communication protocol. The serial bus compatibility of the developed sensor permits a fundamental extension toward the networked node concept: a sensor can be, up to now, added in any part of the bus without changing the entire communication strategies.

## *7.4.4 Software Architecture*

As introduced before both a Virtual Reality environment and a Robot Control library have been implemented for simple user interface development.

A low-count vertex mesh (see Fig. 7.12) has been realized, through a 3D Simulink-ready model, in order to analyze the movement of the manipulator in a kinematic simulation environment and to implement complex motion algorithms without the real robot. The VRML world is made up of multiple entities, one per each actuated joint. Thanks to the Matlab Virtual Reality toolbox it is possible to animate the

**Fig. 7.11** Developed sensor board with seven DoF and ATMEGA8 microcontroller

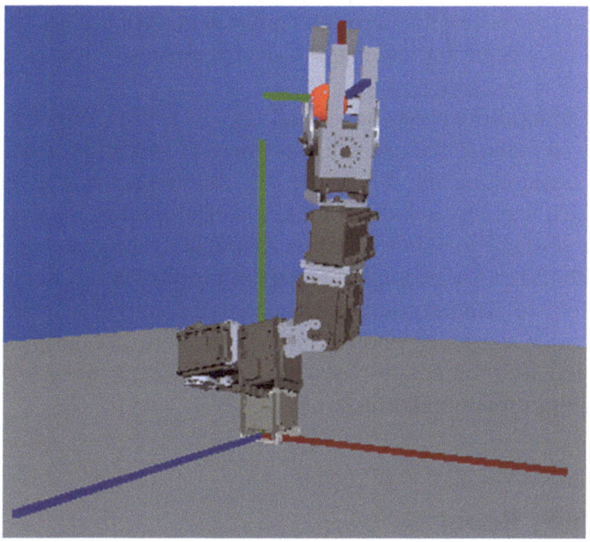

**Fig. 7.12** VRML model of the MiniARM

manipulator respecting the parents relationships and the real physical constrains. This kind of meshed model can be simply used by other external tools, so it is possible to use it in any simulation and control environment, as for example Visual Studio (with multiple languages e.g. C, C++ and C#), Microsoft Robotic Studio and Matlab.

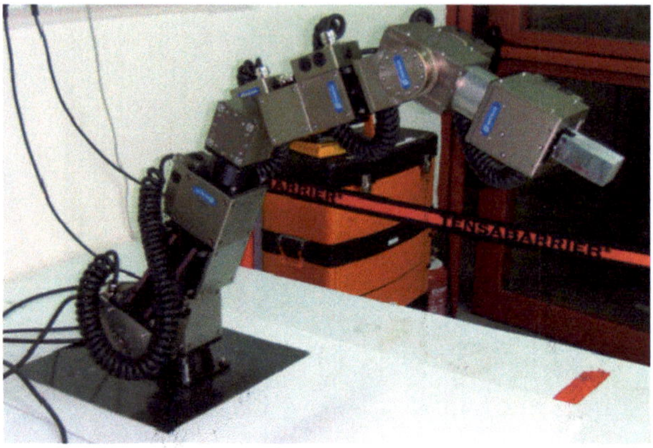

**Fig. 7.13** Picture of PowerCube prototype assembled in our laboratories

The described virtual reality simulation environment has been joined to a Matlab Graphical User Interface (*GUI*) in order to provide real-time configuration representation of the entire manipulator: torque, position, speed and all the other output of connected sensors are accessible through the serial control library. In this way high-level command have been implemented to send sequence of low-level command to the entire robot.

Multiple IK algorithms (both novel and classic approaches) have been implemented in order to manage the end-effector position and orientation. Thanks to the redundant kinematic chain the manipulator is able to reach a given ($x$, $y$ and $z$) position in various way through different kinds of minimization algorithms to realize for example obstacle avoidance. Either Jacobian Transpose and Pesudo-inverse Jacobian algorithms have been implemented both in simulation and in real hardware control. The interface also introduce the image processing of an external USB camera. Real-time video streaming and a simple blob centroid filtering algorithm have been implemented. Reinforcement-based algorithms were applied to the manipulator for solving minimum-path problems [15].

## 7.5 PowerCube Robot

*PowerCube* is a serial manipulator composed by six degrees of freedom as shown in Fig. 7.13. The main characteristic of the PowerCube robot is the modularity. In fact, it consists of a number of rigid links, the module, connected with joints. In order to evaluate cognitive algorithms in this platform, each module has been equipped with a series of sonar sensors (SRF08) that can be used for obstacle avoidance path planning [16].

**Fig. 7.14** High definition rendering of the realized CAD model for the PowerCube

## *7.5.1 Mechanical Design*

The PowerCube serial kinematic chain is designed to be redundant in space, within its dextrous operating space.

The geometry of the manipulator and the assembly of the entire structure have been studied through a CAD project using a standard commercial module [10], whose shaded rendered model is shown if Fig. 7.14. The configuration shows the subsequence of orthogonal four rotational modules, one pan-tilt and the gripper modules (six degrees of freedom). The structure interconnection frames are made with specific parts. The robot mechanical structure involves two different kind of rotatory modules: PR-90, PR-70. The pan-tilt module is called PW-70 and the gripper one is a PG-70 model.The PowerCube robot is equipped with two PR-90 and two PR-70 modules. Both module models use a brushless DC servo-motor as drive controlled in position, speed and torque. The modules can be controlled via Profibus DP, CAN bus or RS-232.

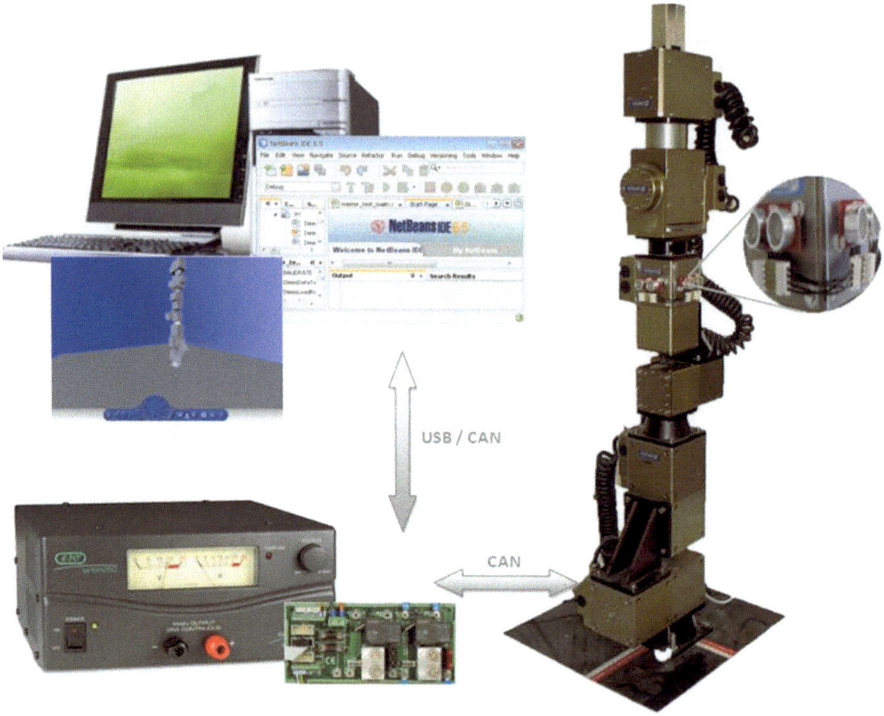


**Fig. 7.15** The powerCube hardware and software architecture

## *7.5.2 Software and Hardware Architecture*

As shown in Fig. 7.15, the robot is controlled by a CAN bus for the module and, as said before, a series of sonar sensors has been added and controlled by I2C bus. A separate module, called driver, permits the interface between the robot and the controller, a standard PC in this case. Moreover, it provides 24 V to the robot using a standard power supply. Thanks to the modular structure of the robot and the features integrated on each module (the module is able to control its position independently), the controller is only a coordinator. It provides the position command to each module according to the global control strategy. The control algorithm has been developed in C code using an open source IDE called NetBeans. Finally, a kinematic simulator has been developed using the Matlab Virtual Reality toolbox. The simulator has been implemented to simplify the debugging phase during the development of cognitive strategies for the manipulator.

Both PowerCube and miniARM robots show a really simple way to implement a distributed control system merged with a sensor network thanks to the modular structure. In fact, both the robots, are based on commercial modules that are able to control the main characteristics (position, speed and torque) autonomously and independently of the other modules of the robot. The communication between them could be very useful to overcome more difficult tasks in which each single unit could be ineffective. Under this point of view the use of a kinematic simulation environment could be of primary importance to operate multiple subsequent strategies in simulation before executing the best one with the real robot or robots (in case of cooperation among different units). Finally, forward and inverse kinematic algorithms have been implemented both in a classical and in a novel recurrent network approach. In particular the RNN called MMC (Mean of Multiple Computation) was implemented to control on the fly the forward and inverse kinematics of the Power Cube and MiniArm in order to reach a given position reference in a redundant space in front of moving obstacles within the operative space [17].

## 7.6 Hybrid Robots

Robotics is a field in continuous evolution. During the last years a considerable attention was focused on finding new original structures with new mechanics solutions, to face with complex tasks. Sometime it is very interesting to create hybrid robots, taking advantages from the different solutions taken into consideration. These were the bases for the design and realization of a modular hybrid robot named TriBot where new solutions were studied, trying to find improvements and to include innovative solutions to add new capabilities to the robot [18, 19]. The TriBot structure is constituted by two wheel-legs modules, an optimal solution for walking in rough terrains and to overcome obstacles. Moreover, a manipulator was added to improve the capabilities of the system that is able to perform various tasks: environment manipulation, object grasping, obstacle climbing and others. Experimental results proved that this hybrid robot is a very performing solution that allows to deal with different environmental conditions, to move agilely both in smooth and in rough terrains thanks to the particular shape of its wheels and to the help of the manipulator.

There is a significant interest in the development of robots capable of autonomous operations in complex environments. Potential operations for such a robot include for example mine clearing and terrain mapping. Therefore, critical issues to be solved to succeed in this operations are: terrestrial mobility, reduced power consumption, efficient navigation and control strategies, robust communication protocols, and a suitable payload. When different tasks are taken into consideration, the robot should be able to deal with either structured (regular) environments or with unstructured environments where the terrain is not a-priori known.

Mainly, there are two kinds of robots: wheeled and legged, which have different characteristics. Wheels, in fact, are relatively simple to control, and allow a vehicle to move quickly over flat terrains [20]. On the other side, a major advantage of legs over wheels is their ability to gain discontinuous footholds, i.e. they alternate between the stance phase, in which they contact the substrate, and the swing phase, in which they are not in contact. This approach is suitable on irregular, discontinuous terrains found in most real-world missions.

**Fig. 7.16** The MiniHex robot. The low level locomotion control is based on the CPG paradigm

Besides, to make a robot as much autonomous as possible, power consumption is of notable importance. This task is more easily attainable for wheeled robots, because of their relatively low energy consumption, while, this problem is mostly present in legged robots, that, instead, introduce a notable energy consumption. On the other hand, multi-legged robots are more robust, in fact they can continue moving also with the loss of a single leg. This characteristic is absent in wheeled vehicles, where a damaged wheel could cause the end of mobility. Up to now, more than half of the earth landmass is inaccessible to wheeled vehicles [21]. The same problem is associated to planetary explorations. Moreover, walking robot performance can be improved taking inspiration from nature, and in particular from insect, that can run stably over rough terrains at high enough speeds to challenge the ability of proprioceptive sensing and neural feedback to respond to perturbations within a stride [22].

An example of bio-inspired mini robot controlled using a VLSI chip that implements a CPG through a CNN-based structure is shown in Fig. 7.16 [23].

To have both advantages of wheeled and legged robots it is possible to design an hybrid robot that uses a strategy of locomotion which combines the simplicity of wheels with the obstacle climbing advantages of legs. Examples of wheel-legs robots are Prolero, Asguard, RHex and Whegs [20].

PROLERO (PROtotype of LEgged ROver) was designed in 1996 by A. Martin Alvarez at the European Space Agency. It uses six drive motors (one for each leg). Each leg consists of a spoke, which rotates in a circular motion. A picture of PROLERO is showed Fig. 7.17a.

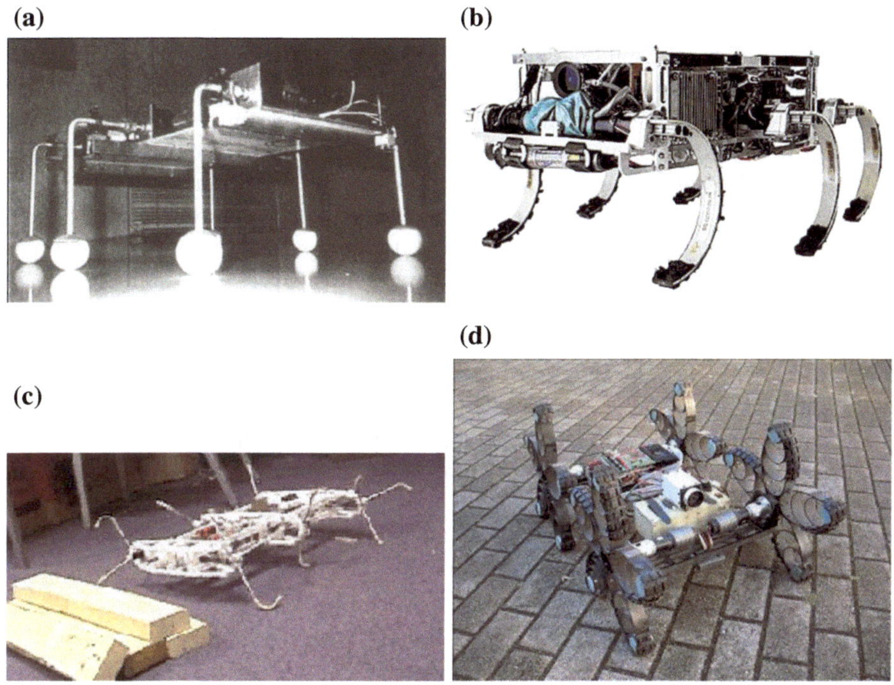

Fig. 7.17 Example of whegs robots. **a** Robot prolero, prototype of Whegs; **b** Robot RHex, predecessor of Whegs; **c** Robot Whegs I; **d** Asguard Robot

Moreover, a cockroach-like robot, named RHex, is the starting point for a major project to understand animals' most distinguishing traits, how they move without falling over. RHex, showed in Fig. 7.17b, also preceded Whegs and was a source of inspiration for Whegs. It also uses a total of six motors and spoke-like legs. During the swing phase of walking, its feet rapidly rotate to return to the stance phase, when the feet rotate much more slowly.

RHex is a man-portable robot with extraordinary rough terrain mobility. It climbs over rock fields, mud, sand, vegetation, railroad tracks and up steep slopes and stairways. RHex has a sealed body, making it fully operational in wet weather, in muddy and swampy conditions, and it can swim on the surface or dive underwater. RHex's remarkable terrain capabilities have been validated in independent testing at US Government Labs.

Finally, Whegs are a series of robots developed at the Biologically Inspired Robotics Laboratory at Case Western Reserve University. They utilize a strategy of locomotion which combines the simplicity of wheels with the obstacle clearing advantages of legs. Therefore, the Whegs vehicles' three-spoke appendages are called "whegs", because they combine the advantages of both wheels and legs.The first robot designed to use abstracted principles of cockroach locomotion was Whegs I showed in Fig. 7.17c.

Another robot, similar to Whegs, that results to be very interesting, is Asguard (Fig. 7.17d), a fast and highly agile quadruped robot [24].

This robot was designed to be used in harsh out-door environments with a focus on security and outdoor surveillance as well as on disaster mitigation missions. For those applications, the requirements for such an in-situ system are that it has to deal with obstacles or uneven and difficult outdoor terrains. Additionally, the robot has to move fast where the ground is leveled and easier to cross. In order to cope with those two requirements, the quadruped robot Asguard was designed, using multiple rotating legs along one hip shaft.

Finally, the bio-inspired robot proposed in this work is named TriBot, due to the inspiration from the Trinacria (from Greek trinacrios, tries that means three and *àkra* that means promoter), the symbol of the Sicily region. The mechanical structure of the robot joins the advantages both of legs and wheels, exactly as wheel-legs robots previously mentioned but, in addiction, it is equipped with a manipulator that allows to add new features to the overall structure. Using this additional module, the robot is not only able to overcome obstacles otherwise unreachable with the only use of wheel-legs, but it is also possible to grasp and carry light objects and to do small and precise movements that, for example, can help the robot in the grasping phase.

## 7.6.1 Mechanical Design

In this section, we discuss about the mechanical characteristics of TriBot, an autonomous mobile hybrid robot.

Up to now two prototypes of the robot TriBot have been developed; the main differences between the two structures refer to the motors used to actuate the wheel-legs modules, the joint that connects these modules with the manipulator and the manipulator configuration.

The mechanical design of the robotic structure and the first prototype are shown in Fig. 7.18. The robot has a modular structure, in particular it consists of two wheel-legs modules and a two-arms manipulator. The two wheel-legs modules are inter-connected by a passive joint, whereas an actuated joint connects these modules with the manipulator, that consists of two legs with three degrees of freedom.

To connect the two modules, a passive joint with a spring have been used. It allows only the pitching movement and facilitates the robot during climbing, in fact the body flexion easily allows, in a passive way, to adapt the robot posture to obstacles. Besides, the actuated joint between the two modules and the manipulator allows the last one to assume two different configurations, therefore it can be useful to improve locomotion capabilities when it is moved down (i.e. used as legs), while when it is moved up it can make manipulation and grasp objects (i.e. used as arms). The whole mechanical structure is realized in aluminium and plexiglass; both materials have been selected for their characteristics of cheapness and lightness. The first prototype uses five Hitec HS-985MG and the manipulator presents a configuration similar to the legs of a bioinspired six legged robot named Gregor [25].

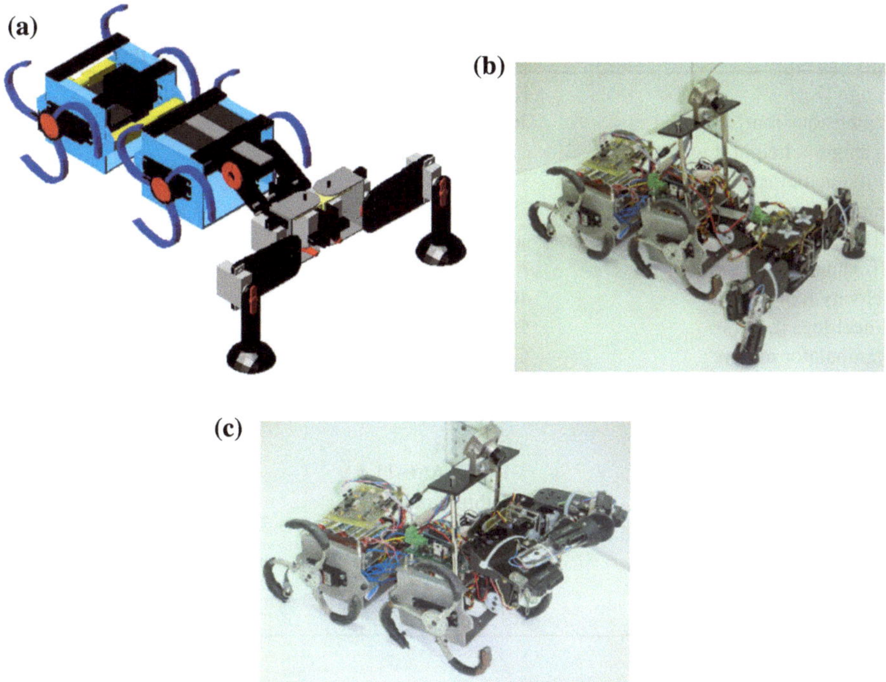


**Fig. 7.18** First prototype of the robot TriBot. **a** AutoCAD design; **b** Manipulator down; **c** Manipulator up

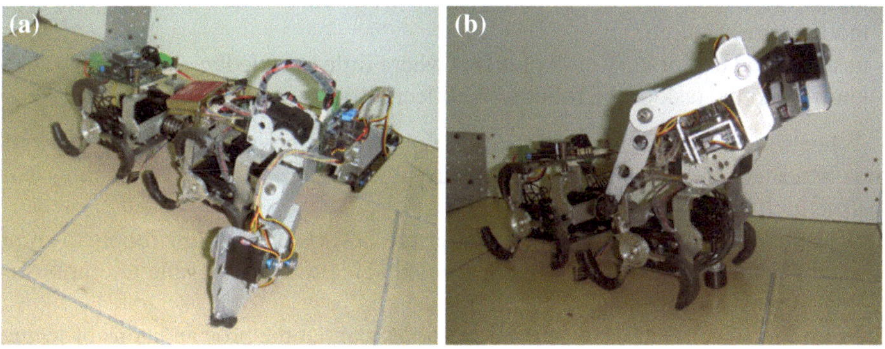


**Fig. 7.19** Second prototype of the robot TriBot. **a** Manipulator down position; **b** Manipulator up position

The second prototype, showed in Fig. 7.19, uses five Dynamixel RX-10 motors and the manipulator is inspired to the robot Tarry, a six-legged walking machine developed at the University of Duisburg [26]. Main technical and mechanical characteristics of both prototype of the robot are shown in Table 7.3.

**Table 7.3** Technical characteristics of TriBot I and TriBot II

| | TriBot I | TriBot II |
|---|---|---|
| Weigh [Kg] | 1,95 | 1.7 kg |
| Dimensions [cm] (length × height × width) (manipulator up) | 36 × 23 × 13 | 32 × 22.5 × 25 |
| Dimensions [cm] (length × height × width) (manipulator down) | 28 × 12 × 25 | 34 × 16.5 × 30 cm |
| Velocity [cm/s] | 46 | 43.5 |
| Wheel-legs motors | 5 × Hitec HS-985MG | 5 × Dynamixel RX-10 |
| Manipulator motors | 6 × Hitec HS-82MG | 6 × Hitec HS-82MG |
| Batteries | 10 × 3000 mAh@1.2 V stylus AA (motors) 1600 mAh@11.1 V Lithium polymers (control boards) | 1600 mAh@11.1 V Lithium polymers |
| Higher obstacle overcoming (using only wheels) | 1.42 | 1.4 times wheel radius |
| Higher obstacle overcoming | 1.8 | 1.85 times wheel radius |

Typically, robotic machines are employed in two types of environments:

- highly structured (regular) industrial environments where precise, repetitive motions are required;
- unstructured exploratory environments where little knowledge about the terrain is known and the task of the robot is to collect information.

Wheels are easy to control, as they act very predictably when running on a well-known surface. Whereas legged robots are much better suited when dealing with unstructured environments.

Wheels allow a vehicle to move over terrain quickly. Legs allow robots to climb obstacles that are higher than what a wheeled vehicle would be able to climb over [27], as it is possible to note in Fig. 7.20.

From the considerations discussed previously about the pros and cons of legged and wheeled machines, legged robots cannot be viewed as competing with wheeled machines: they are rather complementary and are employed in environments otherwise inaccessible.

All these reasons led up to realize each leg of the robot TriBot with a peculiar shape. This is an hybrid solution, the result of a study on the efficiency of a wheel-leg hybrid structure; in this way the robot can have the advantages of using legs, in fact it can easily overcome obstacles and face with rough terrains. On the other hand, wheel-legs have the shape of wheels, therefore the robot TriBot is able to have a quite smooth movement in regular terrains and so to reach high speed. Figure 7.21 shows

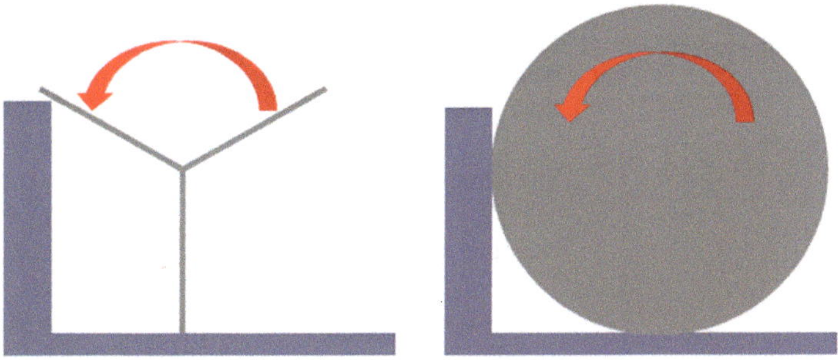

**Fig. 7.20** A wheel-leg allows to overcome obstacles higher than using a wheel. While, a wheel with the same radius can not overcome an obstacle of the same height

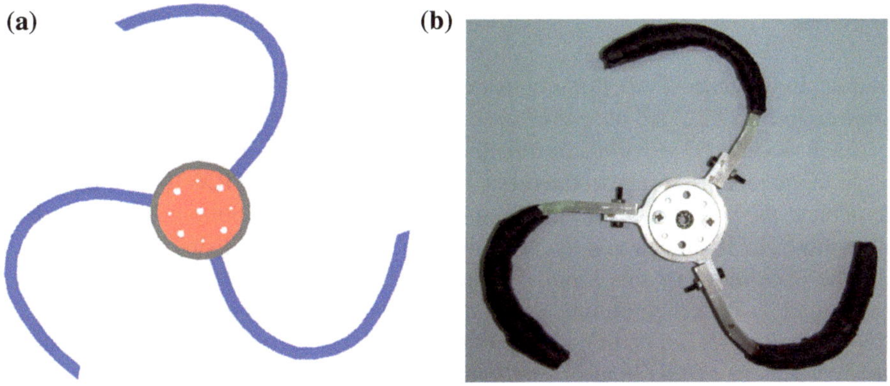

**Fig. 7.21** Wheel-legs of TriBot. **a** AutoCAD design; **b** Physical realization

the wheel-legs design. Each wheel-leg is composed by three spokes that describe a circle of 12 cm of diameter and they are 6 mm thick.

The spokes can be moved in two different directions; if each spoke faces the convex part toward the motion direction, the movement results to be more smooth because the wheel-leg has a quasi continuous contact with the terrain. Whereas, the other configuration is better in overcoming obstacles because it increases the grip with the terrain.

This kind of locomotion is used both in the first and in the second prototype of the robot TriBot, that, as previously said, differs only in terms of motors used to actuate wheel-legs.

In the first prototype of the robot TriBot five Hitec HS-985MG have been used to actuate wheel-legs and the joint connecting the manipulator with the rest of the robot. The main characteristics of this kind of motor are listed in Table 7.4.

During the planning phase, different solutions were analyzed and studies about the characteristics that the robot should have, like to be overturnable and light, were carried out. Moreover, the module should be symmetric, and it has to contain control

**Table 7.4** Technical characteristic of the motor Hitec HS-945MG

| Class | High Torque |
|---|---|
| Applied voltage | 4.8 or 6.0 V |
| Stall Torque | 8.8 Kg·cm @ 4.8 V |
| | 11 Kg·cm @ 4.8 V |
| Idle current | 5.3 mA @ 4.8 V |
| | 6.6 mA @ 6V |
| Running current | 44 mA @ 4.8 V |
| | 500 mA @ 6V |
| Dimension | 39.4 × 20 × 37.8 mm |
| Weight | 56 g |
| Speed 60° | 16 S @ 4.8 V |
| | 12 S @ 6 V |

boards and batteries. All these considerations led up to the design of the chassis showed in Fig. 7.22. All measures are expressed in millimeters.

Each module, 57 mm long, 68 mm thick and 54 mm tall, was realized in aluminum.

In the second prototype of the robot TriBot five Dynamixel RX-10 have been used to actuate the wheel-legs and the joint connecting the manipulator with the rest of the robot. The main characteristics of this kind of motors are listed in Table 7.5.

The peculiarity of this kind of motors is that Dynamixel with a unique ID is controlled by Packet communication on a BUS and supports networks such as TTL, RS485, and CAN depending on the type of model.The main advantage is that, the main processor of this kind of motors can set speed, position, compliance, torque, etc. simultaneously with a single command packet, it can control several Dynamixels with a little resource and in a simpler way than Hitec HS-945MG motors that need PWM signals to be actuated. The whegs module is very similar to the module used in the first prototype, but it was modified in order to position the Dynamixel RX-10 motors. Figure 7.23 shows the design of the module.

The central part of both prototypes of the robot TriBot is connected to a manipulator through an actuated joint. This joint allows the manipulator to move in different configurations. When it is moved down, the legs are able to touch the ground. This configuration is useful in rough terrain to increase the robot stability and to improve climbing capabilities.

Another function that the manipulator adds to the robot is the possibility to perform small and precise movements otherwise impossible with the only use of whegs. Furthermore, it is possible to move the manipulator up to use it for manipulation purposes. In this configuration the robot can push obstacles or bottoms, move sliding doors, grasp and carry objects.

In the two prototypes of the robot two different configurations of the manipulator have been used. The first one is inspired by an hexapod robot called Gregor [25], while the second one uses the configuration of legs used in another hexapod robot, called Tarry [26].

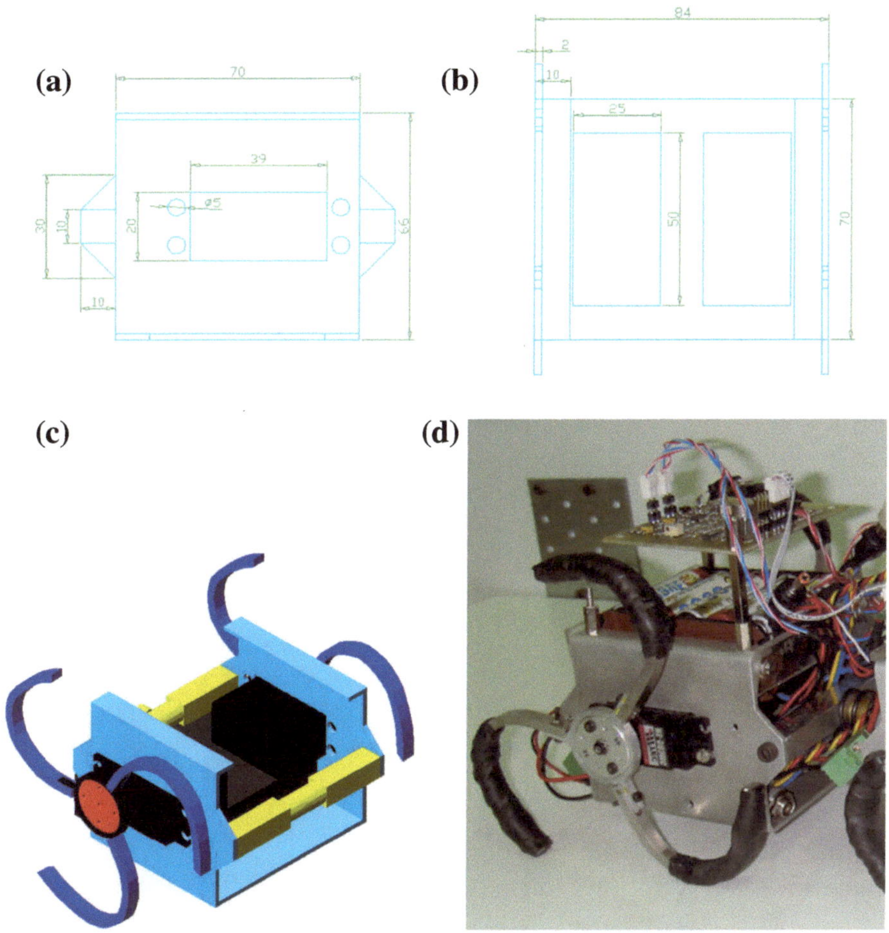


**Fig. 7.22** Three-spoke module I. **a** AutoCAD design, *side view*; **b** AutoCAD design, *top view*; **c** AutoCAD design, *3D view*; **d** Physics realization

**Table 7.5** Technical characteristic of the motor dynamixel RX-10

| Class | High Torque |
|---|---|
| Applied voltage | 10 or 12 V |
| Final reduction stopping Torque | 10.1 Kg·cm @ 10 V |
| | 12.1 Kg·cm @ 12 V |
| Dimension | 35.6 × 50.6 × 35.5 mm |
| Weight | 64.5 g |
| Speed | 0.143 S/60° |

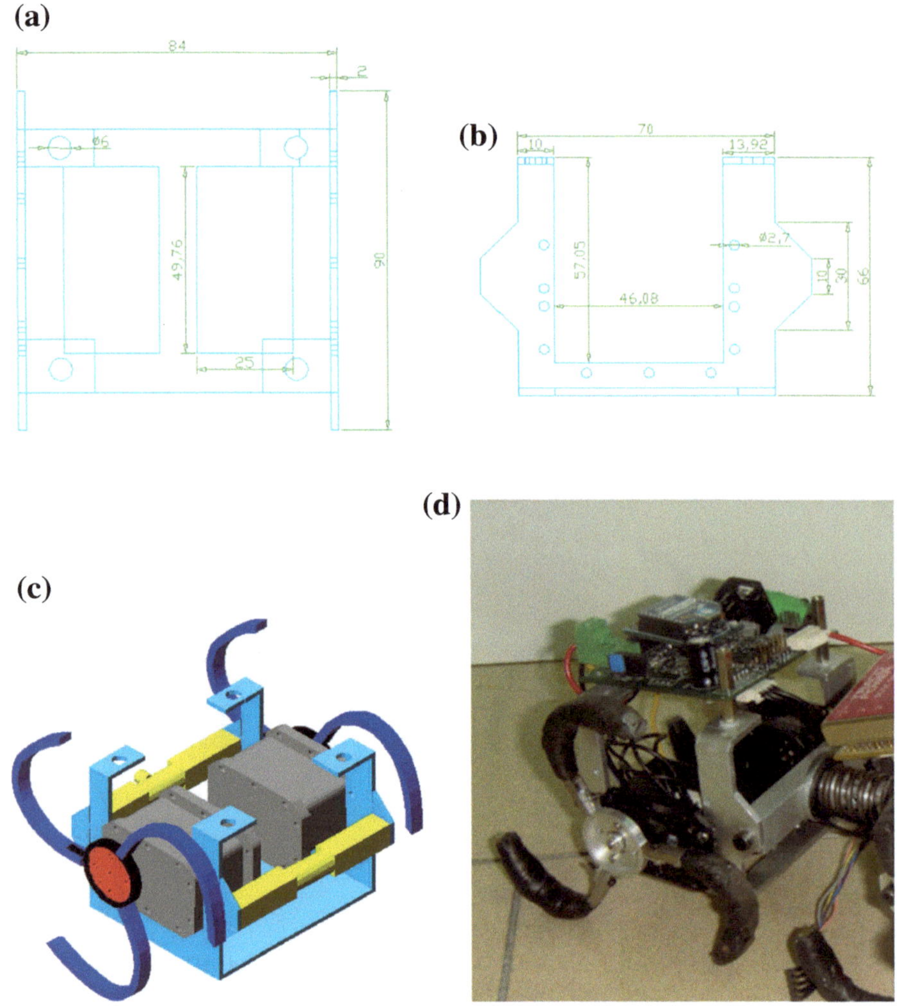


**Fig. 7.23** Three-spoke module II. **a** AutoCAD design, *side view*; **b** AutoCAD design, *top view*; **c** AutoCAD design, *3D view*; **d** final prototype

Both manipulators have been actuated using Hitec HS-82MG motors, whose characteristics are shown in Table 7.6.

In this case the module structure consists of two legs each one with three degrees of freedom (Fig. 7.24).

For this first prototype, plexiglass, aluminium and brass have been selected to realize the structure. Each leg is divided into three segments: coxa, femur and tibia articulated through three rotational joints.

Plexiglass has been used to realize tibias and femurs. It has been selected for its availability and costs, extreme facility of workmanship as well as lightness.

**Table 7.6** Technical characteristic of the motor Hitec HS-82MG

| Motor type | 3 Pole Ferrite |
|---|---|
| Applied voltage | 4.8 V or 6.0 V |
| Stall Torque | 2.8 Kg·cm @ 4.8 V |
| | 3.4 Kg·cm @ 4.8 V |
| Idle current | 8.8 mA @ 4.8 V |
| | 9.1 mA @ 6V |
| Running current | 220 mA @ 4.8 V |
| | 280 mA @ 6V |
| Dimension | 29.8 × 12 × 29.6 mm |
| Weight | 19g |
| Speed | 0.12 S / 60° @ 4.8 V |
| | 0.10 S / 60° @ 6 V |

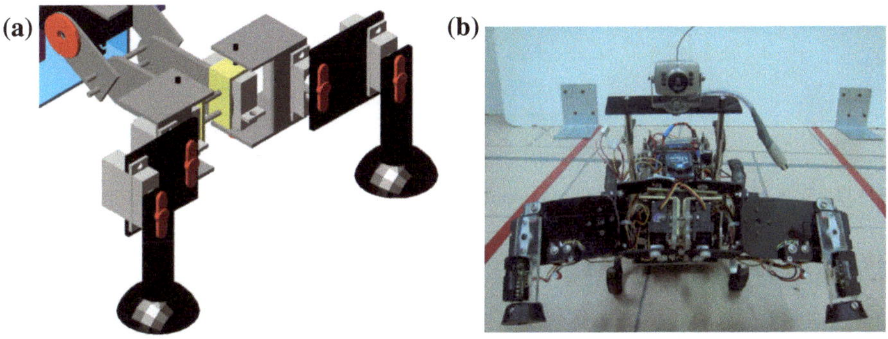


**Fig. 7.24** Manipulator module. **a** AutoCAD design; **b** Physic realization

In particular, the tibia is a rectangular piece of plexiglass $20 \times 70\,\text{mm}^2$. Figure 7.25a shows the AutoCAD design, while Fig. 7.25b shows its physical realization.

Also the femur is completely built in plexiglas. It is a rectangular piece 50 × 50 mm showed in Fig. 7.26.

The last segment of the manipulator, the coxa (Fig.7.27), has to support notable mechanical stress especially during the extension of the arm. For this reason, it was realized in aluminum. The peculiarity of this structure, in comparison to the previous ones, is that the rotation axis of the two motors (the first one used to actuate the coxa while the other one the femur) are orthogonal each other.

Also in this case the manipulator consists of three segments: tibia, femur and coxa like the manipulator of the first prototype, but the configuration of the coxa is different.

Figure 7.28 shows the CAD design and the physical realization of this new structure of manipulator.

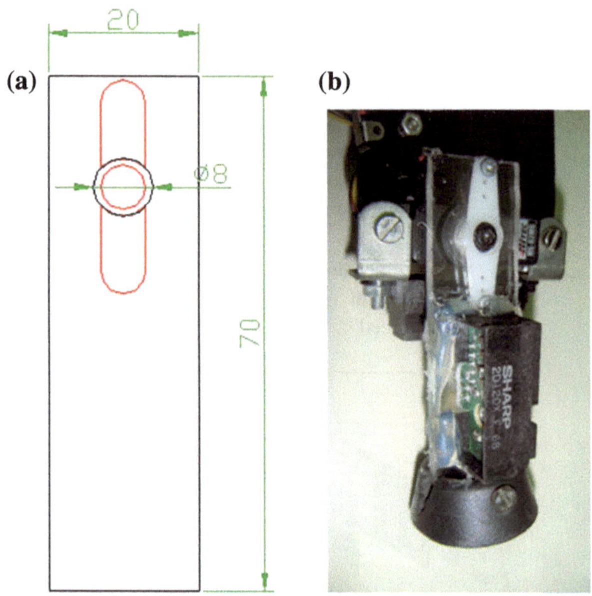


**Fig. 7.25** Tibia of the Gregor style manipulator. **a** AutoCAD design; **b** Physics realization

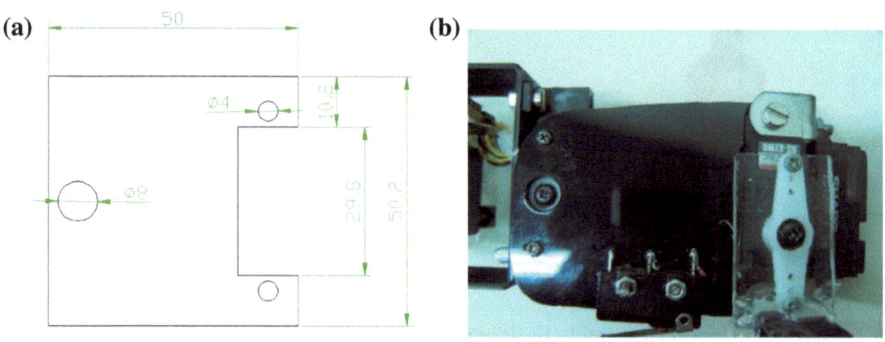


**Fig. 7.26** Femur of the Gregor style manipulator. **a** AutoCAD design; **b** Physics realization

The issue is to improve the performance of the robot during climbing and to increase the reachable space of the manipulator. The inspiration was taken from the structure of a robot called Tarry.

All segments have been realized in aluminium in order to increase the robustness of the structure.

The peculiarity of this manipulator, that makes it more robust than the other one, is the configuration of the coxa segment, showed in Fig. 7.29.

Also the other two segments, femur and tibia, of this structure are quite different than the Gregor style manipulator and they are showed, respectively, in Fig. 7.29d and f.

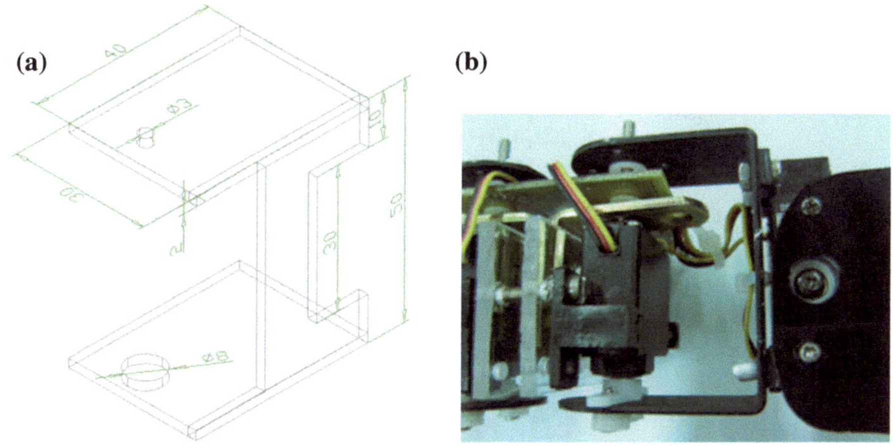


**Fig. 7.27** Coxa of the Gregor style manipulator. **a** AutoCAD design; **b** Physics realization

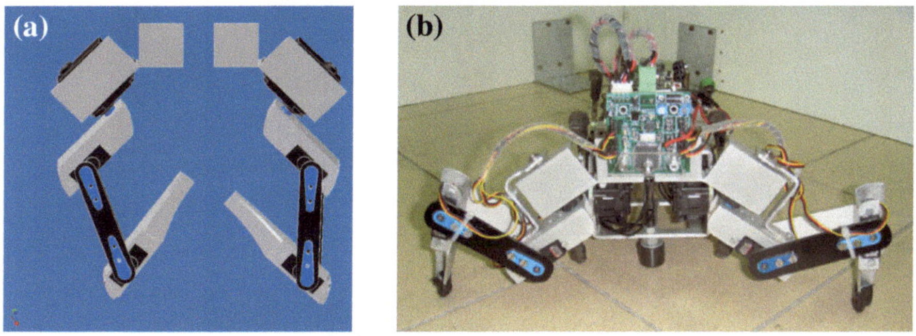


**Fig. 7.28** Tarry style manipulator. **a** CAD design; **b** Physics realization

Different tests have been performed with the TriBot II by using different whegs. The spokes of the whegs have the same shape and the same radius of curvature but differ for the support surface with the terrain. Figure 7.30 shows the three kinds of spokes used.

Tests show that bigger is the support surface with the terrain, higher is the stability of the robot without any decrease in terms of speed.

## *7.6.2 Tribot Hardware Architecture*

The hardware architecture of the robot TriBot follows the modularity of the structure. A block diagram is shown in Fig. 7.31.

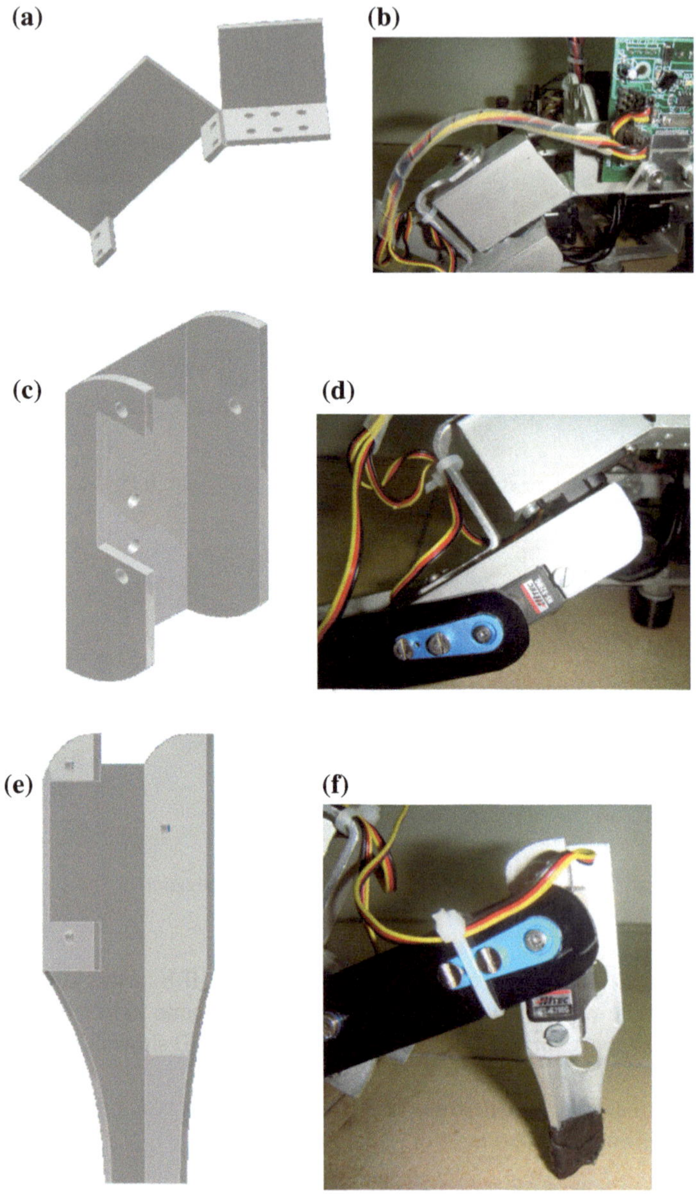


**Fig. 7.29** Segments of the Tarry style manipulator. **a** CAD design of coxa; **b** Coxa physics realization; **c** CAD design of femur; **d** Femur physics realization; **e** CAD design of tibia; **f** Tibia physics realization

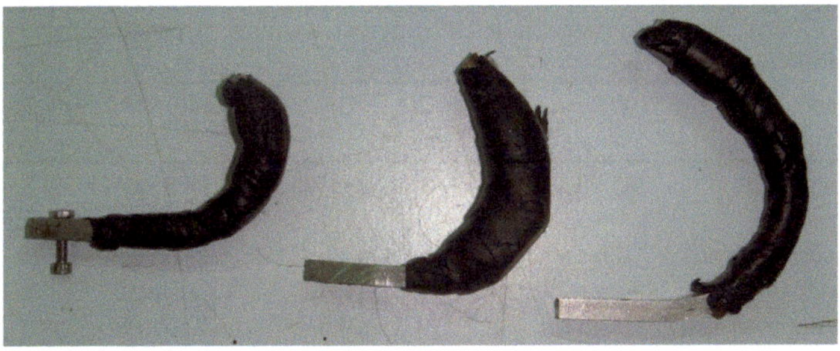

**Fig. 7.30** Three kind of spokes used for the experiments

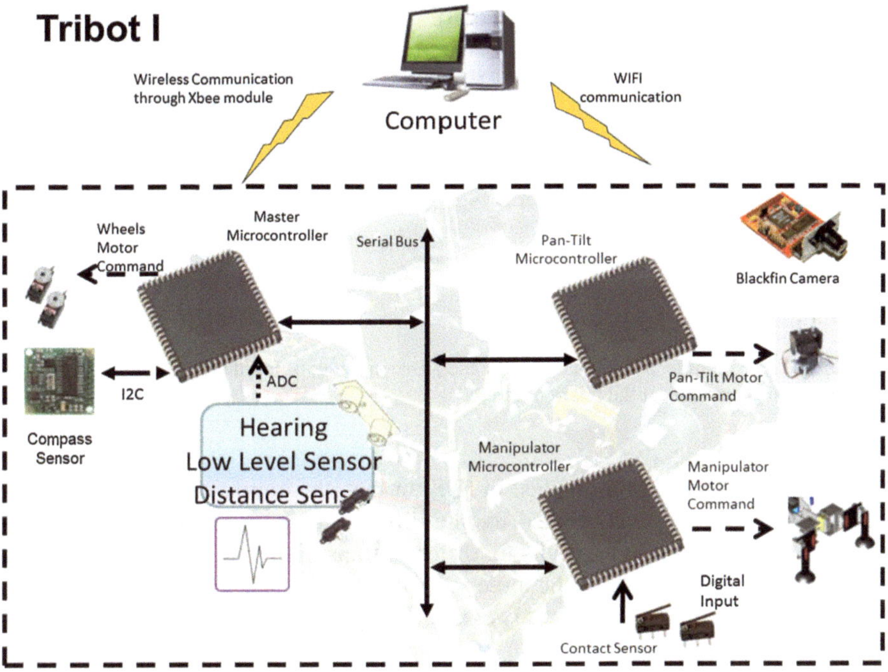


**Fig. 7.31** Functional scheme of the communication architecture

The hardware structure of TriBot is managed by two microcontrollers that handle motors and sensors distributed on the structure. Interesting applications are envisaged with the introduction of active antennae in the robot for object detection and classification [28, 29]. Furthermore through a computer, using a wireless connection, it is possible to acquire data and send commands generated by a high level control algorithm. The robot is controlled by a system of two microcontroller-based boards connected through a master/slave bus. The computer supervises and controls the robot through a RF wireless XBee module, that uses the standard ZigBee. This communication technology was chosen because of its much less energy demanding

**Table 7.7** Hourly energy consumption of different systems depending on the transmission/acquisition time [30]

| Transmission/acquisition time (s) | 10 | 20 | 30 | 40 |
|---|---|---|---|---|
| Hourly consumption of GPRS | 283.05 mWh | 141.53 mWh | 94.35 mWh | 47.18 mWh |
| Hourly consumption of ZigBee | 2.03 mWh | 1.04 mWh | 0.7 mWh | 0.37 mWh |
| Hourly consumption of RF | 1.29 mWh | 0.65 mWh | 0.43 mWh | 0.22 mWh |

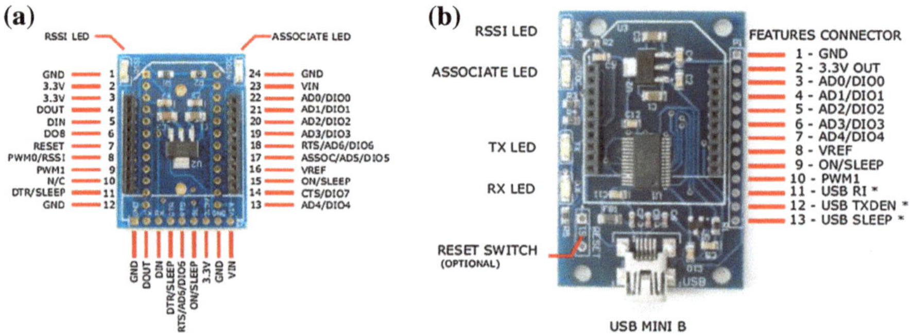


**Fig. 7.32** XBee module. **a** Board containing the XBee module; **b** Board for the module XBee with a MiniUSB port to connect the board to the PC

compared with other technologies. In [30], in fact, the hourly energy consumption of the system has been computed for the different communications technologies. Table 7.7 summarizes these results for different transmission/acquisition times, considering the bit rate equal for the three systems to 9600bit/s.

By comparing these results, it is evident that GPRS transmission technology is much more energy demanding than Zigbee and RF. The Xbee module is connected to a small board (Fig. 7.32a), which was designed to be simply connected to the other boards. Likewise the computer, that controls the robot, needs a similar board, that in this case has also an USB interface (Fig. 7.32). The two XBee modules can be programmed through serial port, using a software, called X-CTU, easily available on the web. A module can be simply programmed through AT commands. The two XBee modules have been programmed in order to have a point to point communication, or the destination address of one module is the address of the other one, and vice versa.

The master control board, showed in Fig. 7.33 is positioned in the central wheel-legs module. Its main role is to control the servomotors that actuate the four wheel-legs and it is also used as a bridge between the PC and the other board mounted on the manipulator. It is obvious that it represents a crucial knot for the communication and

**Fig. 7.33** Three-spoke board

the general management of the robot. It uses an ATmega128, a low-power CMOS 8-bit microcontroller based on the AVR enhanced RISC architecture.

The same board is used to control the second prototype of the robot, the difference is that in this case it is not needed to generate PWM signals since RX-10 motors are driven by receiving binary data, therefore they can be controlled using a little resource.

The manipulator is controlled by a similar board, showed in Fig. 7.34, configured as slave; also in this case an ATmega128 microcontroller is used to generate PWM signals to actuate the six servomotors used to move the manipulator and to the servomotor that actuates the joint connecting the manipulator with the rest of the robot. This board has also the important task to read data from the distributed sensory system embedded in the manipulator.

The sensory system is needed for autonomous navigation and to use the robot as test bed for perceptual algorithms. For these reasons on the manipulator of the first prototype of the robot, four distance sensors have been distributed, for obstacle detection and a series of micro-switches are used to detect collisions and to grasp objects.To implement targeting and homing strategies, a hearing circuit inspired by phonotaxis in crickets has been also included. The aim is to give the robot the ability to approach sound sources, reproducing the behaviour shown by the female cricket to follow a particular sound chirp emitted by a male cricket. Furthermore, one of the most useful and rich sense that guides animal's actions is the visual system. Therefore the first prototype of the robot is equipped with a wireless camera that can be used for landmark identification, following moving objects, and other higher level tasks. Although vision is one of the most important sense, the other ones, first of all hearing but also touch, become essential particularly in cluttered conditions, where visual percepts are somehow obscured by the environment conditions. All these sensor modalities have been included in the robot in order to make the system

**Fig. 7.34** Manipulator board

**Fig. 7.35** ST lis3lv02dq accelerometer

able to safely move in unknown environments. In the second prototype of the robot an accelerometer have been mounted in order to test the stability of the structure. An ST Microelectronics device has been used. As shown in Fig.7.35, this accelerometer was positioned in the anterior three-spoke module, near the manipulator, in order to better perceive the acceleration on the x, y and z axis.

This accelerometer can be supplied with a voltage between 2.16 and 3.6 V and can communicate with a PC through SPI or $I^2C$ connection. In this case the $I^2C$ was chosen and a wireless communication through ZigBee was used. The accelerometer

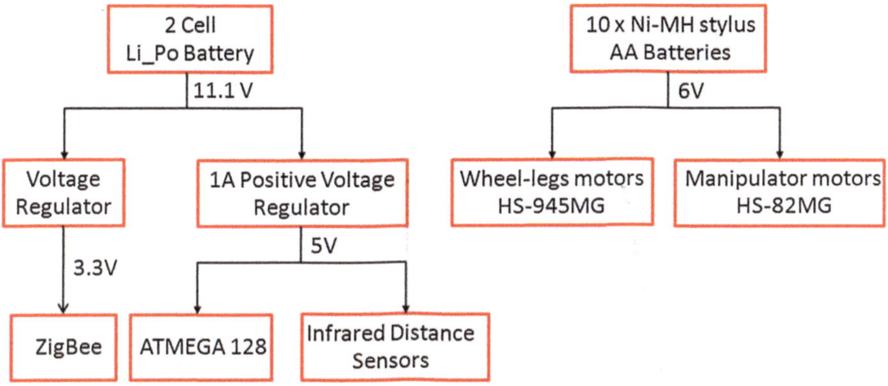


**Fig. 7.36** Flow chart of the first prototype of TriBot power system

gives an acceleration between −2 and 2 g with a sampling frequency of 650 Hz. Moreover, the aim is to mount on the second prototype all sensors mounted in the first one in order to test the new structure.

In the first prototype Hitec HS-945MG servomotors have been used to actuate the wheel-legs modules. To obtain the maximum torque and so better performance, these motors need a 6 V power supply. A NiMH battery pack is equipped on the robot to supply motors. It is important to take into account that, while in most applications the need is to have batteries as much light as possible, but at the same time with a high capacity, in this application the weight of batteries allows to increase the stability of the whole structure. As a matter of fact, at the moment, the first prototype of the robot is equipped with two different battery packs:

- 10 rechargeable 1.2 V, 3000 mAh type Ni-MH AA stylus batteries;
- a rechargeable 1600 mAh @ 11.1 V, Lithium-Polymers battery.

A scheme of the power system of the first prototype of the robot TriBot is shown in Fig.7.36. The flow chart shows as the Ni-MH battery pack power supplies all the motors that actuate the robot, while the Lithium-Polymers battery is regulated to 3.3 V and to 5 V to power supply the wireless module and the control board together with the robot sensory system (i.e. tactile sensors and infrared distance sensors). Experiments show that, using these kinds of batteries, the lifetime of the robot is about two hours.

Regarding the Tribot II, the control board is the same of that used in the first one, therefore also in this case a rechargeable 1600 mAh @ 11.1 V, Lithium-Polymers battery is used to power supply it. Moreover, this battery is used also to supply Dynamixel RX-10 motors since they need a voltage between 10 and 12 V. Then a DC-DC converter, showed in Fig.7.37 is used to give 5 V to motors used to actuate the manipulator module.

Therefore, the scheme of the power system of the second prototype of the robot TriBot is shown in Fig. 7.38.

**Fig. 7.37** DC-DC converter

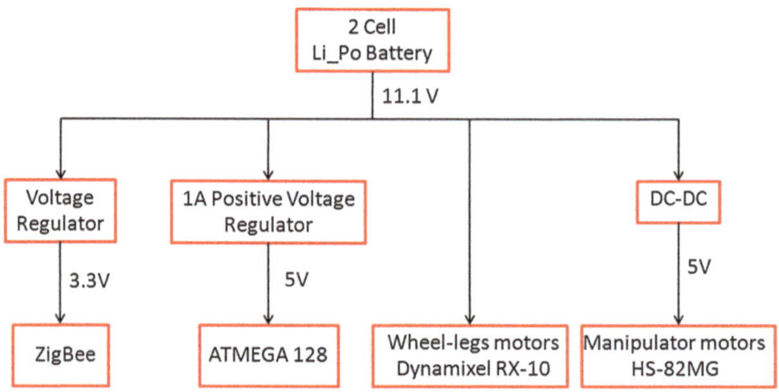


**Fig. 7.38** Flow chart of the second prototype of TriBot power system

## *7.6.3 Experimental Results*

First of all some tests have been done on the first prototype of the robot TriBot in order to underline its main characteristics. Experiments to test the maximum velocity of the robot and its climbing capabilities were performed. Results show that

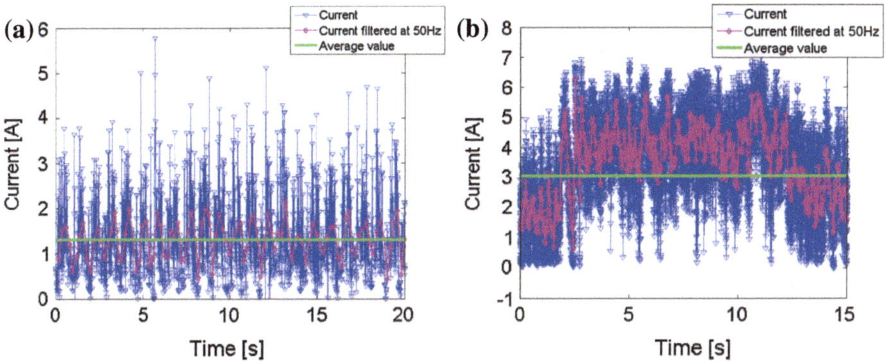


**Fig. 7.39** Behaviour of the instantaneous current of motors. Data have been acquired at 100 Hz (*blue line*) and filtered at 50 Hz (*violet line*). The *green line* represents the average value. **a** Movement at maximum speed; **b** Climbing an obstacle

the robot can reach the velocity of 46 cm/s and the higher obstacle that it is able to overcome is 1.42 times its wheel radius using only wheels and 1.8 times its wheel radius using also the manipulator. Besides, studies about power consumptions have been done. We measured the instantaneous current absorbed by the motor system in two different typical scenarios: movement at maximum speed on a flat terrain and obstacle climbing.

Figure 7.39a shows the behavior of the current absorbed by the motor system, during the movement at maximum speed in a flat terrain, while Fig. 7.39b shows as, when both the whegs module and the manipulator are enrolled to overcome an obstacle of 6 cm at a medium speed (32 cm/s) the power consumption increases.

These results are extremely useful because they allow to evaluate robot performances. One interesting parameter, in fact, that can be used to compare TriBot with other robots, is the Energetic Performance. To measure energy efficiency an increasingly accepted parameter is the "Specific Resistance" [8] that for the Tribot, have been obtained during the movement at maximum speed on a flat terrain ($\varepsilon = 0.9$) and during obstacle climbing using both wheels and manipulator ($\varepsilon = 3.2$). Moreover, on the first prototype of the robot TriBot a correlation-based navigation learning algorithm, based on Spike Timing Dependent Plasticity (STDP) was implemented for learning in order to increase its behavioral capabilities. In fact correlation based algorithms have been found to explain many basic behaviors in simple animals. The main interesting consequence of STDP is that the system is able to learn high-level sensor features, based on a set of basic reflexes, depending on some low-level sensory inputs. Robot experiments demonstrate the potentiality and the effectiveness of the approach. During the learning phase the robot navigates in the environment showing an attractive behaviour. When an object is detected the robot tries firstly to grab it. If the action does not succeed the robot tries to climb the obstacle or eventually to avoid it. At the end of the procedure the synapses subject to the STDP learning rule are updated [19, 31].

**Table 7.8** Minimum, maximum and average values of the y acceleration when the torque is 60, 80 and 100 % of the maximum torque

| Spokes | 60 % | | | 80 % | | | 100 % | | |
|---|---|---|---|---|---|---|---|---|---|
| | Min | Max | Average | Min | Max | Average | Min | Max | Average |
| Small | −0.58 g | 0.495 g | 0.041 g | −0.408 g | 0.688 g | 0.086 g | −0.391 g | 0.544 g | 0.132 g |
| Medium | −0.25 g | 0.99 g | 0.42 g | −0.95 g | 0.54 g | 0.37 g | −0.14 g | 0.23 g | 0.24 g |
| Big | −0.151 | 0.277 | 0.027 | −0.209 | 0.27 | 0.026 | −0.368 | 0.544 | 0.018 |

Some tests on the second prototype of the robot have been done using the three different spokes realized, in order to test the stability of the structure when these different kinds of spokes are mounted on the robot. To study the stability of the robot, testes have been done on a flat terrain varying the velocity, in particular, since the motors used are torque controlled, tests have been performed at 60, 80 and 100 % of the maximum torque. The accelerometer previous described was used to acquire the variation of the acceleration on the x, y and z axis. Theoretically, the acceleration on the z axis should be constant and equal to the gravity acceleration, while the acceleration on the x axis is due to the movements of the robot and finally the acceleration on the y axis is due to the oscillation of the robot structure. Therefore, measuring the last one value, it is possible to determine which is the set of spokes that allows more stability. Table 7.8 shows the minimum, average and maximum values of the acceleration on the y axis using the three sets of spokes when the torque is respectively 60, 80 and 100 % of the maximum torque.

These results show as the acceleration on the y axis decreases if the big spokes are used, therefore these configuration is the best solution to increase stability.

## 7.7 Conclusions

In this Chapter the details about the robotic platforms used as testbed for the Insect Brain Computational Model are reported. In particular four different classes of robots have been considered: wheeled, legged, manipulators, hybrid. For each structure the main characteristics of the Software/Hardware architecture realized are commented.

## References

1. S.L. Anafocus, Eye-ris producer home page, http://www.anafocus.com
2. B. Webb, T. Scutt, A simple latency dependent spiking neuron model of cricket phonotaxis. Biol. Cybern. **82**(3), 247–269 (2000)
3. R. Reeve, B. Webb, New neural circuits for robot phonotaxis. Philos. Trans. Roy. Soc. A **361**, 2245–2266 (2002)
4. Mobile Robots. P3-at vendor, http://www.mobilerobots.com/

5. L. Alba, P. Arena, S. De Fiore, L. Patané, R. Strauss, G. Vagliasindi, Implementation of a drosophila-inspired model on the eye-ris platform. in *Proceedings of the IEEE International Conference of CNNA*, Berkley
6. P. Arena, S. De fiore, L. Patané, Cellular nonlinear networks for the emergence of perceptual states: Application to robot navigation control. Neural Netw **22**(5–6), 801–811 (2009)
7. P. Arena, L. Fortuna, M. Frasca, L. Patané, M. Pollino, An autonomous mini-hexapod robot controlled through a CNN-based CPG VLSI chip. in *10th IEEE International Workshop on Cellular Neural Networks and their Applications* (Istanbul, 2006), pp. 401–406
8. G. Gabrielli, T.H.von Karman, What price speed?. Mech. Eng. **72**(10), 775–781 (1950)
9. P. Arena, L. Patané (Eds.), *Spatial Temporal Patterns for Action-Oriented Perception in Roving Robots*, Vol. 1, (Series: Cognitive Systems Monographs, Springer 2009)
10. SCHUNK GmbH and Co. Schunk home page, http://www.schunk.com
11. DLR Institute. Institute of robotics and mechatronics home page, http://www.dlr.de
12. Lynxmotion, Inc. Lynxmotion home page, http://www.lynxmotion.com
13. Robotis, Corporation. Robotis home page, http://www.robotis.com
14. Dassault Systemes SolidWorks Corp. Solidworks home page, http://www.solidworks.com
15. I. Aleo, P. Arena, L. Patané, SARSA-based reinforcement learning for motion planning in serial manipulators, *WCCI 2010 IEEE World Congress on Computational Intelligence* (Barcelona, Spain, 2007), pp. 3605–3610
16. P. Arena, H. Cruse, L. Fortuna, L. Patané, An obstacle avoidance method for a redundant manipulator controlled through a recurrent neural network. in *Proceedings of Microtechnologies for the New Millennium (SPIE 07)* (Gran Canaria (SPAIN), 2007)
17. I. Aleo, P. Arena, L. Patane, Parallel central pattern generators for locomotion control in a humanoid robot model. in *18th IEEE Workshop on Nonlinear Dynamics of Electronic Systems (NDES 2010)* (Dresden, 2010)
18. P. Arena, S. De Fiore, L. Patané, M. Pollino, C. Ventura, Insect inspired unsupervised learning for tactic and phobic behavior enhancement in a hybrid robot, *WCCI 2010 IEEE World Congress on Computational Intelligence* (Barcelona, Spain, 2010), pp. 2417–2424
19. P. Arena, L. Patané, M. Pollino, C. Ventura, Tribot: a hybrid robot for cognitive algorithm implementation. in *18th IEEE Workshop on Nonlinear Dynamics of Electronic Systems (NDES 2010)*, pp. 1–6
20. Case Biorobotics Lab. Whegs series robots home page, http://biorobots.cwru.edu/projects/whegs/whegs.html
21. Robovolc project. Robovolc project home page, http://www.robovolc.dees.unict.it/
22. J.E. Seipel, P.J. Holmes, R.J. Full, Dynamics and stability of insect locomotion: a hexapedal model for horizontal plane motions. Biol. Cybern. **91**(2), 76–90 (2004)
23. M. Eich, F. Grimminger, S. Bosse, D. Spenneberg, F. Kirchner, Asguard: A hybrid legged-wheel security and sar-robot using bio-inspired locomotion for rough terrain, in *Proceedings of the EURON/IARP International Workshop on Robotics for Risky Interventions and Surveillance of the Environment, Benicassim* Spain, January, 2008
24. P Arena, L Fortuna, M Frasca, L Patané, M. Pavone, Implementation and experimental validation of an autonomous hexapod robot. in *Proceedings of IEEE International Symposium on Circuits and Systems* (Kos, May, 2006), pp. 401–406
25. Tarry project. Tarry robot home page, http://www.tarry.de/
26. S. Baglio, S. Gagliano, D. Neri, N. Savalli, G.M. Tina, Optimal design of photovoltaic systems for wireless sensor networks, *IEEE International Conference on Symposium on Industrial Electronics* (Cambridge, 2008), pp. 2108–2113
27. L. Patané, S. Hellbach, F. André, P. Durr, V. Arena, An insect-inspired bionic sensor for tactile localisation and material classification with state-dependent modulation. Front. Neurorobotics **6**(8), 1–18 (2012)
28. P. Arena, L. Patané, A spiking network for object and ego-motion detection in roving robots. in *International Joint Conference on Neural Networks (IJCNN 2012)* (Brisbane, 2012)
29. P. Arena, S. De Fiore L Patané, M. Pollino, C. Ventura, Stdp-based behavior learning on tribot robot. in *Proceedings of IEEE/RSJ International Conference SPIE* (2009), pp. 1–12

30. R.T. Schroer, M.J. Boggess, R.J. Bachmann, R.D. Quinn, R.E. Ritzmann, Comparing cockroach and whegs robot body motions, *IEEE International Conference on Robotics and Automation* (New Orleans, May, 2004), pp. 3288–3293
31. P. Arena, L. Fortuna, M. Frasca, L. Patané, Learning anticipation via spiking networks: application to navigation control. IEEE Trans. Neural Netw. **20**(2), 202–216 (2009)

# Chapter 8
# Compact Internal Representation of Dynamic Environments: Implementation on FPGA

**L. Salas-Paracuellos and L. Alba-Soto**

**Abstract** Animals for surviving have developed cognitive abilities allowing them an abstract representation of the environment. This internal representation (IR) may contain a huge amount of information concerning the evolution and interactions of the animal and its surroundings. The temporal information is needed for Internal Representations of dynamic environments and is one of the most subtle points in its implementation as the information needed to generate the IR may eventually increase dramatically. Chapter 3 in the book proposed the compression of the spatio-temporal information into only space, leading to a stable structure suitable to be the base for complex cognitive processes in what has been called Compact Internal Representation (CIR). The Compact Internal Representation is especially suited to be implemented in autonomous robots as it provides global strategies for the interaction with real environments. This chapter describes an FPGA implementation of a Causal Neural Network based on a modified FitzHugh-Nagumo neuron to generate a Compact Internal Representation of dynamic environments for roving robots to avoid dynamic and static obstacles.

## 8.1 Introduction

Internal representations are abstract representations of the environments generated by many living beings to interact with the real world. Recently the Compact Internal Representation (CIR) has been proposed as a new biologically-inspired cognitive process allowing the decision making by creating an internal representation of

L. Salas-Paracuellos · L. Alba-Soto (✉)
Innovaciones Microelectrónicas S. L.-Anafocus, Avda. Isaac Newton 4,
Planta 7, E-41092 Seville, Spain
e-mail: luisalba@anafocus.com

L. Salas-Paracuellos
e-mail: luis.salas@anafocus.com

P. Arena and L. Patanè (eds.), *Spatial Temporal Patterns for Action-Oriented Perception in Roving Robots II*, Cognitive Systems Monographs 21,
DOI: 10.1007/978-3-319-02362-5_8, 

dynamic environments. The inclusion of a temporal dimension could dramatically increase the information required to properly represent internally the environment. Nonetheless the CIR minimizes this information by compacting the spatiotemporal representation into a spatial (no time-dependent) stable structure, by means of reaction-diffusion system that models the possible interactions between agent and its environment, and extracts the critical events projecting them univocally onto spatial points.

Compact Internal Representation is especially suitable to be implemented in robots interacting with real environments (roving robots, manipulators, etc) because it provides infinite global solutions (possible decisions to be made by the robots) by investing minimum resources. Moreover, as this is a process based on a physical system it is perfect to be implemented in a neural network simply by reproducing the numerical discretization scheme. One of the most versatile options for this goal is the use of FPGAs [1–3], but it forces to simplify the numeric approach to the generation of the CIR. The CIR reaction-diffusion system is based on a set of coupled partial differential equations usually numerically solved by the Runge-Kutta method, but a feasible scheme of implementation by FPGAs requires a simplified numeric scheme, preferably based on a simple algebraic structure. In this chapter, an FPGA-implementation is proposed and implemented by stating the initial partial differential equations and approaching them by algebraic equations by means of a proper combination of the 'forward' and 'backward' Euler algorithms.

## 8.2 FPGA Design Overview

The FPGA design comprises the mentioned CIR implementation as well as a small microprocessor to control, configure and read back the data generated by the CIR. The microprocessor is the well known Nios II soft core from Altera, which is a general-purpose RISC processor core with the following features:

- Full 32-bit instruction set, data path, and address space
- 32 general-purpose registers
- Optional shadow register sets
- 32 interrupt sources
- External interrupt controller interface for more interrupt sources
- Dedicated instructions for computing 64 and 128-bit products of multiplication
- Floating-point instructions for single-precision floating-point operations
- Single-instruction barrel shifter
- Access to a variety of on-chip peripherals, and interfaces to off-chip memories and peripherals
- Hardware-assisted debug module enabling processor start, stop, step and trace under control of the Nios II software development tools
- Optional memory management unit (MMU) to support operating systems that require MMUs

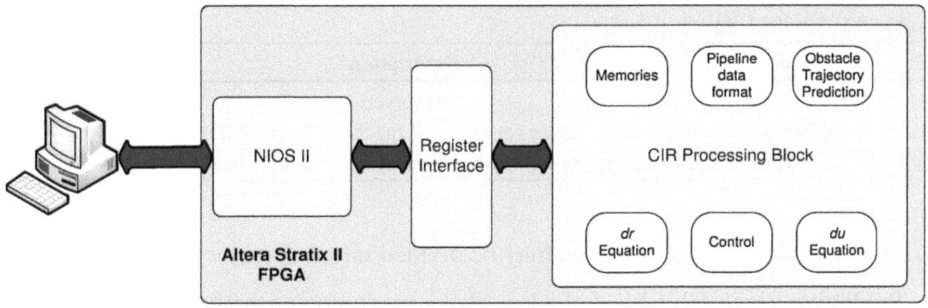


**Fig. 8.1** System overview

- Optional memory protection unit (MPU)
- Software development environment based on the GNU C/C++ tool chain and the Nios II Software Build Tools (SBT) for Eclipse
- Integration with Altera's SignalTap II Embedded Logic Analyzer, enabling real-time analysis of instructions and data along with other signals in the FPGA design
- Instruction set architecture (ISA) compatible across all Nios II processor systems
- Performance up to 250 DMIPS.

In our design the main purpose of the Nios II microprocessor is to simply act as a command server between the PC and the CIR custom logic. The custom logic presents a register memory map to Nios II, trough an Altera proprietary slave interface known as Avalon bus, so once the PC sends the right command to the FPGA Nios II the corresponding configuration registers are updated. The configuration registers will be further explained in the next subsection (Fig. 8.1).

As the communication between the PC and Nios II doesn't require a high bandwidth real-time interface, the implementation of an UART peripheral on Nios II will satisfy bandwidth requirements. Not to mention that the implementation and configuration of an UART in Nios II processors just requires a few mouse clicks.

Upon a power-on sequence, if we want to run the Nios II application as standalone, the program code will be looked for in an external Flash. So another peripheral needed in our Nios II is a Flash memory controller. The main features of our design Nios II microprocessor are listed below:

- CPU in fast mode.
- Embedded multipliers.
- 4 Kbytes of Instruction Cache.
- 2 Kbytes of Data Cache with a line size of 32 Bytes.
- Flash Memory Controller.
- 64Kbytes On-chip RAM.
- General Purpose I/Os.
- 125 MHz clock frequency.

**Table 8.1** Control register description

| Bits | Name | Description | Reset |
|---|---|---|---|
| 31..2 | – | Reserved | – |
| 1 | RESET | Reset CIR. When set to 1 CIR machine block will be resetted | 0 |
| 0 | START | CIR process starts. A 0-to-1 transition will start the CIR process | 0 |

The hardware implementation may be divided into two major blocks:

- Predictive Block. This block deals with the prediction of the future trajectories of the obstacles in the environment. It may be implemented in many diverse and simple ways, but essentially it must take the initial visual information (preprocessed or not) about the movement of an obstacle and return its trajectory. In this work only straight trajectories are considered.
- Internal Representation block. This block implements a reaction-diffusion system responsible for generating the internal representation. The system is modelled by means of an $N \times N$ Causal Neural Network, which expects three different inputs:
  - **Position of the agent**, which could be considered as the origin of the coordinates and will be introduced in the reaction-diffusion system as a source, whose position does not change in time.
  - **Position of the target** (initially assumed as static) which is included in the reaction-diffusion process as a sink.
  - **Predicted position of the obstacles** in each time step and obtained from the previous predictive block.

Once the external inputs are introduced in the Internal Representation Block two diffusive phenomena appear:

- First, an active, non-passive, wavefront propagates maintaining constant velocity. This wave represents in each time the possible positions of the robot, where it is assumed that moves with constant velocity. When in a time step during the course of its propagation the wave finds a cell occupied by a moving or static obstacle the cell is 'frozen'. By 'frozen cell' we mean that its value is set constant and equal to its initial value during all the remaining process. These cells will make up the **effective obstacles**, where the latter passive diffusion will be affected in the same way as with static obstacles. A zero-flux (Neumann) boundary condition is assumed in the boundary of the (effective and static) obstacles, modelling the fact that diffusion does not penetrate into the obstacle. It is worth to mention that if and obstacle does not move, it will be its own **effective obstacle**.
- Second, a diffusive process takes place inside the region enclosed by the propagating wavefront and simultaneously to this propagation, the cells in this area relax to a value given by a standard diffusion. At the end of the process and after the wave has disappeared, the values of the cells in the Causal Neural Network shape a profile defining all the possible trajectories for the agent to reach the target. These trajectories are computed simply by following the gradient of the profile from the agent position (source) to the target position (sink).

**Table 8.2** Status register description

| Bits | Name | Description | Reset |
|---|---|---|---|
| 31..1 | – | Reserved | – |
| 0 | END | Process CIR end. When set, CIR Processing machine finished | 0 |

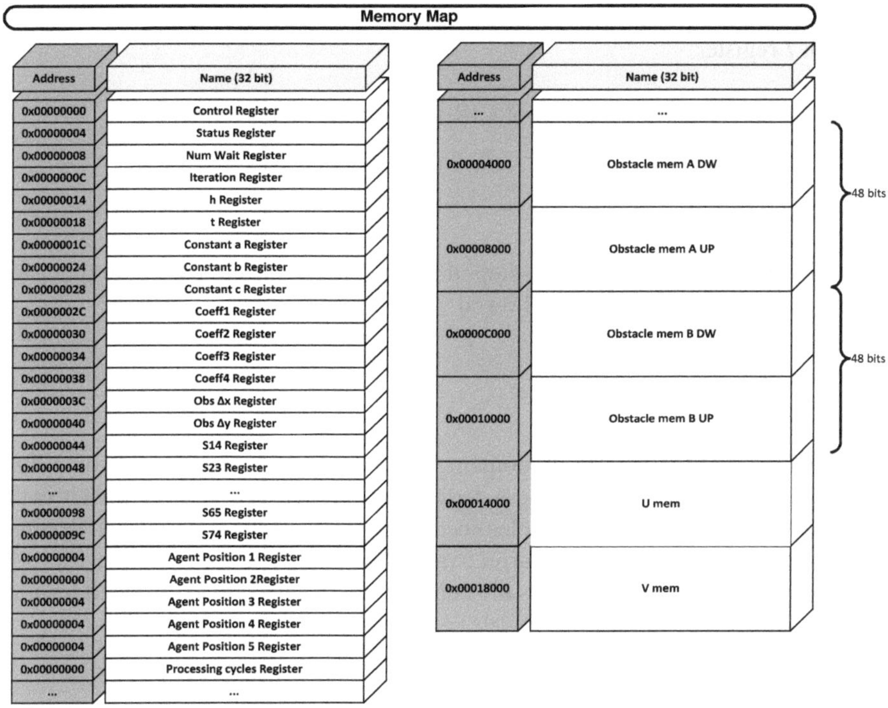


**Fig. 8.2** System memory map

## 8.2.1 System Memory Map

As already mentioned registers are used to control and initialize some parameters needed by the CIR processing block. In the following an overview of them is presented (Fig. 8.2).

Control register

This register allows controlling the CIR block via Nios II software commands. Table 8.1 shows the register fields.

Status register

As its name states, Status register reports the CIR status during and after a CIR execution has been launched. This register is read-only. Table 8.2 shows the register fields.

Constant *I* register

32-bit register storing the number of CIR iterations to be performed.

Constant h register

Register containing the ODE time-step of the iterations. Due to the 20 bit precision of the data, a minimum of 0.1 is allowed.

Numwait register

CIR block uses a different clock domain than Nios II does. For this reason a kind of synchronization is needed using a dedicated block for this task. Numwait is the number of waiting Nios II clock cycles needed to sync both clock domains. It depends on the relation between clocks. Current values are 125 MHz for Nios and 100–140 MHz to CIR. For these frequencies the constant doesn't need to be tweaked from its default value.

Constant *a* egister

This register contains a constant used in the computation of *v* state cell variable.This constant equals to

$$a = \frac{1}{25} timestep$$

Constant *b* register

This register contains a constant used in the computation of *v* state cell variable. This constant equals to

$$b = \frac{7}{25} timestep$$

Constant $c$ register

This register contains a constant used in the computation of $v$ state cell variable. This constant equals to

$$c = \frac{2}{25} timestep$$

Constant *Coeff1* register

Third degree polynomial coefficient of $r^3$ term.

Constant *Coeff2* register

Third degree polynomial coefficient of $r^2$ term.

Constant *Coeff3* register

Third degree polynomial coefficient of $r$ term.

Constant *Coeff4* register

Third degree polynomial coefficient of the independent term.

X displacement register

This register contains the displacement increment along X axis according to rectilinear movement equation. This constant is used in the obstacle evolution block.

Y displacement register

This register contains the displacement increment along Y axis according to rectilinear movement equation. This constant is used in the obstacle evolution block.

$S$ Matrix parameter register

It stores the parameters of the $S$ Matrix needed to calculate the CIR Equation:

$$R(t_n + 1) = SR(t_n)$$

Processing cycles register

This register stores the number of clock cycles that the CIR has spent in processing the number of iterations set by the user.

Memory

Matrix memory representing the state variables $R$, $V$ in every cell of the 60×60 arena as well as the obstacle position.

## 8.3 Mathematical Issues of FPGA CIR Implementation

To ease the Compact Internal Representation (CIR) hardware implementation, the differential equations describing the CIR spatial and time evolution have been resolved by means of backward and forward Euler.

The equations describing the CIR are

$$\dot{r}_{ij} = q_{ij}\{H(r_{ij})(f(r_{ij}) - v_{ij}) + d\Delta r_{ij} - r_{ij}p_{ij}\} \quad (8.1)$$

$$\dot{v}_{ij} = \frac{u_{ij} - 7v_{ij} - 2}{25} \quad (8.2)$$

being

$$f(r) = \frac{-r^3 + 4r^2 - 2r - 2}{7} \quad (8.3)$$

The differential equation contains a diffusive term which makes the ODE resolution rather complex. The issue is that our system has two discrete dimensions, space and time, and numerically resolving the displacements along both discretized dimensions imposes that both dimensions must be coupled in such a way that we do not move forward more quickly along spatial dimension than along the temporal one.

The Euler Algorithm may be used to solve the system although with some minor modifications and maintaining a time step small enough.There are two kinds of Euler Algorithm, Backward and Forward, but Forward Euler fails to solve a diffusion situation forcing us to use Backward Euler.

$$\frac{x(t_n) - x(t_{n-1})}{h} = g(x(t_n)) => x(t_n) = x(t_{n-1}) + hg(x(t_n)) \quad (8.4)$$

Although moving from Forward to Backward Euler may seem a minor change; we have to note that now to obtain $x(t_{n+1})$ it must be solved from $g(x(t_{n+1}))$ which means that now our equation has an implicit dependency on $x(t_{n+1})$. In the specific case of the system intending to generate a CIR this could be a drawback due to now $g$ function contains a 3° polynomial from which may be difficult to solve for $x(t_{n+1})$ and therefore numerically solve the system. In order to avoid the mentioned drawback let us suppose that there is neither target nor active diffusion. In these conditions only $r$ variable exists, therefore the equations now stay for

$$\dot{r}_{ij}(t) = d\Delta r_{ij} = d(r_{i+1j}(t) + r_{i-1j}(t) + r_{ij+1}(t) + r_{ij-1}(t) - 4r_{ij}(t)) \quad (8.5)$$

Time discretizing and applying Backward Euler method yields

$$\frac{r_{ij}(t_n + 1) - r_{ij}(t_n)}{h} = d(r_{i+1j}(t_{n+1}) + r_{i-1j}(t_{n+1}) + r_{ij+1}(t_{n+1}) + r_{ij-1}(t_{n+1}) - 4r_{ij}(t_{n+1})) \quad (8.6)$$

Grouping $t_{n+1}$ and $t_n$ terms yields

$$(1+4hd)r_{ij}(t_{n+1}) - hd(r_{i+1j}(t_{n+1}) + r_{i-1j}(t_{n+1}) + r_{ij+1}(t_{n+1}) + r_{ij-1}(t_{n+1})) = r_{ij}(t_n) \quad (8.7)$$

The best option to represent this system of linear equations is by matrix notation being:

$$R = [r_{11}; \ldots; r_{N1}; r_{12}; \ldots; r_{N2}; \ldots; r_{1N}; \ldots; r_{NN}] \quad (8.8)$$

The system (1.7) now can be written as

$$S'R(t_{n+1}) = R(t_n) \quad (8.9)$$

Where $S'$ matrix would be for a 3×3 neuron grid as follows

$$S' = \begin{pmatrix} 1+4hd & -hd & 0 & -hd & 0 & 0 & 0 & 0 & 0 \\ -hd & 1+4hd & -hd & 0 & -hd & 0 & 0 & 0 & 0 \\ 0 & -hd & 1+4hd & 0 & 0 & -hd & 0 & 0 & 0 \\ -hd & 0 & 0 & 1+4hd & -hd & 0 & -hd & 0 & 0 \\ 0 & -hd & 0 & -hd & 1+4hd & -hd & 0 & -hd & 0 \\ 0 & 0 & -hd & 0 & -hd & 1+4hd & 0 & 0 & -hd \\ 0 & 0 & 0 & -hd & 0 & 0 & 1+4hd & -hd & 0 \\ 0 & 0 & 0 & 0 & -hd & 0 & -hd & 1+4hd & -hd \\ 0 & 0 & 0 & 0 & 0 & -hd & 0 & -hd & 1+4hd \end{pmatrix}$$

Thus defining $S = S'^{-1}$, the simplified system resolution will be given by

$$R(t_n + 1) = SR(t_n) \quad (8.10)$$

The problem arises when we introduce the non lineal term. On that case it should be possible to solve $(t_{n+1})$ variable from the non lineal term. However this is not an option because by definition we wont be able anymore to express the algorithm in matricial terms due to now it is non lineal. To solve this issue we assume the Backward Euler method is applied only to the diffusive term, keeping Forward Euler for the reactive terms. So having the system

$$\dot{r}_{ij} = q_{ij}\{H(r_{ij})(f(r_{ij}) - v_{ij}) + d\Delta r_{ij} - r_{ij}p_{ij}\} \tag{8.11}$$

$$\dot{v}_{ij} = \frac{(r_{ij} - 7r_{ij} - 2)}{25} \tag{8.12}$$

With $d = 0.2$ and $f(r) = \frac{(-r^3 + 4r^2 - 2r - 2)}{7}$
We can discretize the system as follows

$$R(t_n) = SQ(t_{n-1})(R(t_{n-1}) + hH(R(t_{n-1}))(f(R(t_{n-1})) - V(t_{n-1})) - hR(t_{n-1})P(t_{n-1}))) \tag{8.13}$$

$$V(t_n) = V(t_{n-1}) + h\frac{(R(t_{n-1}) - 7V(t_{n-1}) - 2)}{25} \tag{8.14}$$

Finally the Von Neumann boundary conditions are imposed, having each cell in the obstacle boundary the same value than the neighbor cell perpendicular to its boundary. Combining forward and backward Euler for an arena, the place where the robots wander, of 60×60 cells we would have an S matrix of 3600×3600 elements resulting in at least 12,960,000 multiplications per iteration.

## 8.4 Hardware Architecture

The original Causal Neural Network mapping the CIR is comprised from a set of neurons based on the FitzHugh-Nagumo model and disposed like a grid with local coupling among them.

Cellular neural networks modelling space-invariant systems have been successfully implemented on FPGAs for different tasks like image processing using different architectures and techniques, reducing the operations to multiplications between the feedback, input synaptic operator and a 3×3 or 5×5 template of coefficients. The modified CIR model to be implemented in hardware platform for a grid of 60×60 neurons or cells needs an S matrix of 3600 elements which should be multiplied element by element by the *r* matrix for each time step in order to emulate analog behavior. Even the most powerful FPGAs don't have such amount of multipliers, not to mention the additional resources needed to carry out additional operations so accommodating this is not feasible in practice. To solve this issue a kind of pipeline and parallel architecture has been developed. Obstacle prediction, *dr* and *dv* are computed in parallel, while other auxiliary operations and time step evolution are computed sequentially.

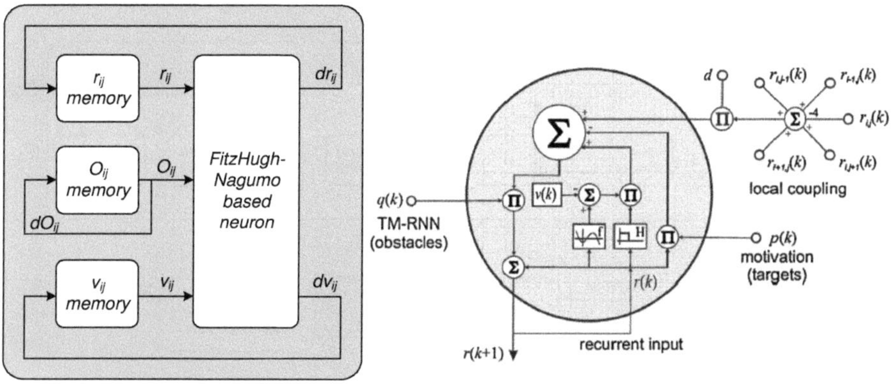


**Fig. 8.3** CIR simplified model and circuit implementation of the modified FitzHugh-Nagumo neuron used in the Causal Neural Network. The blocks *f*, *H*, $\sum$ and $\pi$ stand for cubic nonlinearity, Heaviside step function, sum, and product respectively. The block *v(k)* updates the state of the corresponding recovery variable (linear sum of *r(k)* and *v(k))*

Even with this approximation the direct implementation of the CIR model in the FPGA needs minor changes due to the number of neurons needed and the number of iterations to be performed to obtain a representative CIR. For example with an arena mapped to a grid of $60 \times 60$ neurons and no less than 5000 iterations the FPGA will need to allocate 3600 neurons and perform 3600 multiplications per neuron and per time iteration, among other operations, which is not feasible in practice.

Thus our proposal is to reduce the $S$ matrix to the minimal neuron neighborhood coupling needed for diffusion regime and generate this minimal neighborhood for each neuron on each iteration. Simulation results with the MATLAB CIR model showed that a minimum neighborhood of $7 \times 7$ neurons is needed, checking that operating with a $5 \times 5$ neighborhood wave and diffusion regimes are affected. This reduced version of $S$ coefficients is computed offline and stored in registers reducing the local coupling operation to a maximum of 25 parallel multiplications and its subsequent sums, taking into account that for a $7 \times 7$ neighborhood and a timestep of $0.1$, $S$ coefficients are reduced to 25 different values (Fig. 8.3).

The final approach is to create a single fully pipelined neuron which upon an initial offset outputs valid data per cycle, thus an N-cell CIR is compressed in just one-recurrent-neuron. This single recurrent neuron has its inputs and outputs connected to three dual port memories for *r*, *v* and *obstacles* respectively. The memories are initialized in MATLAB and then written to the FPGA through the command of NIOS II microcontroller and once the CIR is signaled to start, these memories go on storing the new iteration updated values. As the neuron has the capability to store all the *r* state variable of its $7 \times 7$ immediate neighbors the origin and destination memory can be the same, reducing the number of memories to just the mentioned independently of the iterations. The neuron is based on the FitzHugh-Nagumo model.

Besides the implementation of the modified neuron additional tasks have to be carried out. These tasks consist on update the effective obstacles supplied to the

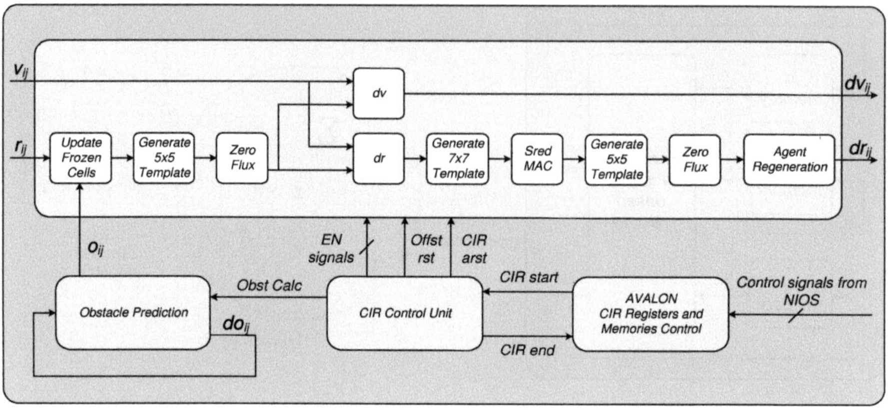


**Fig. 8.4** Simplified CIR model

neuron, apply Von-Neumman zero flux boundary conditions to the current cell based on the surrounding effective obstacles, create the right neighborhood for applying the mentioned zero flux and another one for the operation with the reduced version of $S$ coefficients, apply Von-Neuman zero flux boundary conditions to the neuron outputted $r$ variable and regenerate the agent with its initial value for the next iteration (Fig. 8.4).

In order to achieve an optimal CIR using this approach a small timestep must be used. The smaller the time step used the more accurate the CIR will be. Whether using small or larger timesteps, effective obstacles will be obtained, but larger timesteps will severely affect the passive diffusion profile leading to poor navigation trajectories. An optimal timestep value could be 0.001 which would require a precision above 32 bits to represent the derived reduced $S$ coefficients. In our case we have chosen a precision of 20 bits which derives into a timestep of 0.1, which will be showed latter that is an acceptable value to obtain good compromise between calculation time and performance.

As $S$ coefficients have a 20-bit precision, state variables $r$ and $v$ will be Q3.20 signed fixed point numbers. Memory positions storing $r$ values expand from 24 to 28 bit words in order to store additional information of the cell like for example if it is a target to be reached or is a frozen cell.

Summarizing we can say that the CNN has three main layers, $r$, $v$ and $O$ (which stands for Obstacles) as well as $T$ (Target) and $Q$ (effective obstacles) that evolve in every timestep to end up with the CIR generation at the last timestep. Once explained the modifications made to the CIR in order to be implemented on an FPGA and its overall scheme, we go on describing how one iteration is performed.

**Table 8.3** Qupdate VHDL entity template

| Qupdate.vhd | |
|---|---|
| Description | Updates Q checking if U is between 1 and 2 and if there is obst |
| Entity | Qupdate is<br>generic (Ubits : integer:= 28);<br>port (clk : in std_logic;<br>nreset : in std_logic;<br>en_in : in std_logic;<br>ncol : in unsigned (5 downto 0);<br>obst : in std_logic;<br>data_in : in std_logic_vector (Ubits-1 downto 0);<br>ulimit : in signed (23 downto 0);<br>dlimit : in signed (23 downto 0);<br>data_out : out std_logic_vector (Ubits-1 downto 0));<br>end Qupdate; |

### *8.4.1 r Data Preprocessing*

There are three main steps before the incoming *rij* data can be computed according to *dr* and *dv* equations.

Update frozen cells

The first step is to update if the cell becomes into a frozen cell or not, this is to check if the cell will be an effective obstacle by comparing *rij* with a value between an upper (ulimit in Table 8.3) and lower threshold (dlimit in Table 8.3) and if *Oij* (obst in Table 8.3) equals 1. The resulting value is stored into the bit $Q$, which is bit 25 of $r$ word. Down and up limits are stored in registers for quick access. Moreover the previous value of $Q$ is added to the just calculated $Q$ to maintain the effective obstacles frozen along the total number of iterations. This operation takes one cycle being the first stage of the neuron pipeline (Fig. 8.5).

The Qupdate VHDL entity template is shown in Table 8.3.

Current cell neighborhood generation

The next block needs to be fed with the current cell *rij* state variable as well as with the *rij* state variable of its 5×5 neighborhood at the same time.

This block takes as input data the serial data output from the updated frozen cells block, and with the aid of as much FIFO's as the neighborhood order minus one and a ladder of registers builds up a neighborhood of 5×5 cells. Each FIFO must have at least the depth of a whole row of cells. For example, in our case 60 cells must

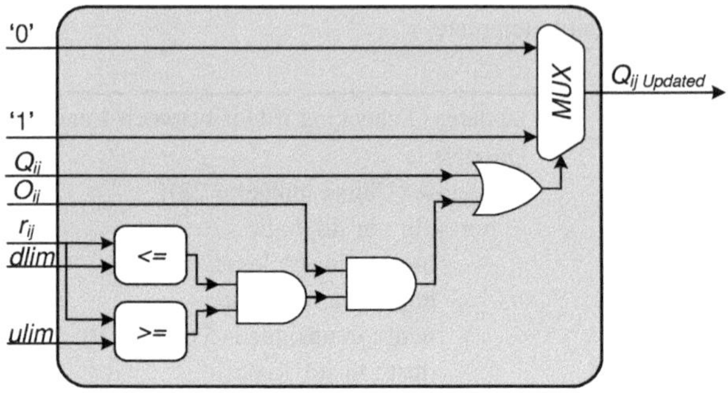


**Fig. 8.5** Q update

be stored in each FIFO. The data format 5×5 block takes care of the corners and boundaries, needing an initial offset of 128 cycles to output the first valid data with its neighborhood (Fig. 8.6).

The block diagram shown in Fig. 8.6 is for a 5×5 neighborhood generation, for a 7×7 data format it must be expanded following the rules:

$$Number of FIFOs = Order of neighborhood - 1$$

$$Number of stages = (Order of neighborhood \times 2) - 1$$

The data_format5×5 VHDL entity template is shown in Table 8.4.

## Zero-Flux Boundary Conditions

The next step is to impose Von Neumann boundary conditions over the current cell. This means to check if the current cell is an effective obstacle and in case it is, to set each cell *rij* in the obstacle boundary to the same value as its perpendicular neighbor cell *rij* generating a kind of toroid around the effective obstacle. In practice we need two rings of cells around the current cell, the inner one will be the toroid and the outer will be the one copied into the toroid cells, in the case where appropriate.

This operation is carried out by the block called zero flux. As the computation of whether the current cell and each one of its up, right, down and left neighbor cells are already frozen or toroid is performed in parallel, this block needs to be fed with a neighborhood of 5×5 cells.The toroid is not stored between timesteps, instead it is computed each time as explained before updating thus the value of *rij* as appropriate. This block consumes two cycles and only updates *rij* value when is part of a toroid and not an effective obstacle.

The zero_flux VHDL entity template is shown in Table 8.5

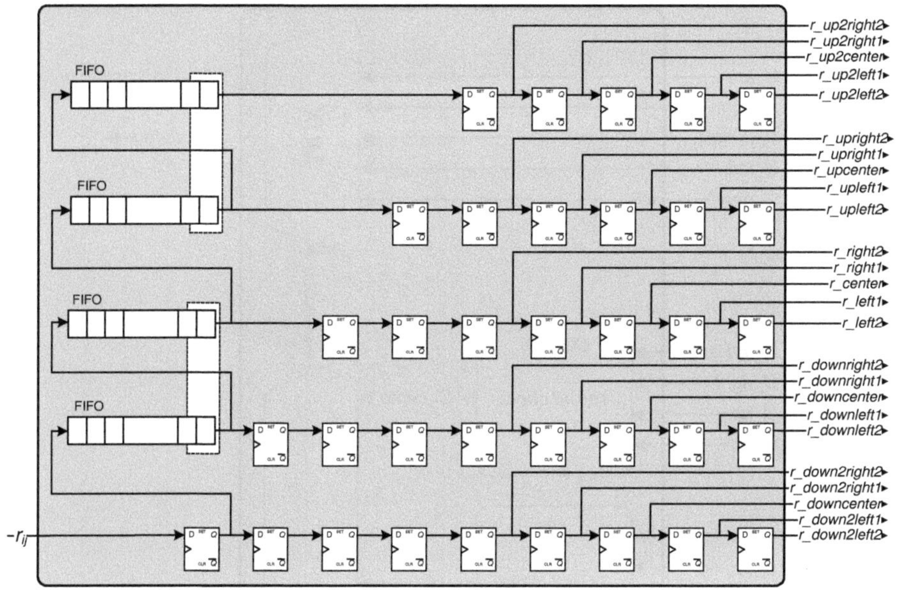


**Fig. 8.6** Block diagram of a 5×5 neighborhood generation

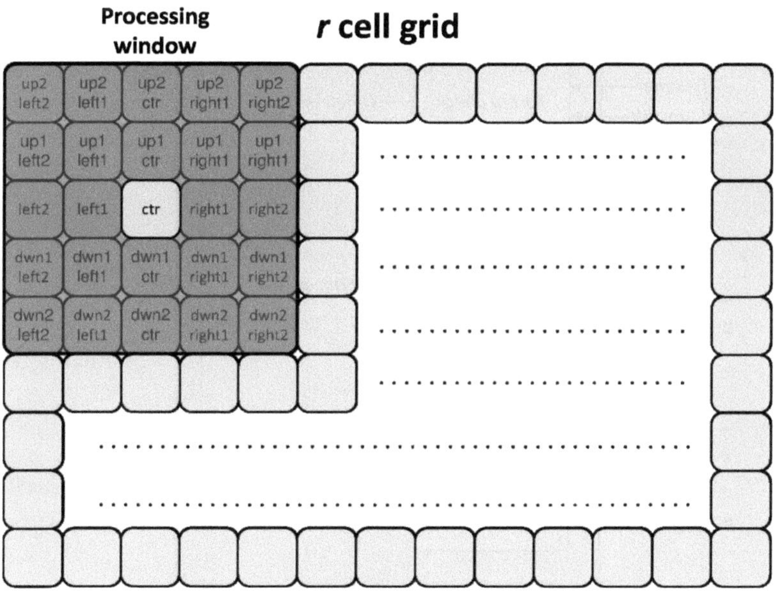


**Fig. 8.7** Current cell and its 5×5 neighborhood

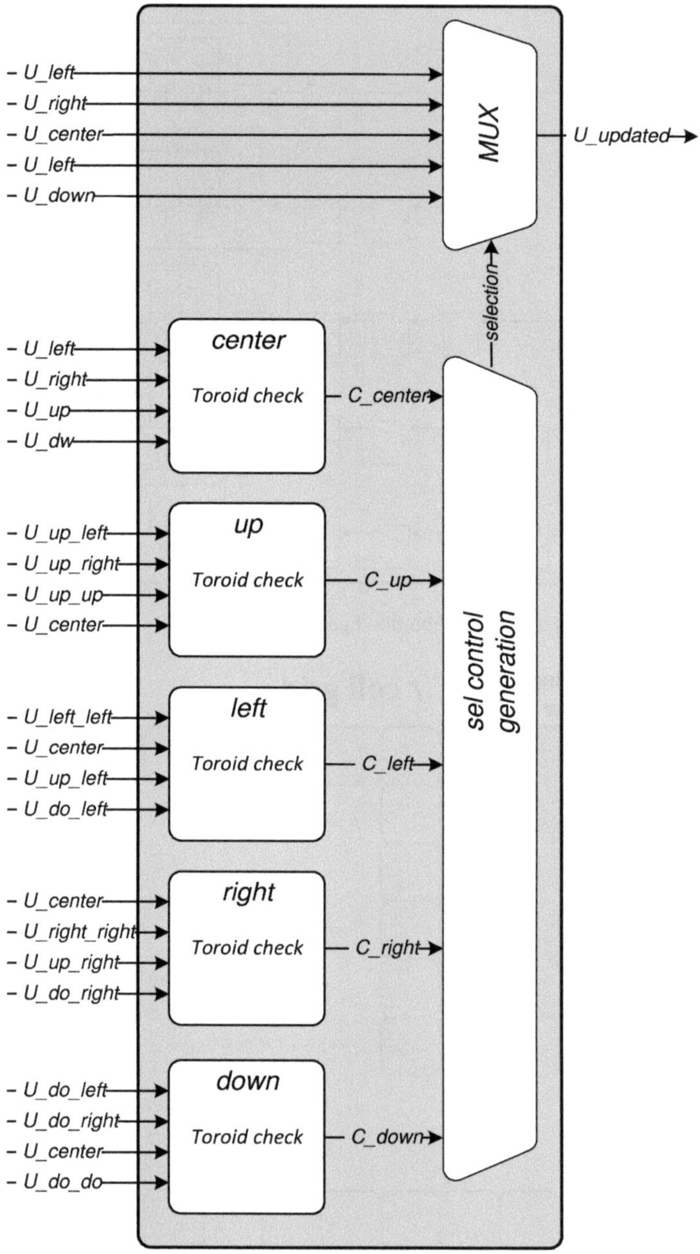


**Fig. 8.8** Zero flux block diagram

**Table 8.4** Data_format 5×5 VHDL entity template

| data_format 5×5.vhd | |
|---|---|
| Description | 5×5 cell neighborhood generation |
| Entity | data_format 5×5 is<br>generic (dbits : integer:= 28);<br>port (clk : in std_logic;<br>nreset : in std_logic;<br>ncol : in unsigned (5 downto 0);<br>en_in : in std_logic;<br>data_in : in std_logic_vector (dbits−1 downto 0);<br>do_upn_leftn : out std_logic_vector (dbits−1 downto 0);<br>do_upn_left : out std_logic_vector (dbits−1 downto 0);<br>do_upn_mid : out std_logic_vector (dbits−1 downto 0);<br>do_upn_right : out std_logic_vector (dbits−1 downto 0);<br>do_upn_rightn : out std_logic_vector (dbits−1 downto 0);<br>do_up_leftn : out std_logic_vector (dbits−1 downto 0);<br>do_up_left : out std_logic_vector (dbits−1 downto 0);<br>do_up_mid : out std_logic_vector (dbits−1 downto 0);<br>do_up_right : out std_logic_vector (dbits−1 downto 0);<br>do_up_rightn : out std_logic_vector (dbits−1 downto 0);<br>do_leftn : out std_logic_vector (dbits−1 downto 0);<br>do_left : out std_logic_vector (dbits−1 downto 0);<br>do_mid : out std_logic_vector (dbits−1 downto 0);<br>do_right : out std_logic_vector (dbits−1 downto 0);<br>do_rightn : out std_logic_vector (dbits−1 downto 0);<br>do_down_leftn : out std_logic_vector (dbits−1 downto 0);<br>do_down_left : out std_logic_vector (dbits−1 downto 0);<br>do_down_mid : out std_logic_vector (dbits−1 downto 0);<br>do_down_right : out std_logic_vector (dbits−1 downto 0);<br>do_down_rightn : out std_logic_vector (dbits−1 downto 0);<br>do_downn_leftn : out std_logic_vector (dbits−1 downto 0);<br>do_downn_left : out std_logic_vector (dbits−1 downto 0);<br>do_downn_mid : out std_logic_vector (dbits−1 downto 0);<br>do_downn_right : out std_logic_vector (dbits−1 downto 0);<br>do_downn_rightn: out std_logic_vector (dbits−1 downto 0));<br>end data_format 5×5; |

### 8.4.2 Evolution of r and v Cell State Variables

Once the input data $r_{ij}$ has been preprocessed it is ready to be fed directly to the blocks *dv* and *dr* which implements the differential equations. Both blocks perform its computations in parallel. Block *dv* takes less cycles to output the next timestep $v_{ij}$ value than *dr* does with $r_{ij}$, so internal data in *dv* must be delayed the right amount of cycles in order to both blocks output valid data synchronously.

**Table 8.5** Zero_flux VHDL entity template

| zero_flux.vhd | |
|---|---|
| Description | Von-Neumann Zero-Flux Boundary Conditions |
| Entity | zero_flux is<br>generic (dbits : integer:= 28);<br>port (clk : in std_logic;<br>nreset : in std_logic;<br>en_in : in std_logic;<br>Uupn_leftn : in std_logic_vector (dbits-1 downto 0);<br>Uupn_left : in std_logic_vector (dbits-1 downto 0);<br>Uupn_center : in std_logic_vector (dbits-1 downto 0);<br>Uupn_right : in std_logic_vector (dbits-1 downto 0);<br>Uupn_rightn : in std_logic_vector (dbits-1 downto 0);<br>Uup_leftn : in std_logic_vector (dbits-1 downto 0);<br>Uup_left : in std_logic_vector (dbits-1 downto 0);<br>Uup_center : in std_logic_vector (dbits-1 downto 0);<br>Uup_right : in std_logic_vector (dbits-1 downto 0);<br>Uup_rightn : in std_logic_vector (dbits-1 downto 0);<br>U_leftn : in std_logic_vector (dbits-1 downto 0);<br>U_left : in std_logic_vector (dbits-1 downto 0);<br>U_center : in std_logic_vector (dbits-1 downto 0);<br>U_right : in std_logic_vector (dbits-1 downto 0);<br>U_rightn : in std_logic_vector (dbits-1 downto 0);<br>Udo_leftn : in std_logic_vector (dbits-1 downto 0);<br>Udo_left : in std_logic_vector (dbits-1 downto 0);<br>Udo_center : in std_logic_vector (dbits-1 downto 0);<br>Udo_right : in std_logic_vector (dbits-1 downto 0);<br>Udo_rightn : in std_logic_vector (dbits-1 downto 0);<br>Udon_leftn : in std_logic_vector (dbits-1 downto 0);<br>Udon_left : in std_logic_vector (dbits-1 downto 0);<br>Udon_center : in std_logic_vector (dbits-1 downto 0);<br>Udon_right : in std_logic_vector (dbits-1 downto 0);<br>Udon_rightn : in std_logic_vector (dbits-1 downto 0);<br>U_out : out std_logic_vector (dbits-1 downto 0);<br>en_out : out std_logic);<br>end zero_flux; |

Figure 8.9 shows how *dv* is a simple block computing the evolution of *v* cell state variable according to the discretized CIR equation and needs only 4 cycles to output valid data. The *dr* block is responsible for the computation of the evolution of *r*. Inside this block it must be also evaluated a three degree polynomial that models a cubic non-linearity and controls the evolution of the wavefront exploring the grid. It is worth to mention that the robot or agent is modelled as a 5 cell object.

*dv* block VHDL entity template is shown in Table 8.6

*dr* block VHDL entity template is shown in Table 8.7. *dU* stands for *dr*.

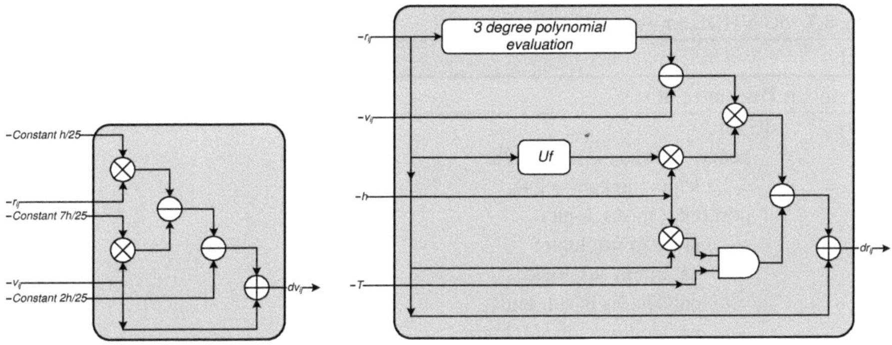


**Fig. 8.9** From left to right variable blocks *dv* and *dr* respectively

**Table 8.6** dV VHDL entity template

| dV.vhd | |
|---|---|
| Description | Derivative of *v* |
| Entity | dV is<br>generic (Ubits : integer:= 24);<br>Vbits : integer:= 24);<br>port (clk : in std_logic;<br>nrst : in std_logic;<br>en_in : in std_logic;<br>U : in signed (Ubits-1 downto 0);<br>V : in signed (Vbits-1 downto 0);<br>coef1 : in signed (Vbits-1 downto 0);<br>coef2 : in signed (Vbits-1 downto 0);<br>coef3 : in signed (Vbits-1 downto 0);<br>dv : out signed (Vbits-1 downto 0));<br>end dV; |

Horner Algorithm

This block evaluates the three degree polynomial using the Horner Algorithm which minimizes the arithmetic and can be pipelined in 6 stages, obtaining the first valid data at the seventh cycle.

$$f(r) = \frac{(-r^3 + 4r^2 - 2r - 2)}{7}$$

In our case as the polynomial coefficients are fixed, they are stored in registers as a signed fractional Q1.15 number saving a division operation by seven. So we end up with four coefficients which derive in the following equation

**Table 8.7** dU VHDL entity template

| dU.vhd | |
|---|---|
| Description | Derivative of *r* |
| Entity | dU is<br>generic (Ubits : integer:= 28;<br>Vbits : integer:= 24);<br>port (clk : in std_logic;<br>nrst : in std_logic;<br>en_in : in std_logic;<br>en_d7×7 : in std_logic;<br>en_sredmac : in std_logic;<br>ncol : in unsigned (5 downto 0);<br>cir_agent_position1 : std_logic_vector (31 downto 0);<br>cir_agent_position2 : std_logic_vector (31 downto 0);<br>cir_agent_position3 : std_logic_vector (31 downto 0);<br>cir_agent_position4 : std_logic_vector (31 downto 0);<br>cir_agent_position5 : std_logic_vector (31 downto 0);<br>U : in std_logic_vector (Ubits-1 downto 0);<br>V : in signed (Vbits-1 downto 0);<br>Obst : in std_logic;<br>s14 : in signed (23 downto 0);<br>s23,s24,s25 : in signed (23 downto 0);<br>s32,s33,s34,s35,s36 : in signed (23 downto 0);<br>s41,s42,s43,s44,s45,s46,s47 : in signed (23 downto 0);<br>s52,s53,s54,s55,s56 : in signed (23 downto 0);<br>s63,s64,s65 : in signed (23 downto 0);<br>s74 : in signed (23 downto 0);<br>tstep : in signed (Ubits-5 downto 0);<br>pco1 : in signed (Ubits-5 downto 0);<br>pco2 : in signed (Ubits-5 downto 0);<br>pco3 : in signed (Ubits-5 downto 0);<br>pco4 : in signed (Ubits-5 downto 0);<br>dU : out std_logic_vector (Ubits-1 downto 0));<br>end dU; |

$$f(r) = ((-\frac{1}{7}r + \frac{4}{7})r - \frac{2}{7})r - \frac{2}{7}$$

$$Coeff1 = -0.142857$$

$$Coeff2 = -0.571428$$

$$Coeff3 = -0.285714$$

$$Coeff4 = -0.285714$$

**Table 8.8** Pol_eval VHDL entity template

| pol_eval.vhd | |
|---|---|
| Description | Evaluates a third degree polynomial using Horner Algorithm |
| Entity | pol_eval is<br>generic (Ubits : integer:= 24);<br>port (clk : in std_logic;<br>nrst : in std_logic;<br>en_in : in std_logic;<br>coef1 : in signed (Ubits-1 downto 0);<br>coef2 : in signed (Ubits-1 downto 0);<br>coef3 : in signed (Ubits-1 downto 0);<br>coef4 : in signed (Ubits-1 downto 0);<br>U : in signed (Ubits-1 downto 0);<br>Ueval : out signed (Ubits-1 downto 0));<br>end pol_eval; |

Below is a block diagram of the Horner Solver showing the multiplier, truncation and adder operation and how they are arranged in pipeline stages.

Pol_eval block VHDL entity template is shown in Table 8.8

*Uf* represents the Heaviside function controlling the transition between Wave and Diffusion Regimes. When the state of the current neuron is under a threshold value, the block outputs a logic '1' and that neuron is subject to the active diffusion, but when is above the threshold then the block outputs a logic '0' and the neuron state evolves according to passive diffusion.

The target to be reached by the robot is represented by $T$ which is the bit 24 of the 28 bit $r$ word. The logic 'and' operation between $T$ and $r$ represents the reaction term, where the binary result of the operation is set to '1' if the neuron is occupied by a target and clear to '0' if not. As *dr* runs in parallel with *dv* and takes more cycles than *dv* to calculate its output, the control unit is responsible for controlling when to enable the writing of the updated $v$ value to its memory and when to do the same with the updated $r$ value. Regarding the delay *dv* takes 4 cycles to output first valid data while *dr* takes 9 cycles.

The local coupling between $r_{ij}$ cell state variable and its neighbors is modelled by the sred_mac block, which is not strictly a multiply and accumulate unit. This block consumes the majority of the FPGA DSP resources invested in the design, as this block performs 25 parallel multiplications and then partial sums in order to obtain a unique $r_{ij}$ updated value.

Sred_mac block VHDL entity template is shown in Table 8.9

The MAC unit needs to be supplied with $r_{ij}$ and its $7 \times 7$ immediate neighbors so another block to create a $7 \times 7$ neighborhood is needed.The $7 \times 7$ block is implemented in the same way the $5 \times 5$ one with the same kind of FIFO's but increasing its number up to 6. The $7 \times 7$ neighborhood generation block and MAC unit take an offset of 191 and 6 cycles respectively to output valid data (Figs. 8.7, 8.8 and 8.11)

**Table 8.9** Sred_mac VHDL entity template

| sred_mac.vhd | |
|---|---|
| Description | Computes parallel multiplication element by element and accumulates each sub-product |
| Entity | sred_mac is<br>generic (Ubits : integer:= 24;<br>dbits : integer:= 24;<br>intbits : integer:=3;<br>fracbits : integer:=20);<br>port (cir_clk : in std_logic;<br>cir_en : in std_logic;<br>cir_reset : in std_logic;<br>cir_ncol : in unsigned (5 downto 0);<br>Ust : in std_logic_vector (3 downto 0);<br>U_upnn_mid : in signed (Ubits-1 downto 0);<br>U_upn_left : in signed (Ubits-1 downto 0);<br>U_upn_mid : in signed (Ubits-1 downto 0);<br>U_upn_right : in signed (Ubits-1 downto 0);<br>U_up_leftn : in signed (Ubits-1 downto 0);<br>U_up_left : in signed (Ubits-1 downto 0);<br>U_up_mid : in signed (Ubits-1 downto 0);<br>U_up_right : in signed (Ubits-1 downto 0);<br>U_up_rightn : in signed (Ubits-1 downto 0);<br>U_leftnn : in signed (Ubits-1 downto 0);<br>U_leftn : in signed (Ubits-1 downto 0);<br>U_left : in signed (Ubits-1 downto 0);<br>U_mid : in signed (Ubits-1 downto 0);<br>U_right : in signed (Ubits-1 downto 0);<br>U_rightn : in signed (Ubits-1 downto 0);<br>U_rightnn : in signed (Ubits-1 downto 0);<br>U_down_leftn : in signed (Ubits-1 downto 0);<br>U_down_left : in signed (Ubits-1 downto 0);<br>U_down_mid : in signed (Ubits-1 downto 0);<br>U_down_right : in signed (Ubits-1 downto 0);<br>U_down_rightn : in signed (Ubits-1 downto 0);<br>U_downn_left : in signed (Ubits-1 downto 0);<br>U_downn_mid : in signed (Ubits-1 downto 0);<br>U_downn_right : in signed (Ubits-1 downto 0);<br>U_downnn_mid : in signed (Ubits-1 downto 0);<br>s14 : in signed (23 downto 0);<br>s23 : in signed (23 downto 0);<br>s24 : in signed (23 downto 0);<br>s25 : in signed (23 downto 0);<br>s32 : in signed (23 downto 0);<br>s33 : in signed (23 downto 0);<br>s34 : in signed (23 downto 0);<br>s35 : in signed (23 downto 0); |

**Table 8.9** continued

| sred_mac.vhd |
|---|
| DescriptionComputes parallel multiplication element by element and accumulates each sub-product |

```
            s36 : in signed (23 downto 0);
            s41 : in signed (23 downto 0);
            s42 : in signed (23 downto 0);
            s43 : in signed (23 downto 0);
            s44 : in signed (23 downto 0);
            s45 : in signed (23 downto 0);
            s46 : in signed (23 downto 0);
            s47 : in signed (23 downto 0);
            s52 : in signed (23 downto 0);
            s53 : in signed (23 downto 0);
            s54 : in signed (23 downto 0);
            s55 : in signed (23 downto 0);
            s56 : in signed (23 downto 0);
            s63 : in signed (23 downto 0);
            s64 : in signed (23 downto 0);
            s65 : in signed (23 downto 0);
            s74 : in signed (23 downto 0);
            en_out : out std_logic;
            U_out : out std_logic_vector (27 downto 0));
    end sred_mac;
```

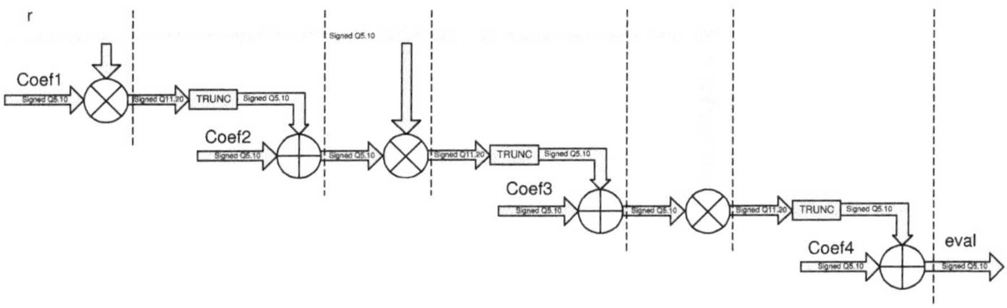


**Fig. 8.10** MAC simplified block diagram

## 8.4.3 r Data Postprocessing

Finally we need to apply once again zero flux conditions over the output of the MAC unit; in other words over the evolution of $r_{ij}$, to model the boundary conditions of the grid. As with the first zero flux application we need to generate a neighborhood of $5\times5$ cells.

The last step is to regenerate the agent for the next iteration where the diffusion takes place. The agent regeneration block simply checks the position of the current cell and in case it matches the position of the agent $r_{ij}$ value is rewritten with its

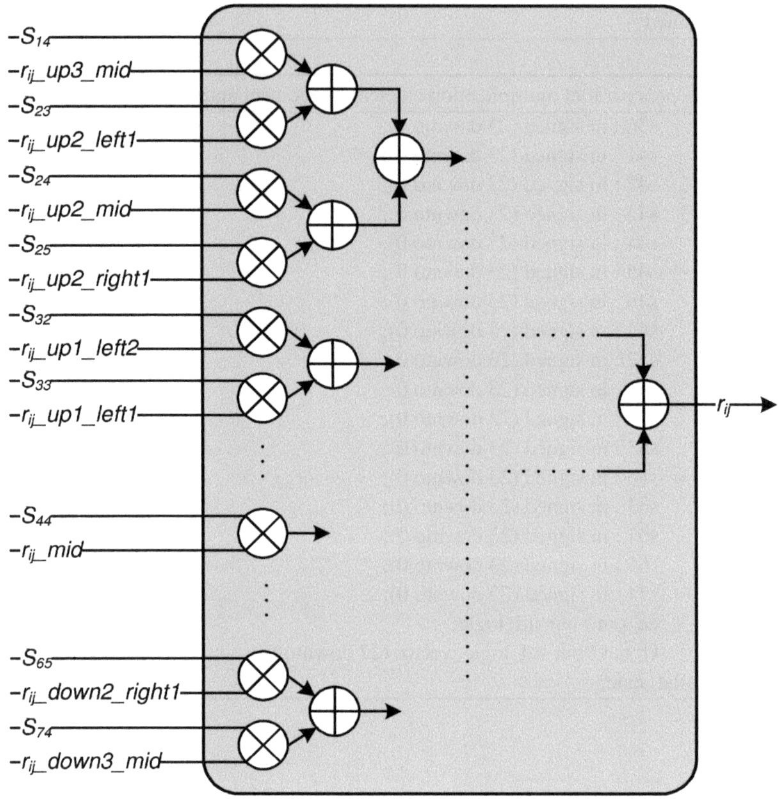


**Fig. 8.11** Horner solver diagram

initial value, in this implementation the value 5,and in case it is not the agent position the $r_{ij}$ value just passes through this block being written to $r$ memory. This block takes 1 cycle to be performed.

Now the CIR is ready to perform another iteration and so on until the total number of iterations has been reached. The total initial offset is 470 cycles. This is the number of cycles needed to full all the pipeline stages and thus to obtain the output of the first neuron of the cell grid for the first iteration. After this number of cycles a new pair of valid updated $r$ and $v$ values is obtained per cycle. Each iteration takes as much cycles to end as the number of neurons, in our case 3600 cycles.

Besides the mentioned blocks which comprise the CIR neuron there is a control unit which is basically a finite state machine that controls the synchronization among the blocks, the memories, the start and end of the CIR responding to NIOS II commands and the obstacle evolution block. The NIOS II can access the memories to initialize and read them through the Altera Avalon Bus so there is also a block who acts as an arbiter over the memory data and address buses (Figs. 8.10, 8.12 and 8.15).

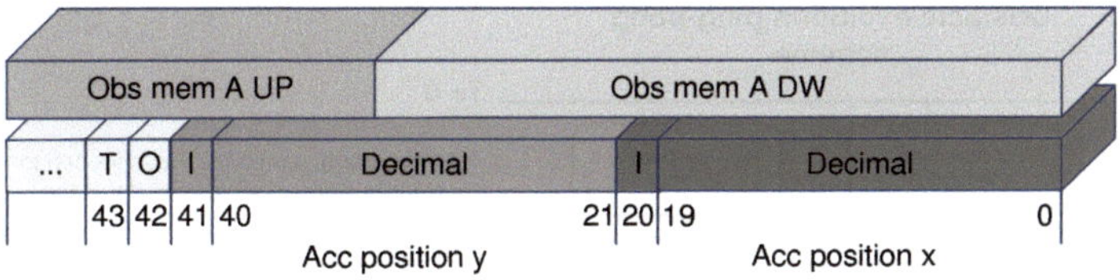


**Fig. 8.12** Obstacle format

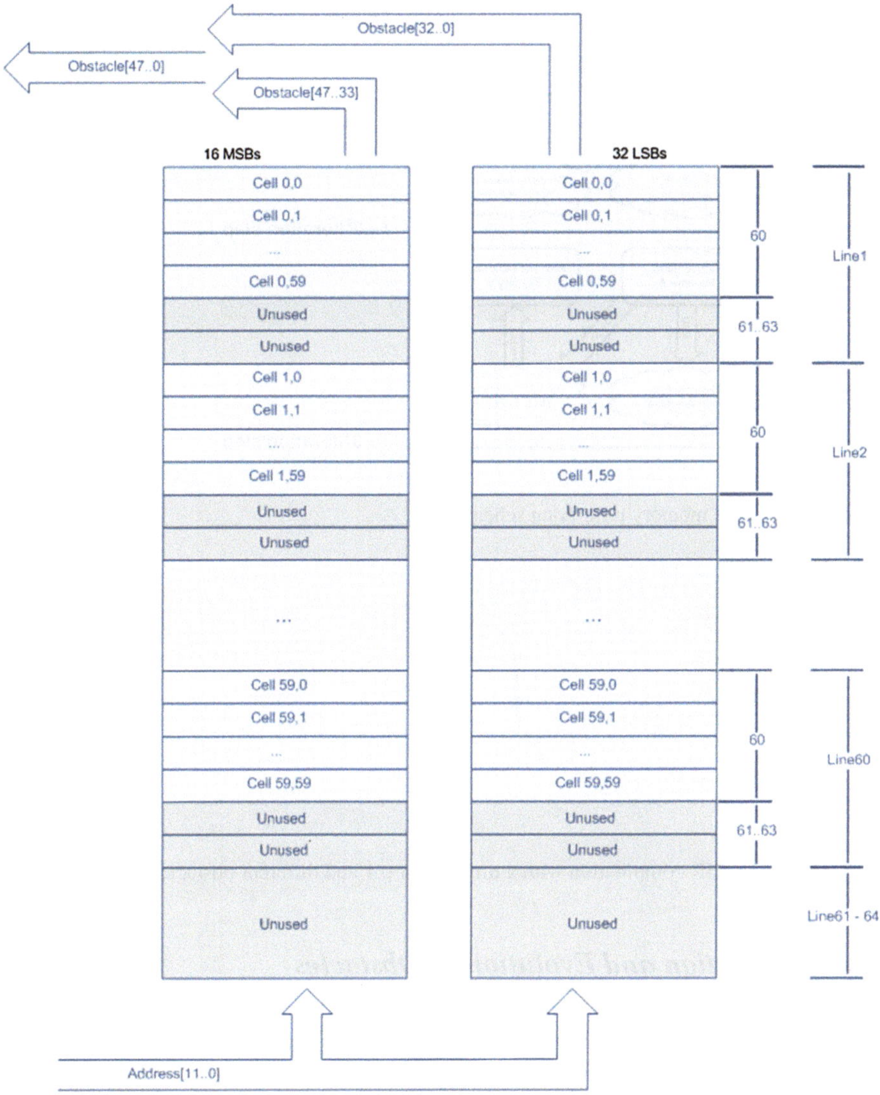


**Fig. 8.13** Obstacle memory arrangement

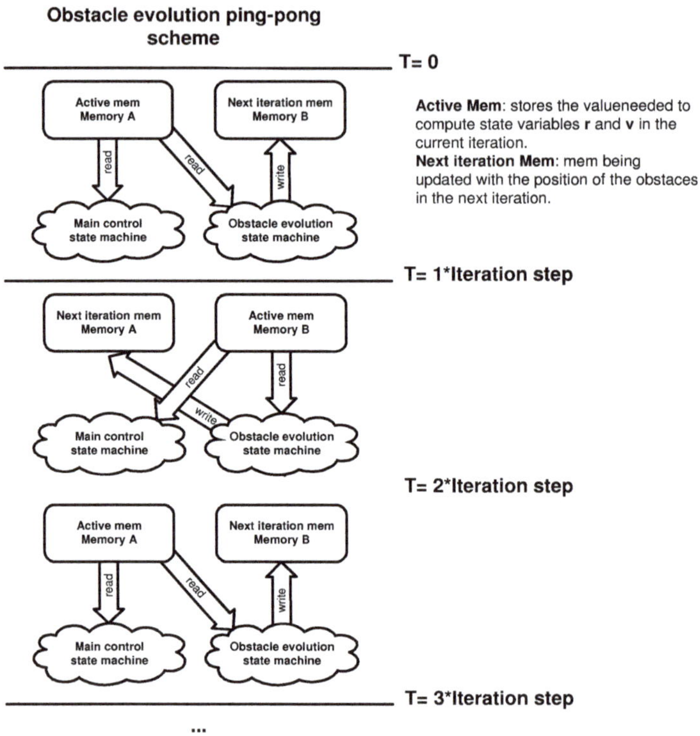


Fig. 8.14 Obstacle memory ping-pong scheme

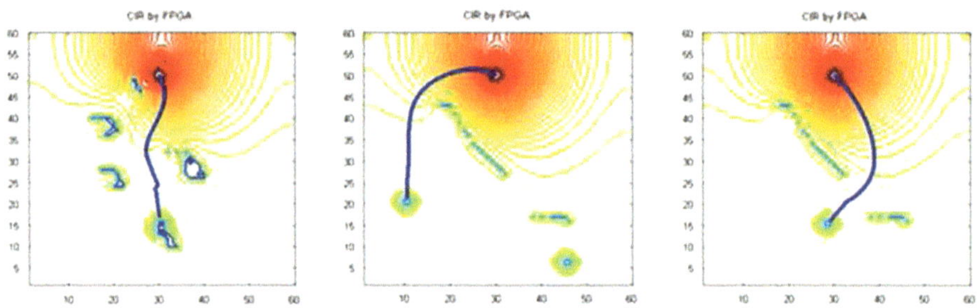

Fig. 8.15 FPGA CIR computation with a timestep of 0.1 and different obstacle scenarios

## 8.4.4 Generation and Evolution of Obstacles

The obstacle evolution block updates the position of the obstacle in every iteration according to a rectilinear movement (in both, x and y axis). It implements a ping-pong scheme with a couple of memories. This means that while the CIR Processing block is reading from one memory the current position (it is, in the current iteration) of the obstacle, the obstacle evolution block is in parallel updating the other memory with the obstacle position of the next iteration. This scheme allows the CIR processing to always have at its disposal the current obstacle position while in parallel the obstacle position is being updated without affecting its current position.Once the iteration finishes, both memories swap, and so on for every iteration.

As said before there are two types of obstacles, static and dynamic ones. The static ones are frozen whereas the dynamic ones can move along the x and y axis with $V_x$ and $V_y$ speed.

In order to deal with obstacles, as with $r$ and $v$ state variables,a $60\times60$ matrix representing a grid version of the arena is used but in this case each obstacle cell or matrix position will have the capability for retaining an obstacle for more than one iteration and determining how close is to its neighbors cells. For the sake of simplicity let's assume the obstacle is smaller than a grid cell, that is, an obstacle can move within the same cell several steps before changing to another cell. It can also be seen as a large arena where cells are big enough.

From the implementation point of view, a memory will store the obstacle information of every cell within the matrix. The format will be as follows:

Where:

- **T**: defines the type of obstacle. '1' for static ones and '0' for dynamic ones.
- **O**: defines the obstacle. '1' if the cell is occupied by an obstacle.
- **X position**: position of the obstacle within a cell along the x axis.
- **Y position**: position of the obstacle within a cell along the x axis.

The position of a dynamic obstacle will be updated in every iteration step according to the $V_x$ and $V_y$ speed of the object.This will be further explained.

It can be seen that due to the 48-bit length of the obstacle word, we need a $60\times60$ memory of 48-bit words. In order to ease the implementation of such memory, two memories will be created. One storing the 32 LSBs and the other storing the 16 MSBs. The outputs will be concatenated to conform the 48-bit word. Finally we would like to remark that due to the binary nature of the digital design, the memory size will be 4096 words instead of $60\times60 = 3600$ and it will be arranged as depicted in Fig. 8.13.

The **Obstacle Memory** is initialized at the beginning with the positions of the static and dynamic obstacles. Likewise, first and last row/column, it is the arena boundaries, will be marked as static obstacles.

Ping-Pong memories are configured as three-port memories using Altera Quartus II tool, in order to be accessed at the same time by the obstacle evolution block and the CIR block.

- One writing port: to write the updated values.
- One read port: will be managed by the main control to know if the cell contains an obstacle.
- A second read port: managed by the obstacle evolution state machine. Needed to compute the evolution of the obstacle.

Static obstacles typically mark the arena perimeter but may also be used to mark walls or whatever static obstacle is placed between the robot position and its target.

Dynamic obstacles are wandering robots which may interfere the trajectory of the robot, also known as agent,to its target or not. The issue here is to predict the trajectory of the dynamic obstacles to avoid them.

As stated before, a memory will store the obstacle info per cell. A ping-pong mechanism computes the evolution of the obstacle(new values of the cell matrix in the next frame step). The drawback of the ping-pong mechanism is that it implies that the memory must be duplicated. One will be marked as the active one, which holds the current value of the obstacle matrix, while the other one will be written with the updated values. Every iteration step, once the $r$ and $v$ updated values have been calculated, both memories swap their functionality. It can be seen in Fig. 8.14.

Considering that the obstacles position is mapped into the arena or cells grid as x and y coordinates, we should be able to compute or predict the new coordinates in each time-step, that is, a new $\Delta$x and $\Delta$y should be calculated.

The computing is quite simple. At a first step, the speed as well as the direction (in x and y axes)of the robot is calculated (by means of an Eye-RIS system for instance). Such speed will be used as a constant displacement in the computing phase.

The processing itself starts using a counter to read the content of every memory cell. In every step,Three possibilities arise depending on its value:

- No obstacle or static obstacle: the value of the cell remains unchanged so no write to the next time frame memory is required.
- Dynamic obstacle found: update the position of the obstacle in x and y axis by adding the normalized displacement:

$$Newposition\_x(cell_{ij}) = \Delta x + currentposition$$
$$Newposition\_y(cell_{ij}) = \Delta y + currentposition$$

  This new value can make the robot:

  - Stay in the same cell: in this case we update the value of the obstacle within the cell in the non-active memory.
  - Move to another cell: if the accumulator saturates the current cell is cleared and we increment  decrement by one the address x or y depending on the direction of the movement.

Once the complete memory has been updated, the system waits until the main control machine warns that the iteration has finished so we swap memories for the next frame step.

Obstacle_evolution VHDL entity template is shown in Table 8.10

In a real time implementation an MLP could be implemented to replace this block and predict complicated obstacle trajectories.

**Table 8.10** Obstacle_evolution VHDL entity template

| obstacle_evolution.vhd | |
|---|---|
| Description | Computation of obstacles trajectories |
| Entity | entity_obstacle_evolution is<br>port (clk : in std_logic;<br>nreset : in std_logic;<br>slave_address : in std_logic_vector (11 downto 0);<br>slave_writedata : in std_logic_vector (31 downto 0);<br>cir_obs_a_dw_mem_data : out std_logic_vector (31 downto 0);<br>cir_obs_a_up_mem_data : out std_logic_vector (15 downto 0);<br>cir_obs_b_dw_mem_data : out std_logic_vector (31 downto 0);<br>cir_obs_b_up_mem_data : out std_logic_vector (15 downto 0);<br>obs_a_dw_mem_en : in std_logic;<br>obs_a_up_mem_en : in std_logic;<br>obs_b_dw_mem_en : in std_logic;<br>obs_b_up_mem_en : in std_logic;<br>obs_mem_a_dw_wren : in std_logic;<br>obs_mem_a_up_wren : in std_logic;<br>obs_mem_b_dw_wren : in std_logic;<br>obs_mem_b_up_wren : in std_logic;<br>v_x : in std_logic_vector (20 downto 0);<br>v_y : in std_logic_vector (20 downto 0);<br>rd_addr : in std_logic_vector (11 downto 0);<br>start_calc : in std_logic;<br>obs_out : out std_logic;<br>end_calc : out std_logic;<br>act_mem : out std_logic;<br>wren_obs : out std_logic);<br>end obstacle_evolution; |

## 8.5 Resources and Processing Times

The CIR has been implemented in an Altera Stratix II EP2S60F672C3 FPGA. The blocks have been described using low level VHDL primitives with ieee.numeric_std.all library for arithmetic operations instead of ieee.std_logic_arith.all, while the place and route has been left to Altera synthesis tool Quartus II.

The memories used as FIFO's involved in the creation of a 5×5 neighborhood are 64×28 bit with synchronous write and read operation. The size of the memory is determined by the number of column cells, as this memory has to store one full row, while the width of the data is determined by the width of r data plus 4 bits of additional information about the cell in that position. Four FIFO's of this kind are needed each one with a resource usage of 1792 ram bits. These memories are also used in the creation of the 7×7 neighborhood block, but in this case six FIFO's are needed.

**Table 8.11** FPGA resources usage

| Resources | Units | % Used |
|---|---|---|
| Combinational ALUTs | 6,660 | 14 |
| Dedicated logic registers | 6,387 | 13 |
| Total block memory bits | 902,304 | 35 |
| DSP block 9-bit elements | 264 | 92 |
| Total PLLs | 1 | 17 |

The proposed design also uses three memories to initialize and update the cell state variables, r and v, and the obstacles. State cell variables r and v memories are 28 bit and 24 bit 3600 words respectively. State r variable memory consumes 100800 ram bits while v state variable memory consumes 86400 ram bits. On the other hand obstacle memory is made up of two memories with a total 200704 ram bits consumption.

The design has been synthesized and implemented using Altera Quartus II 9.1 tools. The overall system is completed with a NIOS II 32-bit microcontroller targeted to control the CIR synthesized with Altera SOPC tool.

A summary of the resources consumed by the whole system is depicted in Table 8.11

The total offset to obtain the first cell output valid data, corresponding to the cycles needed to fill all the pipeline stages and output the first iteration of the first cell data, is 470 cycles. Once this initial offset has passed, the performance of the system is to output valid data on each cycle. For valid data we mean the output of the two cell state variables and the position of the obstacle for a specific cell on a specific iteration.

The CIR runs at 125 MHz and so the NIOS II microcontroller who controls it. By removing the NIOS II from the design a huge increment in the working frequency can be expected. The only way to speed up the computation is to augment the running frequency which is severing limited by the NIOS II microcontroller.

A series of tests on the CIR implementation have been carried out using several obstacle schemes. The scenario resembles typical obstacles moving from one side to the other side of the arena, a robot and a target opposite to it to be reached. The objective of these tests is assessing the performance of the CIR, its right creation and if it is valid to compute the optimal trajectory to avoid the crossing obstacles.

To obtain an optimal CIR, at least 7000 iterations are needed which leads to an FPGA processing time of 0.2 s.

## 8.6 Conclusions

A CIR implementation based on an FPGA has been proposed achieving similar results as with MATLAB model but with better performance. This architecture proposes an approach for implementing the CIR on an FPGA suitable for running a large amount of iterations with minimal delay needed to obtain a reliable Compact Internal Representation. The approach is purely sequential except in *dv* and *dr* computation which are performed in parallel. This CIR implementation is better suited for a large number of iterations and a sequential data input which the current case is.

## References

1. F. Javier Toledo-Moreo, J. Manuel Ferrández-Vicente, J. Javier Martínez-Alvarez, F. Javier GarrigÓs-Guerrero, High performance implementation of an fpga-based sequential dt-cnn. Lect. Notes Comput. Sci. **4528**, 1–9 (2007)
2. S. Péter, N. Zoltán, Configurable multi-layer cnn-um emulator on fpga. IEEE transactions. **50**(6), 1057–7122 (2003)
3. G. Vagliasindi P. Arena D. Mazon-A. De Maack S. Palazzo, A. Murari and JET EFDA contributors. Image processing with cellular nonlinear networks implemented on field-programmable gate arrays for real time applications in nuclear fusion. EFDA-JET-PR(10)01
4. M.G. Alba Luis-Velarde, V. Makarov, J.A. Villacorta-Atienza, L. Salas-Paracuellos, Compact internal representation as a protocognitive scheme for robots in dynamic environments. in *Proceedings SPIE, 8068* (2011)

# Chapter 9
# Visual Routines for Cognitive Systems on the Eye-RIS Platform

**D. J. Caballero-Garcia and A. Jimenez-Marrufo**

**Abstract** The purpose of this chapter is to describe how different visual routines can be developed and embedded in the AnaFocus' Eye-RIS Vision System on Chip (VSoC) to close the perception to action loop within the roving robots developed for the testing of the insect brain models. The Eye-RIS VSoC employs a bio-inspired architecture where image acquisition and processing are intermingled and the processing itself is carried out in two steps. At the first step, processing is fully parallel owing to the concourse of dedicated circuit structures which are integrated close to the sensors. At the second step, processing is realized on digitally-coded information data by means of digital processors. The Eye-RIS VSoC systems provide with image-processing capabilities and speed comparable to high-end conventional vision systems without the need for high-density image memory and intensive digital processing. Current perceptual schemes are often based on information derived from visual routines. Since real world images are quite complex to be processed for perceptual needs with traditional approaches, more computationally feasible algorithms are required to extract the desired features from the scene in real time, to efficiently proceed with the consequent action. In this chapter the development of such algorithms and their implementation taking full advantage of the sensing-processing capabilities of the Eye-RIS VSoC are described.

## 9.1 Introduction

The AnaFocus Eye-RIS v2.1 Vision System is a compact vision system that improves the characteristics of previous release already adopted for robotic applications [1]. The Eye-RIS v2.1 includes all the elements needed for acquiring images, enhancing

D. J. Caballero-Garcia · A. Jimenez-Marrufo (✉)
Innovaciones Microelectrónicas S. L. —Anafocus, Avenida Isaac Newton 4, Planta 7, 41092 Seville, Spain
e-mail: amanda.jimenez@anafocus.com

P. Arena and L. Patanè (eds.), *Spatial Temporal Patterns for Action-Oriented Perception in Roving Robots II*, Cognitive Systems Monographs 21,
DOI: 10.1007/978-3-319-02362-5_9, 

sensor operation, processing the image flow in real-time, interpreting the information contained in such image flow, and supporting decision-making based on the outcome of such interpretation. Eye-RIS v2.1 system employs the innovative and proven Smart Image Sensors (SIS) proprietary architecture, in which image-processing is accomplished following a hierarchical approach with two main levels:

- Early-processing: This level comes right after signal acquisition. Inputs are full-resolution images. The basic tasks at this level are meant to extract useful information from the input image flow. Outputs of this level are reduced sets of data comprising image features such as object locations, shapes, edges that are important for visual learning purposes [1].
- Post-processing: Here the amount of data is significantly smaller. Inputs are abstract entities in many cases, and tasks are meant to output complex decisions and to support action-taking. These tasks may involve complex algorithms within long and involved computational flows and may require greater accuracy than early processing.

One unique characteristic of the Eye-RIS vision systems compared to other commercial solutions is that image acquisition and early-processing is performed simultaneously in all pixels of the image sensor. Consequently, images do not need to be downloaded from the sensor for the initial stages of the processing. This concept of concurrent sensing-processing extends the Image Sensor (IS) concept to the Smart Image Sensor (SIS) one. What is more, many of the early-processing tasks in the SIS are accomplished in the analogue domain directly operating with analogue (grey-scale) images. The above has several advantages compared to purely digital processing, but imposes some restrictions in programming that shall be understood and taken into consideration when developing applications in these systems.

In contrast, image post-processing is digital and it is carried out by a RISC microprocessor.

## 9.2 Software Libraries

### *9.2.1 Problem Description*

In robotic applications, the images acquired from a camera contain relevant information that should be extracted to be properly processed. In same architectures used for visual homing the image flow can be used as raw data trying to find correlations with previously stored locations [2] but in general a segmentation algorithm together with a feature extraction process is needed [1, 3].

In the following the segmentation process for a set of objects from the images acquired by a camera mounted on different mobile robots is discussed. In this context, to segment means:

- Determine the presence/absence of an object of interest.

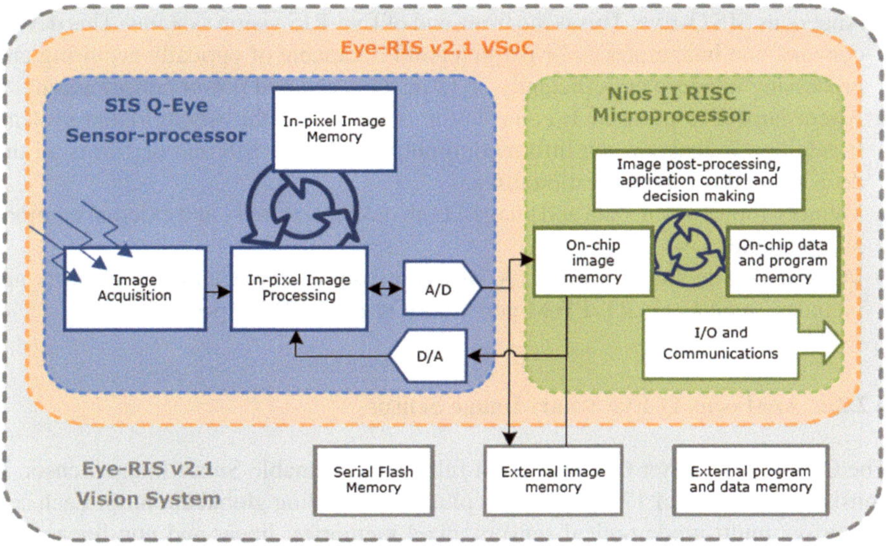


**Fig. 9.1** Eye-RIS Vision System on Chip architecture

- Provide its location.
- Provide a set of features of the object that permits its univocal identification.
- Segmentation must be robust against different illumination conditions, since the algorithms must be able to work both indoors and outdoors.

In order to solve these problems the Eye-RIS Vision Systems provided by AnaFocus are used. In particular, the Eye-RIS Vision System on Chip (VSoC) has been chosen, since its particular features make it very suitable for the kind of application that we are facing.

Next section will describe the main features of the Eye-RIS VSoC, and why they make this system very appropriate for the current project. After that, the algorithms developed under the Eye-RIS VSoC platform are explained in detail. And finally, the results of testing these algorithms under real situations is shown.

### *9.2.2 Introduction to Eye-RIS Vision System on Chip*

Thanks to AnaFocus Vision System on Chip technology, the tasks of image early-processing and post-processing are carried out in the same chip, the Eye-RIS v2.1 VSoC, which is the heart of the Eye-RIS v2.1 Vision System. Figure 9.1 depicts the architecture of the Eye-RIS v2.1 VSoC and the Eye-RIS v2.1 Vision System. In this figure, the arrows represent the images flow.

The main components of the Eye-RIS v2.1 VSoC regarding the implementation of visual routines are:

- AnaFocus SIS Q-Eye. This is the front-end of Eye-RIS vision systems. This device contains one image-processor per pixel and is capable of optically acquiring and processing images very efficiently, in terms of speed and power consumption.
- Altera Nios II processor. It controls the operation of the whole vision system. In addition it analyzes the information output from the SIS and performs all the decision-making and actuation tasks.
- Memory for program, data and images, both on-chip memory and external memory are available
- I/O and communication ports. A variety of digital input and output ports are included, such as UART, PWM ports, GPIOs and USB 2.0 port.

#### 9.2.2.1 AnaFocus Q-Eye Smart Image Sensor

The Q-Eye is a quarter CIF resolution fully-programmable Smart Image Sensor. It consists of an array of 176 × 144 cells plus a surrounding global circuitry. Each cell comprises multi-mode optical sensors, pixel memories, linear and non-linear analogue processors and binary processors. Each cell is interconnected in several ways with its 8 neighboring cells, allowing for highly flexible, programmable, efficient, real-time image acquisition and spatial processing operations.

In Smart Image Sensors, each processor is merged with an optical sensor. This means that each pixel can both sense the corresponding spatial sample of the image and process this data in close interaction and cooperation with other pixels. In fact, the Smart Image Sensor Q-Eye is also called Focal-Plane Processor (FPP) because it processes images at the same physical layer where they are sensed. The incorporation of processing and storage circuitry directly at pixel level avoids the usual bottlenecks in conventional vision systems. Figure 9.2 illustrates this point.

As shown, in a conventional vision system, only the image acquisition takes place in the sensor. Analogue-to-digital conversion is usually performed off-chip, as well as all the processing chain. Thus, the output rate of the sensor (F in figure) needs to be high enough to download full-sized images. However, in the case of a SIS, the sensor not only acquires the images, but additionally processes them, downloading to the external processor only information about images (edges, centers of mass, etc.) and not complete images. As a result, the required output bandwidth of a SIS (f in the figure) is much lower than that required for a conventional vision system, allowing vision systems powered by SIS technology to achieve dramatically high frame rates.

Figure 9.3 shows the internal architecture of the Q-Eye Smart Image Sensor. It consists of four main functional parts:

- Cell array. Consisting of 176 × 144 interconnected sensing-processing cells.
- The Q-Eye Control Unit (QC). It is a simple microcontroller which allows the execution of user code within the SIS Q-Eye.
- I/O Block. Consisting of a set of A/D and D/A converters employed to load and download images to/from the Q-Eye.

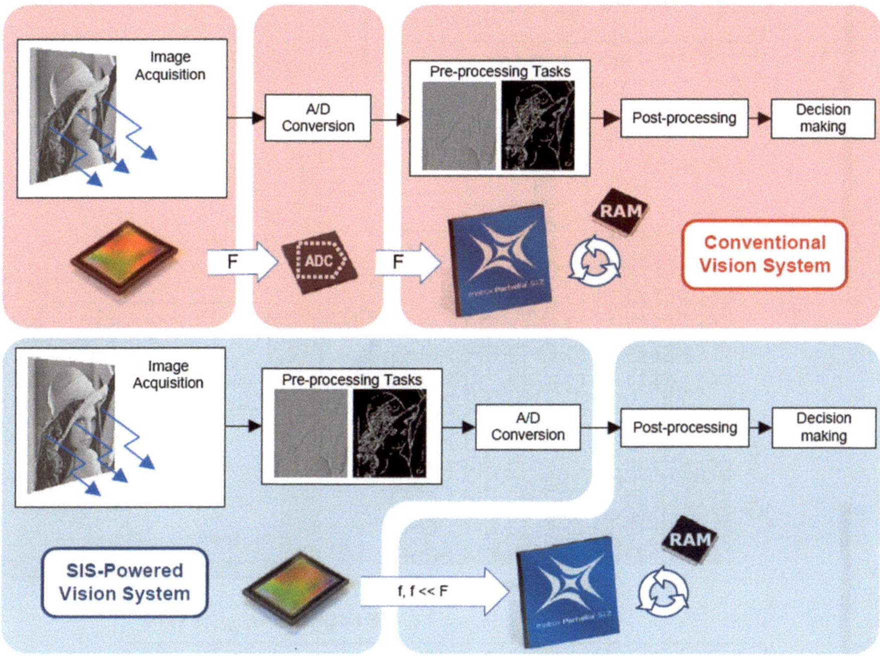


**Fig. 9.2** Comparison between conventional and SIS-powered vision systems

- External interface. Composed by a set of configuration registers allowing for controlling the operation of the Q-Eye. In the Eye-RIS v2.1, such control is accomplished by the Nios II processor. Additionally, general data (like parameters for the Programming Circuitry operation or results of a certain task) can be exchanged between the SIS and an external system through this interface.

### The Cell Array

The cell array is made up of a network of $176 \times 144$ interconnected sensing-processing cells. Each of these cells contains mixed-signal circuitry capable of:

- Sensing images; including adaptive linear and high-dynamic range acquisition
- Storing up to 7 grey scale images and up to 4 binary images
- Performing grey-scale image additions, subtraction and multiplication by a constant
- Performing spatio-temporal image filtering (low-pass, high pass, band pass and stop band) with programmable bandwidth
- Performing grey scale image convolutions
- Performing grey scale to binary, and binary to grey scale conversions
- Performing generic combinational and sequential binary operations

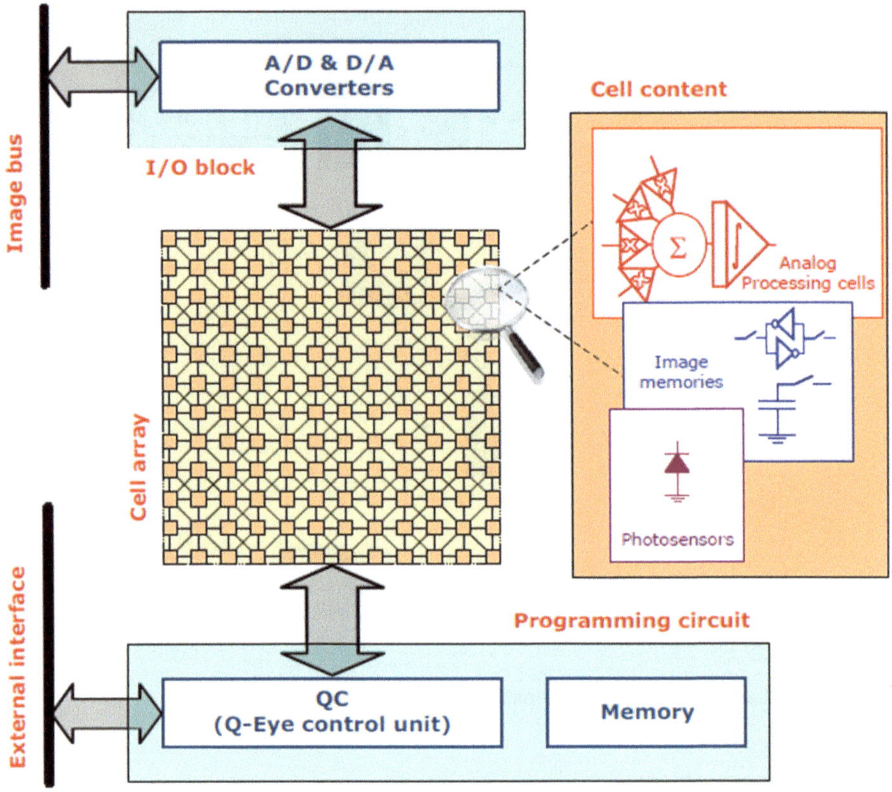


**Fig. 9.3** Conceptual architecture of the Q-Eye chip

- Performing Hit-and-Miss binary operations
- Real-time loading and downloading of images
- Ultra fast calculation and input/output of certain image features such as centroids, mean values, pixel coordinates...

The accomplished tasks are fully programmable. They can be performed in the analogue and digital domains, for grey-scale and binary images respectively. The equivalent processing accuracy is approximately 8-bits.

Figure 9.4 depicts a detailed diagram of a Q-Eye cell. It can be described as an association of fourteen functional blocks connected to two main data buses: one analogue bus called ladn (Local Analogue Data Node) mainly used for grey-level image handling, and another digital bus called lddn (Local Digital Data Node) utilized for binary images. These fourteen blocks and the interaction between them are described next:

- *Optical Sensor*
  The Optical Sensor in the Q-Eye cell consists of a 4T pixel and a voltage buffer. It incorporates a P-well photodiode with an active area of $50\,\mu^2$ rendering 1.0/V(lux

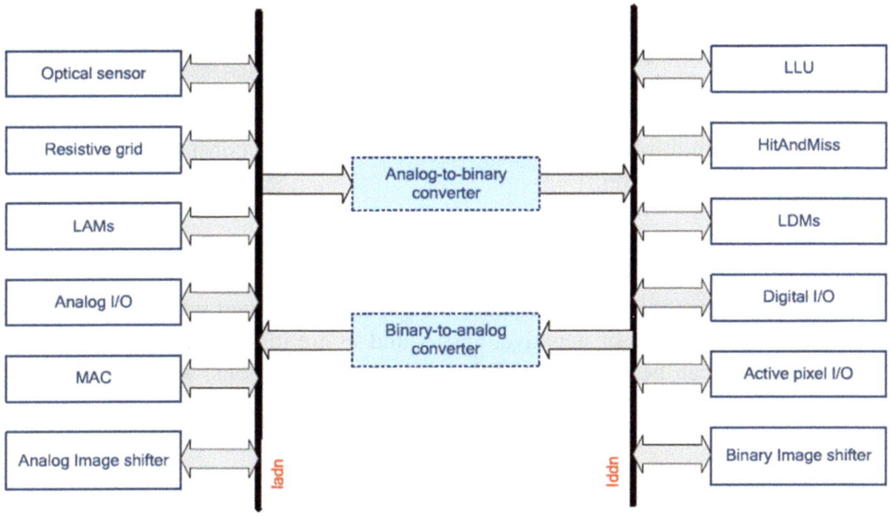


**Fig. 9.4** Block diagram of a Q-Eye cell

sec) sensitivity for green light ($\lambda = 555$ nm). The voltage buffer connects the pixel output with any block in the ladn. For example, the acquired values can be stored in an analogue memory (LAM), or be filtered using the Resistive Grid Module.

- *Local Analogue Memories (LAMs)*
  Local Analogue Memories (LAMs) are included inside every cell to temporarily store grey-level (analogue) images. Each cell includes two banks of LAMs: Bank 1 contains seven memories for general purpose, while Bank 2 contains two memories which are used for internal operation and are not accessible to the user. LAMs are intended for grey-level, analogue image storage with an equivalent resolution of 8 bits. LAMs can store grey-level images coming from any block connected to the ladn, such as the Optical Sensor, the MAC or the resistive grid module. They can also serve to store input operands for the same blocks. The LAMs have a limited retention time due to leakage. They are conceived for retaining an image during several frame-times (the time between two consecutive frames) at 25 fps frame rate.
- *Resistive Grid*
  The Resistive Grid is a very efficient (in area occupation and power consumption) analogue circuit conceived to perform low-pass Gaussian filters with programmable spatial bandwidth over the input images. The Resistive Grid can be combined with the Multiplier-Adder Circuit (MAC) to implement any type of filter: high-pass, band-pass. Similar to what happens with all blocks inside the Q-Eye cell, the operation of the resistive grid is global. This means that all resistive grids inside all the Q-Eye cells perform the same operation at the same time. There is, however, the possibility of selectively deactivating the operation of this block in

certain cells so that the image filtering only takes place in certain region(s) of the image.

- *Multiplier-Accumulator (MAC)*
  The Multiplier-Accumulator (MAC) is a mixed-signal block that serves to sum images and to multiply the difference of two images by a constant. It implements the following expression:

$$I_{out} = I_0 + k(I_1 - I_2)$$

  where $I_0$ is a programmable constant image, k is a constant whose value can be 1, 1/2 and 2, positive or negative, and $I_1$ and $I_2$ are input images. The result of this operation can be accumulated through several iterations, whereby the result is:

$$I_{out} = i_0 + k \sum\nolimits_{i=0}^{N} I_i$$

- *Thresholder*
  This block realizes a comparison between two analogue images outputting a binary image. Its operation is represented by the following equation:

$$I_{out} = sign(I_1 - I_2) = \begin{cases} 1 \text{ if } I_1 > I_2 \\ 0 \text{ in other case} \end{cases}$$

  Where $I_1$ and $I_2$ are the analogue input images and $I_{out}$ is the resulting binary image. Depending on the nature of $I_2$, the threshold process is classified as:

  – Global threshold if $I_2$ is a spatially-invariant image, and
  – Local threshold if $I_2$ is not constant.

  It is important to note that input images are analogue images, which must be stored in LAMs, but the result image is a binary image, and hence must be stored in one of the LDMs. Consequently, this block is the natural "bridge" between the analogue and digital blocks in the Q-Eye cell.
- *Analogue Image Shifter*
  This block shifts an analogue image to any of the eight neighbors of every cell. The source image has to be inside one LAM and the resulting shifted image will be placed in another LAM.
- *Analogue I/O block*
  This module allows for a readout of an analogue image stored inside a LAM. Images are digitized with an 8-bit resolution by a built-in bank of A/D converters and stored in digital memories external to the Q-Eye. The opposite operation is also possible: Images stored in digital memory external to the Q-Eye can be converted to analogue by a built-in bank of D/A converters and loaded in one of the LAMs.
- *Local Digital Memories (LDMs)*
  Local Digital Memories (LDMs) are included within every cell, and are intended for storing binary images. As with the LAMs, each cell of the Q-Eye chip contains

two banks of Local Digital Memories (LDMs). Bank 1 contains eight LDMs, whereas Bank 2 contains one memory. Only four memories in Bank 1 are available to the user; the rest of them are utilized for internal operation. Such LDMs can be used to store input or output data to the different blocks connected to the lddn. Unlike LAMs, LDMs are non-volatile and can store binary images as long as the chip is biased without degradation

- *Local Logic Unit (LLU)*
  The Local Logic Unit (LLU) is a two-input logic block that performs a logic operation between both input binary images. Such operation is defined with a programmable truth table. Both the input and the output operands of this block are data in LDMs.
- *HitAndMiss*
  This block is used to perform a binary Hit-and-Miss morphological operation for a given binary image. HitAndMiss is a well known image-processing operation that is often used to look for particular binary patterns within an image. It is also the basic operation of binary morphology, and most of the binary morphological operators can be derived from it.
  The HitAndMiss operation checks whether the $3 \times 3$ neighborhood of a cell matches a specified pattern or not. This is performed in parallel in all the cells of the Q-Eye chip. The inputs to this block are the image to be analyzed (that must be stored in a LDM) and the pattern to match. The result is an image with white pixels at points whose $3 \times 3$ neighborhood matches the input pattern.
  Three states can be defined as the value of a pixel: 1 (white), 0 (black) and DNC (Do Not Care), meaning that the value of that pixel is irrelevant to the matching.
- *Active Pixel I/O block*
  This block provides a fast way for loading and downloading sparse binary images to/from the Q-Eye (a sparse binary image contains a reduced number of white pixels). The Active Pixel I/O block takes as input the binary image inside a given LDM and outputs the (x,y) coordinates of the active pixels. The adopted criterion is that a cell is "active" if it is white (logic '1'). Therefore, active pixel downloading becomes especially effective when the readout images have few white (or black) pixels. The same operation can be used in the opposite way; isolated pixel in a binary image can be activated by providing its (x,y) coordinates from outside.
- *Binary Image Shifter*
  This block shifts a binary image to any of the eight neighbors of every cell. The source must be an LDM and the destination another LDM.
- *Binary I/O Block*
  This module is used to read-out a binary image stored in an LDM. When a binary image is downloaded via the Binary I/O Block, the value of every cell is codified with one bit. Therefore a binary image is converted into a digital image of 3,168 bytes. The opposite is also possible: digital images stored in memory outside the Q-Eye are converted into binary and loaded in one of the internal LDMs.
- *Binary to Analogue Converter (BAC)*
  This block is used to generate an analogue image from a binary one. The input must be the binary image stored in one LDM and the output the LAM where the

resulting analogue image will be stored. The generated image has the maximum analogue value for pixels that are 1 in the binary input image, and has the minimum analogue value for pixels that are 0 in the input image.

#### 9.2.2.2 The NIOS II Processor

The Altera Nios II is the digital processor controlling the execution flow and post-processing the image flow provided by the Q-Eye. Nios II is a digital microprocessor which has been integrated into the Eye-RIS v2.1 VSoC, together with the SIS Q-Eye. The main features of the Altera Nios II processor are:

- General-purpose RISC processor core.
- Harvard architecture.
- Full 32-bit instruction set, data path, and address space.
- 32 general-purpose registers.
- 32 external interrupt sources.
- Single-instruction 32 × 32 multiply and divide producing a 32-bit result.
- Single-instruction barrel shifter.
- Access to a variety of on-chip peripherals, and interfaces to off-chip memories and peripherals through Avalon Bus.
- Hardware-assisted JTAG debug module enabling processor start, stop, step and trace.
- Instruction and Data Cache Memory.
- JTAG Debug Module.
- Exception and interrupt controller.

#### 9.2.2.3 Eye-RIS v2.1 Image Memory

The Eye-RIS Image Memory (both External and On-chip sections) is used for storing images. This memory is accessed both by the SIS Q-Eye and the Nios II processors although not simultaneously. It is mainly used for two functionalities:

1. Long term images storage: Due to the limited retention time of the LAMs in the SIS Q-Eye, digital non-volatile memory might be needed to store images that do not change (or vary very slowly) during the execution of a certain algorithm.
2. Debugging purposes: The user can store images coming from the PC in this memory so they can be loaded in real time to the SIS Q-Eye through the I/O block. Likewise, the SIS Q-Eye can download images to this memory for displaying them in the screen of the PC or for storing in hard drive for later revision. Such image load/download is useful during the development vision applications since it allows debugging the algorithm with pre-acquired or synthetical images before processing real-time images.

## *9.2.3 Working with Eye-RIS Vision System on Chip*

### 9.2.3.1 Analogue Processing in the SIS Q-Eye

Unlike conventional vision systems, the image processing operations accomplished within the SIS (Smart Image Sensor) Q-Eye take place in the analogue domain. This fact leads to several considerations that must be taken into account when programming the Eye-RIS Vision Systems. Within this section, we will describe how images are internally represented in the SIS Q-Eye and other important issues about analogue image processing.

#### Precision of Analogue Grey-Level Operations

Analogue processing has important advantages that are summarized below:

- Analogue circuitry is less power-hungry than digital,
- The ratio speed vs. silicon area is better in analogue circuitry.

Thus, analogue systems are usually faster and more compact than digital systems. Nevertheless, they also have some drawbacks that are important to have in mind when programming Eye-RIS:

- Precision of grey-level (analogue) operations is mid-low (in the range of 6-8bits, depending on the specific block).
- Analogue data is volatile. For example, a grey-level image stored within a pixel-level memory degrades around 0.8LSBs every 40ms. Least Significant Bit is a measure for accuracy, or resolution. For eight bits, 1 between 256 levels.

Consequently, we can say that Eye-RIS Vision Systems are capable of performing complex image processing in an ultra-efficient manner, in terms of both speed and power consumption. On the other hand, as a generic rule, Eye-RIS Vision System has not been conceived to perform algorithms containing iterative, intensive grey-level image processing operations. This issue must be taken into account when programming Eye-RIS.

#### Image Codification in the SIS Q-Eye

The SIS Q-Eye is the front-end of the Eye-RIS Vision System v2.1, meaning that it acquires and processes an image flow that represents a real scene. The image processing operations that SIS Q-Eye performs over this image flow are carried out in both analogue and binary domains.

An important issue to understand how these operations are achieved, is that in SIS Q-Eye images are signed. Internally, an image is defined in the range [−x, x], where x is a certain analogue signal level. The zero level is the center of the signal range.

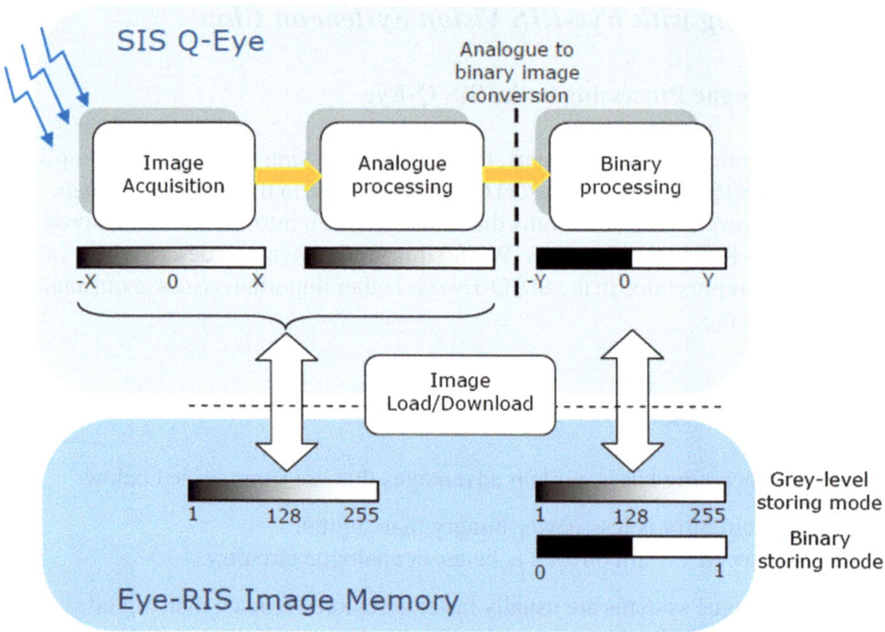


**Fig. 9.5** Execution flow within SIS Q-Eye and image codification

However, when an image is downloaded outside the SIS Q-Eye, it must be converted to digital, and all the signal range is linearly mapped into the positive interval [1, 255]. Likewise, when a digital grey-level image is loaded to the SIS Q-Eye, the interval [1, 255] is mapped into the analogue, signed signal range. This explanation also applies to binary images. This concept is schematically illustrated in Fig. 9.5. It shows how images are codified within every usual step of a vision application, and the correspondence between grey level, signal level and digital number. The sign of analogue images allows the SIS Q-Eye to obtain important information about the image flow, like the sign of a difference image, the direction of the movement of an object present in the scene, the direction of zero crossings, etc.

### Signal Range and Analogue Processing in the SIS Q-Eye

Generally speaking, the signal range in grey-level (analogue) processing must be controlled by the user in the SIS Q-Eye. For example, if we want to subtract two images of range [−X, X], the result will be in the range [−2X, 2X], which is out of range. The solution is pre-scaling both input image multiplying by 1/2 before performing the subtraction operation, so the result is inside the signal range. However, depending on what we want to do with that result, it may be very convenient even to amplify it. One simple case would be if we would like to threshold the difference, in

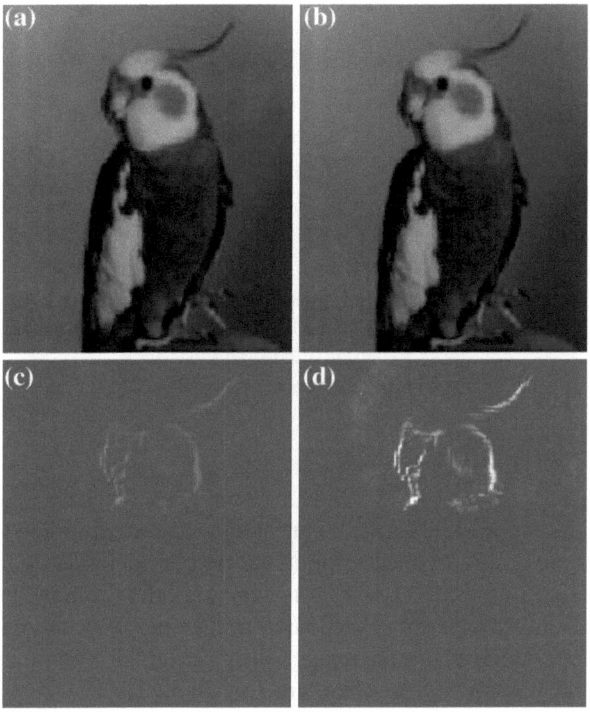


**Fig. 9.6** Analogue grey-level processing in the SIS Q-Eye. **a** and **b** are input images to a subtraction operation. **c** is the result pre-scaling input images **d** is the result amplifying the result by 2

such a case we will be interested in enhancing the contrast of that image. Figure 9.6 illustrates these concepts. Note that image (d), a difference image that has been amplified, has more contrast than image (c), which has been pre-scaled to avoid the over-range. It may look like amplifying is always better; nevertheless, it should be done only if next operation is a threshold. The signal range (0, 255 in digital numbers) must not be overtaken if other grey-level operations are to be performed after. This is not actually a remarkable limitation, as thresholding is often the natural step to be accomplished after grey-level high-contrast operations (image difference, edge enhancement, etc.).

### SIS Q-Eye Local Memories

As described previously, the SIS Q-Eye is a special sensor that contains storing and processing circuitry inside each pixel in addition to the sensing circuitry usual in conventional image sensors. Particularly, those pixel- memories used for storing images are called SIS Q-Eye's Local Memories. There are two types of local memories:

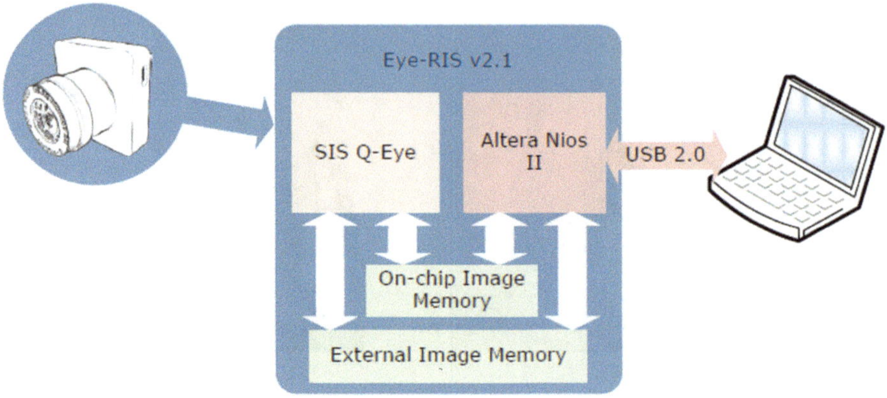


**Fig. 9.7** Eye-RIS image memory

- Local Analogue Memories (LAMs): they store analogue grey-level images. These memories can be configured to be input and output operands of all the analogue operators present in the SIS Q-Eye. It is very important to have in mind that the retention time of LAMs is small compared to the ones of usual digital memories. They have been conceived to store images during several frame-times in 25-frames-per-second vision applications. There are 7 available LAMs.
- Local Digital Memories (LDMs): they store binary images, which can be input and output of all the binary operators present in the SIS Q-Eye. The retention time of LDMs is infinite, meaning that they are capable of maintaining image's values as long as the Eye-RIS Vision System is fed up.

Binary images can be also stored in LAMs, although it is strongly recommended that LAMs are used to store grey-level images.

### The Eye-RIS Image Memory

The Eye-RIS Image Memory is a SDRAM (Synchronous Dynamic Random Access Memory) conceived to store both grey-level and binary images, mainly intended to be used for debugging purposes. This memory is shared between the Nios II Processor and the SIS Q-Eye (see Fig. 9.7). Nios II Processor can send images from this memory to a computer (through the USB port) in order to display or to save them to disk. It is also capable of receiving images from the PC and save them to this memory. Likewise, SIS Q-Eye can move grey-level and binary images between this memory and its local memories (either LAMs or LDMs).

Images can be stored within the Eye-RIS Image Memory in two modes: grey-level mode and binary mode.

*Grey-level mode* each pixel is codified using 8 bits, giving 256 grey levels per pixel. In this case, the size of an image is: $176 \times 144 = 25344$ bytes. Grey-level storing mode usually is used for storing grey-level images, although it is also possible to store binary images with values 0 -> black and 255 -> white.

*Binary mode* each pixel is codified using 1 bit, giving 2 levels (black or white) per pixel. This mode has been conceived as a compact way of storing binary images. In this case, the size of an image is 3168 bytes.

The Eye-RIS Image Memory is divided into pages, so that each image is stored in one page. There are two types of pages, corresponding to both storing modes. The Eye-RIS 2.1 has both external image memory and on-chip image memory. The way to access an image stored in the Eye-RIS Image Memory is to directly refer to the page where it is stored.

## *9.2.4 Vision Algorithms*

### 9.2.4.1 Purpose of the Visual Routines

The purpose of the currently developed vision algorithms is the development of a set of algorithms that allow the implementation of an insect brain in different kind of robots, so that they are able to navigate in random environments basing on visual information [4, 5].

This purpose is concreted in the case of the visual routines in finding and recognizing a set of landmarks present in the scene, and doing this robustly under different lighting conditions, since the robot must work both in indoor and outdoor scenarios.

The main innovation within this development consists on the implementation of these algorithms within the Eye-RIS VSoC, which particular architecture allows a very compact and fast, complete vision system working on a single chip. The chosen algorithms have been adapted to this particular architecture in order to achieve the best robustness-speed compromise.

### 9.2.4.2 Summary of the Vision Algorithms

Figure 9.8 shows the Flow Diagram of the vision application.

The main application steps are:

1. Adaptive acquisition of images: in order to achieve robustness against indoor and outdoor scenarios, the acquisition of images must adapt to temporal and spatial changes of lighting within the scene. This means adapting the integration time of the exposure and also its dynamic range.
2. Segmentation of objects of interest: The Eye-RIS Vision Systems are greyscale sensors, they don't provide color information, therefore alternative methods must be found to the color-based segmentation developed in past works. In this case two main cues are used:

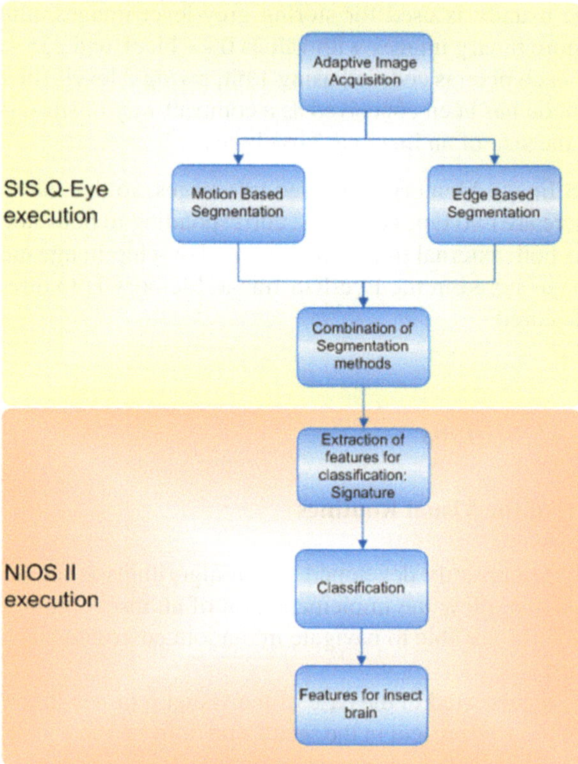


**Fig. 9.8** Application overview

- Motion based segmentation of nearest moving objects.
- Edge based segmentation of closed contours

   Both motion and edges can be very efficiently extracted using SIS Q-Eye processing capabilities.

3. Both segmentation cues are combined properly in order to achieve improved robustness in the segmentation. The result of this step will be the contour of the nearest landmark present in the scene.
4. Signature extraction of the contour of the segmented object. This signature is normalized in order to be able to apply a very simple classifier.
5. The shape classifier is based on distances to models. This classifier although very simple has proven very robust for recognizing the different landmark shapes.
6. A complete set of features is extracted from the recognized shape that will be used by the insect brain to perform the interaction with its environment.

A very important issue that must be taken into account when designing and implementing those algorithms is the distribution of the different processing steps among the two processors that conform the Eye-RIS VSoC, namely the SIS Q-Eye and the

NIOS II digital microprocessor. Due to their very different architecture, each one is more appropriate for certain kind of processing. When working with this kind of systems we can distinguish between:

- Pre-processing: is the processing which input is a complete image (or a Region of Interest, ROI). The processing is carried out at the pixel level. This processing can be done within the SIS Q-EYE very efficiently, since all the pixels can be processed in parallel at the same time. Besides, due to the analogue processing not only a higher speed can be achieved, but also smaller power consumption and smaller processing circuitry size (which means smaller silicon area). The output of this processing is, ideally, only information of interest of the image. Nevertheless also complete images can be obtained from this processing. To summarize, the pre-processing takes an image, which contains a lot of information, and eliminates the information that is of no interest, reducing then considerably the amount of information to be managed by further processing.
- Post-processing: in this processing the input is information of interest about an image, or a pre-processed image. This processing must be done within the NIOS II microprocessor when working with Eye-RIS v2.1. The output of this processing are decisions about the scene, or processed useful information about the same. Since this processor is not very fast and powerful the amount of data to be managed by the processor and/or the complexity of the implemented algorithms should be reduced in order to keep high frame rates. An adequate selection of the pre-processing algorithms running on the SIS Q-Eye enables the use of not very complex post-processing algorithms for very robust and fast vision applications.

In Fig. 9.8 the most appropriate distribution of processing for this application is also displayed.

### Image Acquisition Algorithms

As we explained before, the acquisition of images must be adaptive in order to achieve robustness against indoor and outdoor sceneries. In these scenarios lighting can experience two kind of changes:

1. Inter-frame changes: the level of available lighting varies from one image to the next, that is a temporal variation, so the integration time must adapt from one frame to the next to the average lighting level in the scene.
2. Intra-frame changes: the level of lighting varies within the same image, that is a spatial variation, so the integration time within the image must adapt to these local changes. Intra-frame changes are important when there are very dark and very bright regions in a scene, so that the sensor can not accommodate the complete dynamic range.

### *High Dynamic Range Images*

Scenes from the real world can exhibit a broad range of light variations [6]. This is particularly true for scenes containing both areas of low and high illumination, like a dark room with a sunlit window. Also, the presence of objects with reflective or opaque surfaces contributes to the variety of local light levels. It is clear that in such scenes, the dynamic range, where the term refers to the ratio of the highest and lowest recorded level of light, can be very high. Thus it is impossible to obtain an adequate representation of a scene having a wide dynamic range by using a film or a digital still camera, or in our case a CMOS sensor. There will be an information loss due to the limited dynamic range of the used device. In the case of a CMOS sensor, the dynamic range is usually described in terms of the ratio between the maximum charge that the sensor can collect (full well capacity) and the minimum charge that just overcomes sensor noise (noise floor).

The light values captured by a CCD or CMOS sensor are usually finally quantized (linearly or not) in a range between [0:255] to produce an 8 bit per pixel storage, even if more bits were initially used by the sensor analog to digital converter (ADC) to cover the input signal. This yields a coarse resolution interpretation of the physically, continuously varying values present in the real world. Information loss is observed in highly illuminated areas (where all light variations are mapped to the same value, and thus become saturated) and in dimly illuminated areas (where information is overridden by sensor-noise, producing pixel values floating nearly between 0 and 1 in a random fashion). Once portions of the dynamic range have been definitively lost, there is no way to restore them, even using some sort of postprocessing technique.

Therefore a number of techniques have been developed that allow increasing the dynamic range that can be captured by an image sensor. Next a summary of the most common is included [7].

*Multiple Capture* The idea is to capture several images within normal exposure time:

- Short integration time images will capture the high light regions.
- Long integration time images will capture low light regions.

Then these images are combined into a HDR image, e.g. using Last Sample Before Saturation algorithm.

*Spatially Varying Exposure Technique* This technique implements multiple capture using a conventional sensor by sacrificing spatial resolution. The idea is to deposit an array of neutral density filters on the sensor so that in a single capture pixels with darker filters sample high lights while pixels with lighter filters sample low lights. The High dynamic range image is synthesized using low pass filtering or more sophisticated techniques such as cubic interpolation. This technique requires special filters to be implemented, besides, this blocking light filters reduce sensor sensitivity and SNR. A very high resolution sensor is needed, since spatial resolution is reduced. Therefore this algorithm is not considered to be implemented in the Eye-RIS VSoC.

*Logarithmic sensors* In logarithmic sensors, photocurrent is directly converted to voltage for readout. High dynamic range achieved via logarithmic compression during conversion (to voltage), using exponential I-V curve of MOS transistor in subthreshold:

$$v_{out} = k \, ln \, \frac{i_{ph}}{I_0}$$

A special sensor with specific reading circuitry is needed for this HDR implementation.

*Local Adaptation (Artificial Retina)* The idea is to mimic biological vision systems where local averages are subtracted of each receptor value in order to increase the contrast while maintaining wide dynamic range. An hexagonal resistive network is used to weight closer pixels exponentially higher than distant ones, thereby approximating a Gaussian smoothing.

*Well Capacity Adjustment* The Well-Capacity Adjustment technique modifies the reset voltage of the pixels during the integration time, to achieve a piece-wise linear compression characteristic in the photo-generated charge integrated in the photo-diode. The result of such a compression characteristic is an increase in the intra-frame dynamic range (DR) as compared to linear integration. The amount of increase depends on the number of intermediate levels for the reset voltage and the time interval for which each particular level is maintained.

#### *Implementing High Dynamic Range Acquisition in Eye-RIS VSoC*

From the techniques described in the previous paragraph, two are implementable in the Eye-RIS VSoC, namely, Multiple Acquisition and Well Capacity Adjustment. The implementation of these techniques is described in this paragraph.

*Multiple Acquisition* A set of functions has been created so that the optical sensor in the SIS Q-Eye can be:

- Initialized so that the sensing process is started.
- Several images can be acquired at different configurable intervals without resetting the optical sensor. In other words, the signal that is being integrated can be sampled.
- Finally, each of the acquired images must be conditioned, this means removing noise by a Correlated Double Sampling algorithm and applying a certain gain.

This way a High Dynamic Range image can be acquired by multiple acquisitions without resetting the sensor. This means that the total acquisition time for the HDR image equals the necessary acquisition time for the darkest part of the image.

Having local memories and processing circuitry at pixel level, allows a very efficient implementation of this HDR algorithm, besides it provides some other advantages.

Figure 9.9a shows an image of a HDR scene, as can be seen there are very dark areas and very bright areas where the information is lost. The exposure time has been adjusted so that the mean value of the image equals the signal mid-range,

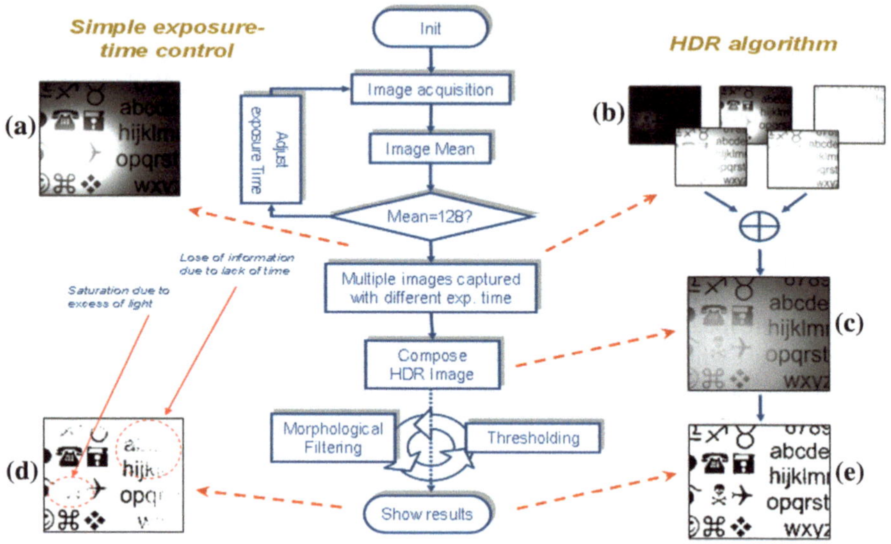


**Fig. 9.9** HDR images with multiple acquisitions

but still information is lost. In Fig. 9.9b we can see how different images can be acquired with increasing exposure times using the previously described functions. These images can be used to compose a HDR image (c) within the same Smart Image Sensor, therefore with a negligible extra processing time. Finally, since each one of the multiple acquired images are available, they can be separately processed and the result combined, leading to no loss of information due to HDR compression (compare results of thresholding the simple exposure image (d) and thresholding and combining the multiple acquired images (e)).

*Well Capacity Adjustment* As explained before, the Well-Capacity Adjustment technique modifies the reset voltage of the pixels during the integration time, see Fig. 9.10a, to achieve a piece-wise linear compression characteristic in the photo-generated charge integrated in the photo-diode, as the one shown in Fig. 9.10b.

The result of such a compression characteristic is an increase in the intra-frame dynamic range (DR) (Fig. 9.10c) as compared to linear integration (Fig. 9.10d). The amount of increase depends on the number of intermediate levels for the reset voltage and the time interval for which each particular level is maintained.

A function for SIS Q-Eye has been implemented that performs HDR acquisition based on this technique.

If it is considered that the well adjustment mode uses N intermediate levels for the reset signal and that those intervals are uniformly distributed in the signal range,

$$V_{reset}^{i+1} - V_{reset}^{i} = \frac{V_{range}}{N+1}$$

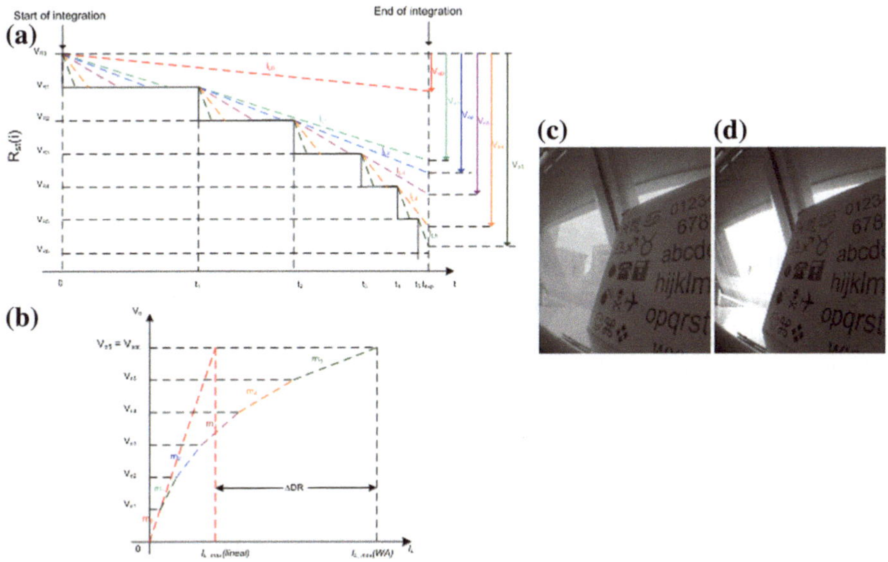


**Fig. 9.10** HDR using the well capacity adjustment technique

And the reset control signal transitions ($t_i$) follow a geometric progression given by the following equation:

$$t_{exp} = k(t_{exp} - t_1) = k^2(t_{exp} - t_2) = k^3(t_{exp} - t_3) = \cdots = k^N(t_{exp} - t_N)$$

where, $t_{exp}$ is the exposure time. Then, the increment of intra-frame dynamic range that we could expect from the well adjustment mode compared to linear integration is given by the equation:

$$\Delta\ \mathrm{DR} = \frac{I_{LWAMax}}{I_{LlinearMax}} = \frac{k^{N+1} - 1}{(N+1)(k-1)}$$

Hence, the increase in the dynamic range depends on the number of intermediate levels (N) for the reset signal of the pixel and on the particular waveform of such signal (k). In the algorithm implemented within Eye-RIS v2.1, N = 6, so the $\Delta$ DR is parameterized with the time each reset voltage is applied (k). The table contains the $\Delta$ DR as a function of k, and the whole intra-frame dynamic range of the image for each single value of k. Total DR is the sum of the well-capacity adjustment DR and the linear integration DR, which is around 50 dB, since the effective resolution of the ADC is 8 bits. Taking this into account, the HDR acquisition function created for the Eye-RIS VSoC takes as input parameters the total exposure time and desired increase of dynamic range, and internally configures the different steps that conform this acquisition.

| k | Well-capacity Adjustment DR ($\Delta$ DR) (dB) | Total DR (dB) |
|---|---|---|
| 2 | 25 | 75 |
| 3 | 44 | 94 |
| 4 | 58 | 108 |
| 5 | 69 | 119 |
| 6 | 78 | 128 |

Implementing Adaptive Acquisition Algorithm in Eye-RIS VSoC

An algorithm has been developed and implemented in the Eye-RIS VSoC that performs fully adaptive image acquisition. Fully, here, means that the acquisition adapts to both temporal and spatial lighting changes (Fig. 9.11).

The algorithm works as follows:

1. Acquire an image, with the set exposure and $\Delta$DR.
2. A condition of High Dynamic Range scene is checked out. In this case, we consider that a High Dynamic Range scene happens if there is more than 5 % clipping at the brightest and darkest values of the image at the same time. That is, more than a 5 % pixels in the image are set to white and more than 5 % pixels in the image are set to black. In this case we consider that the dynamic range of the scene is bigger than the dynamic range of the image, so the algorithm goes to step 3. In other case the algorithm goes to step 4.
3. Increase in 10 dB the dynamic range of the acquired image and go to step 1.
4. In this step a condition of Low Dynamic Range scene is checked up. In this case, we consider that the Dynamic Range of the scene is shorter than the Dynamic Range of the image, if the image range doesn't spread across at least half the signal range. That is, the mean deviation of the image is smaller than 128 (which is half the signal range, being the signal range [0,255]). In this case the algorithm goes to step 5, in other case it continues to step 8.
5. Then we check if $\Delta$DR is bigger than the minimum, which is zero. If this happen the algorithm continues with step 6, otherwise it goes to step 9.
6. Here we have an image of higher dynamic range than the scene that it represents, besides the $\Delta$DR is greater than zero, therefore the $\Delta$DR is decreased and the algorithm goes back to step 1.
7. In this case the image has higher dynamic range than the scene that it represents, but the $\Delta$DR equals zero, therefore the $\Delta$DR cannot be decreased further. So the Dynamic Range adaptation finishes and the Automatic Exposure Control begins (AEC), the algorithm proceeds to step 9.
8. At this step we have an image that correctly covers the dynamic range present in the scene, so the Dynamic Range adaptation finishes an the Automatic Exposure Control begins (AEC), the algorithm proceeds to step 9.
9. The first step of the AEC algorithm consists in checking wether the mean value of the image is around the middle of the signal range, that is if the mean value approximates 128. If this is true then the exposure time is correctly adjusted and

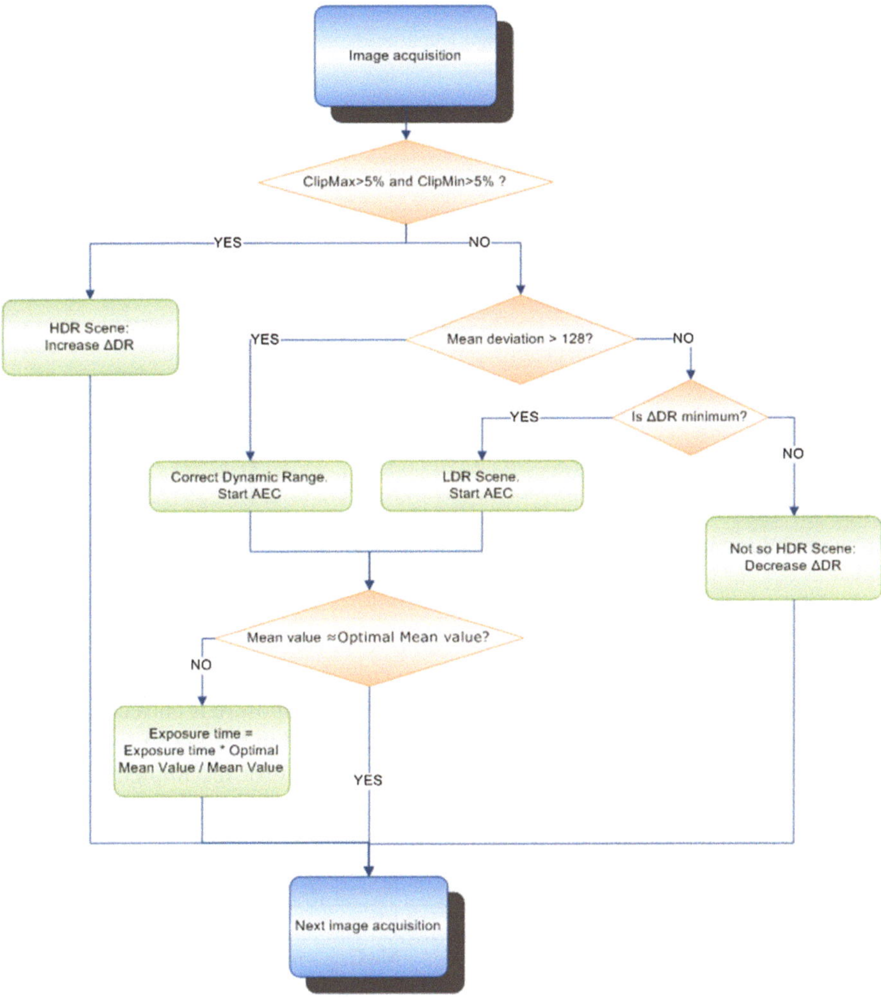


**Fig. 9.11** Image acquisition algorithm flow diagram

the algorithm finishes going back to step 1. In other case the exposure time must be adjusted and the algorithm goes to step 10.

10. In this step the exposure time is adjusted so that the mean value of the image tends to the optimal mean value, which is considered to be the signal mid-range. The update equation for the exposure time is:

$$Exposure\ Time\ (t+1) = Exposure\ Time\ (t)\frac{Optimal\ Mean\ Value}{Mean\ Value\ (t)}$$

Where (t) means the current frame and (t + 1) is the next frame.

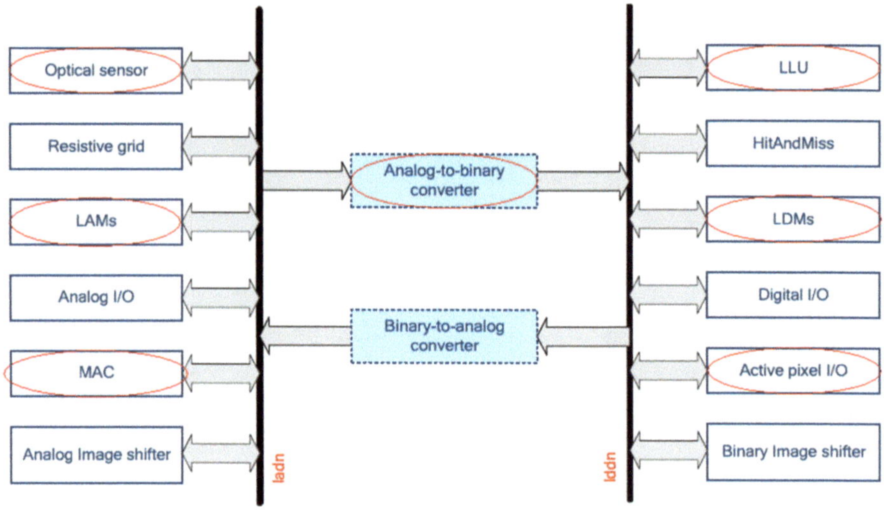


**Fig. 9.12** Blocks used in adaptive image acquisition algorithm

All the flow control of this algorithm can be executed within the SIS Q-Eye, at a negligible processing time. The image features to be extracted in order to determine if the image is High or Low Dynamic Range and if the exposure is correct, have been chosen so that they can be very efficiently extracted within the SIS Q-Eye.

Figure 9.12 shows the blocks of the SIS Q-Eye used for the implementation of this algorithm. Next table shows the execution times of each functionality that is included within the Adaptive Acquisition Algorithm.

| Functionality | Functions | Blocks | Time (μs) |
|---|---|---|---|
| Clipping | Threshold | MAC | 1,7 |
| | | Analog to binary converter | |
| | Logic Not | LLU | 1,8 |
| | Counting white points | Voltage converter | 40 |
| | | Active pixel I/O | |
| | | Current converter | |
| Mean deviation | Mean value | Voltage converter | 14 |
| | Absolute difference | MAC | 9 |
| | | Analog to binary converter | |
| | Mean value | Voltage converter | 14 |
| Mean | Mean value | Voltage converter | 14 |

Taking into account that exposure times are usually in the order of the 10 ms, the extra processing time which in the worst case only amounts 100 μs can be considered negligible.

**Fig. 9.13** Sample images of the adaptive acquisition algorithm

In Fig. 9.13 some sample images of a sequence acquired using the Adaptive Acquisition algorithm are shown. In this sequence the camera changes from outdoor to indoor scenes, and while indoor the lighting of the scene is also modified using a flashlight. As can be seen, when the illumination perturbation happens the image is not correctly exposed, but it quickly adapts thanks to the developed algorithm.

## Segmentation of Objects of Interest

Segmentation in this context means extracting, from the big amount of information contained in the image, those regions that can be considered objects of interest. Segmentation is a crucial step in any vision application, a proper segmentation eases the further processing that is necessary to get the information of interest from the scene, whereas an incorrect segmentation could not only difficult the post-processing but even prevent the extraction of this information.

The Eye-RIS Vision Systems are greyscale sensors, they don't provide color information, therefore greyscale methods that can be implemented in the Eye-RIS Vision systems must be found.

Two main cues have been used for the segmentation of the objects of interest:

- Motion based segmentation of nearest moving objects. Previous results shown that walking fruit flies are attracted by near-by objects whose distance is estimated by the parallax motion of their images on the retina [8, 9]. Therefore we will implement a parallax based method for segmenting the nearest object.
- Edge based segmentation of closed contours The human visual system observes and understands a scene/image by making a series of fixations [10]. Every fixation point lies inside a particular region of arbitrary shape and size in the scene which can either be an object or just a part of it. We define as a basic segmentation problem the task of segmenting that region containing the fixation point. Segmenting the region containing the fixation is equivalent to finding the enclosing contour around the fixation.

Both motion and edges can be very efficiently extracted using SIS Q-Eye processing capabilities.

### *Motion Based Segmentation*

The motion based segmentation developed in [11] has been modified in order to keep information about the location and shape of the objects so that it can contribute to the segmentation of the same. Besides, the Normal Flow Method has been also considered [12]. This method makes use of temporal and spatial derivatives of the acquired images to extract and approximation of the optical flow. We are combining both methods into an approach that can be efficiently implemented using the processing blocks included at the pixel-level in the SIS Q-EYE.

The objective of motion based segmentation part is therefore, detecting the object that is nearest to the camera. This is done this way, since it is known that the "insect" will move first to the object that is nearest.

Next the final Motion Based Segmentation Algorithm is explained in detail.

*Normal Flow Extraction*

Figure 9.14 shows the basis of Normal Flow approach. As can be seen, combining the spatial derivative of an image with its temporal derivative, the orientation of the movement can be obtained. The spatial derivative of an image gives us information about the sign of the edges, combining this edge information with the temporal derivative, that can be obtained by the subtraction of consecutive images, the orientation of the motion is easily obtained.

Spatial and temporal derivatives can be very efficiently obtained within the SIS Q-EYE. Thanks to the signed representation of results, that was previously explained, the signs of the different operations can be easily extracted, thus obtaining the Normal Flow information at really high speeds.

The spatial derivative is extracted making use of the Sobel templates, the different Sobel templates will provide the motion in the different directions: horizontal Sobel will give the vertical motion and vertical Sobel provides the horizontal motion.

Figure 9.15 shows the flow diagram of the implementation of the normal flow approach. As can be seen, first the temporal and spatial derivatives are obtained. The temporal derivative is just a difference between consecutive images (See Fig. 9.17c),

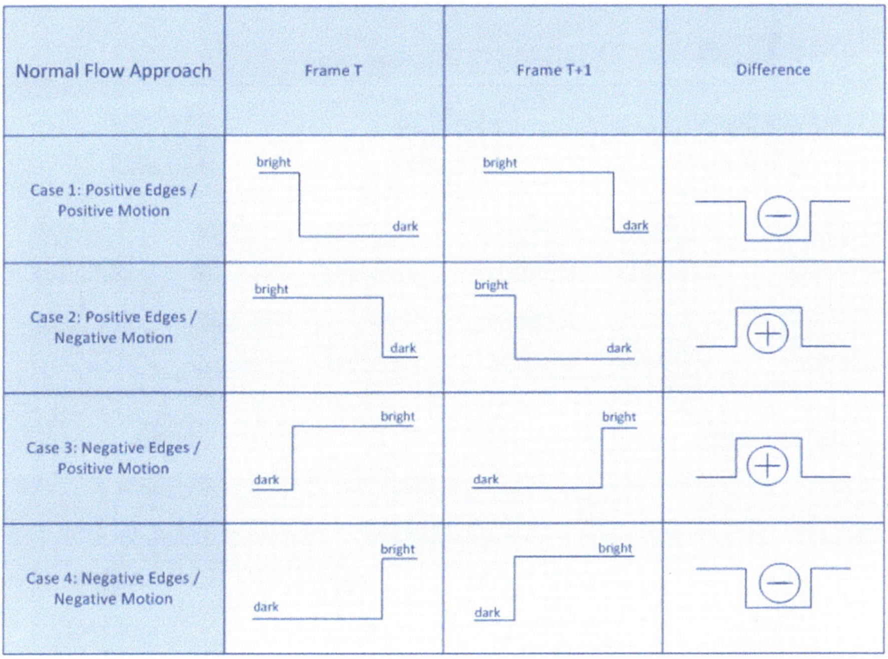


**Fig. 9.14** Normal flow chart

and the spatial derivatives are obtained using the well-known Sobel filters in both the vertical and horizontal directions (See Fig. 9.17b). These operations are carried out very efficiently within the SIS Q-Eye, using the MAC and Analogue Image Shifter blocks as can be seen in Fig. 9.16. After that the derivatives are binarized, extracting both the positive and negative parts of each image (See Fig. 9.17d). Since the SIS Q-Eye works with signed information, this discrimination by sign can be easily implemented. Then the positive and negative results are combined through logic operations (ANDs an ORs) to get the images containing the motion in the different orientations (See Fig. 9.17e and f) (Left-to-right, Right-to-left, Top-to-bottom and Bottom-to-top).

Finally we select the dominant motion direction by comparing the number of active pixels in each direction (See Fig. 9.17e and f). This processing is quite simple but it fits for this particular application, however other different processing can be done to the resulting motion images, for example composing the different directions to determine more complex motions: diagonals, rotations, approaches and others.

Some noise removal processing is also intermingled with the previously explained steps, both in the analogue and in the digital domain.

*Binary Matching*

The Normal Flow approach, as many other Optical Flow methods, can provide very noisy results, therefore, a method has been developed to remove noise from the

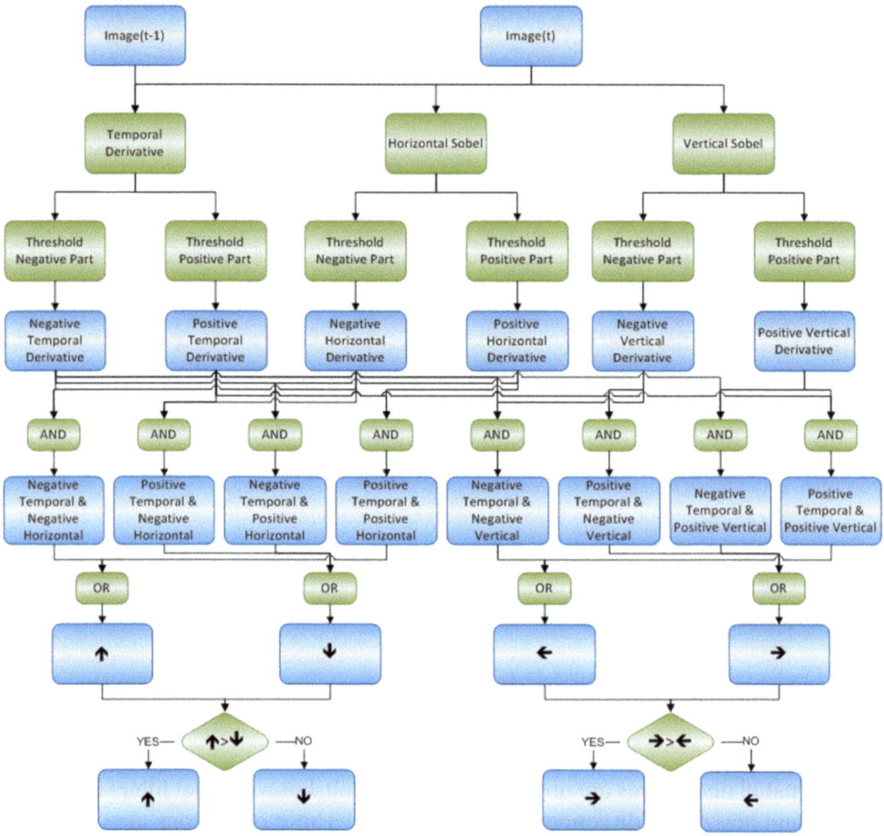


**Fig. 9.15** Normal flow implementation

obtained results. This method is called Binary Matching, and is based on the temporal consistence of correct results compared to noise.

A historical of acquisitions is stored. These images are shifted in the direction of dominant detected motion, and a number of pixels that is as wide as old is the image. Then the next image in the historical record is reconstructed using as seed the previously shifted image. The result will be shifted again and used as seed for reconstructing the next image. This continues until the current image is reached. In this way we are propagating in the time domain the real objects of the image, and eliminating the noise. In Fig. 9.18 the diagram flow of this algorithm is shown, while in Fig. 9.19 a sequence of images for the binary matching and the final result are displayed.

Shifting can be easily implemented in the SIS Q-Eye using the specific block for binary image shifting, but also seed-based reconstruction of objects can be very efficiently carried out in the SIS Q-Eye by combining analogue and digital processing within the pixel.

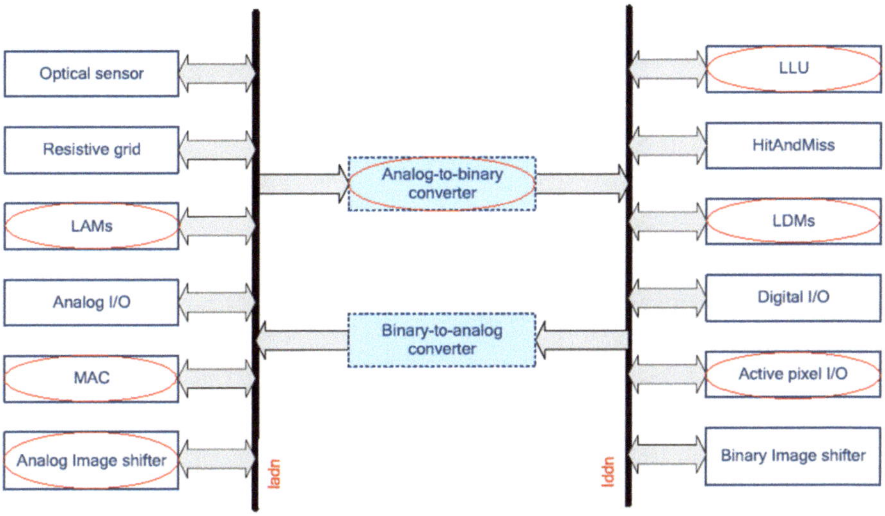


**Fig. 9.16** SIS Q-Eye blocks used in normal flow extraction

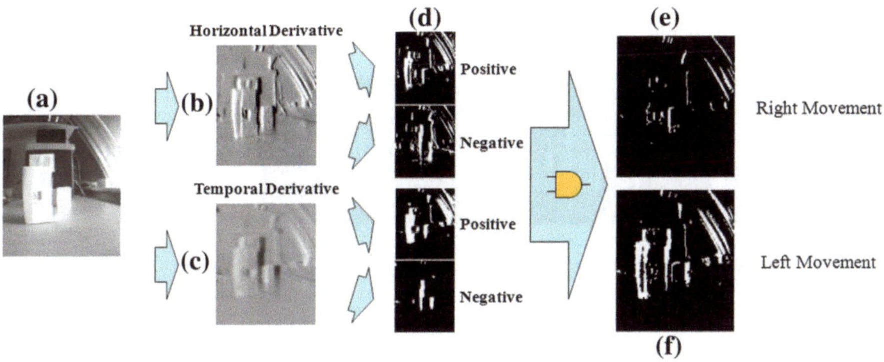


**Fig. 9.17** Normal flow extraction sample images

As can be seen in Fig. 9.20 the Resistive Grid can be used to perform a fast diffusion of an initial point, called the seed. This diffused image can be directly converted to the digital domain to have a binary image. The shape of the object to be reconstructed is obtained by applying a mask. This operation is called FloodFill, and will be intensively used in next processing due to its high efficiency in reconstructing objects. In Fig. 9.21 some intermediate steps in object reconstruction can be seen: the mask, or object to reconstruct is displayed in grey; in white the seed in the first image, and the partially reconstructed object. A time parameter determines the size of the diffusion (and therefore of the reconstruction), these times are in the order of the microseconds, so the reconstruction of even very big objects can be achieved in very short times. Finally, in order to check that the object has been completely

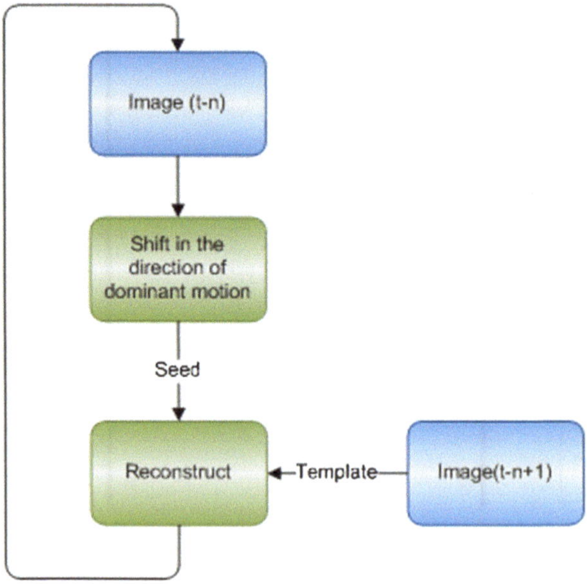


**Fig. 9.18** Binary matching flow diagram

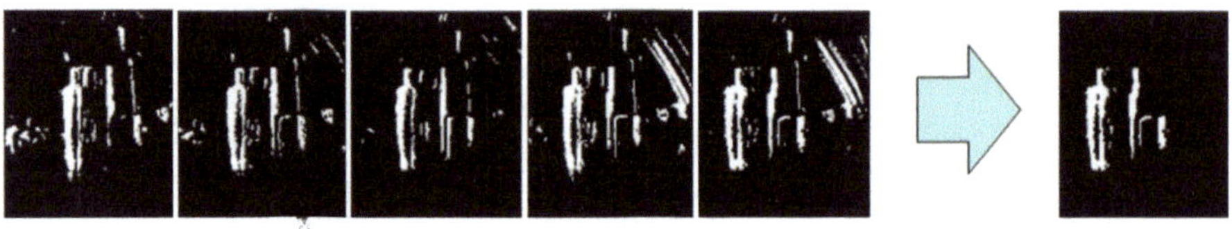

**Fig. 9.19** Binary matching sample images

reconstructed a "no change" condition is checked out using the logic unit and the active points output.

*Nearest Object Detection*

An interesting property of motion is the possibility of extracting the distance to objects similarly to how many animals do, using the parallax. Parallax is an apparent displacement or difference in the apparent position of an object viewed along two different lines of sight [13]. Nearby objects have a larger parallax than more distant objects when observed from different positions, so parallax can be used to determine distances.

Since we are interested in detecting the nearest object to the cameras, we can search for the object with the highest parallax in the image.

The method developed for doing that is displayed in Fig. 9.22, starting from the Binary Motion Image obtained in previous steps of the algorithm, we apply one iteration of directional erosion in the direction of the motion, then an Active Points download is performed.

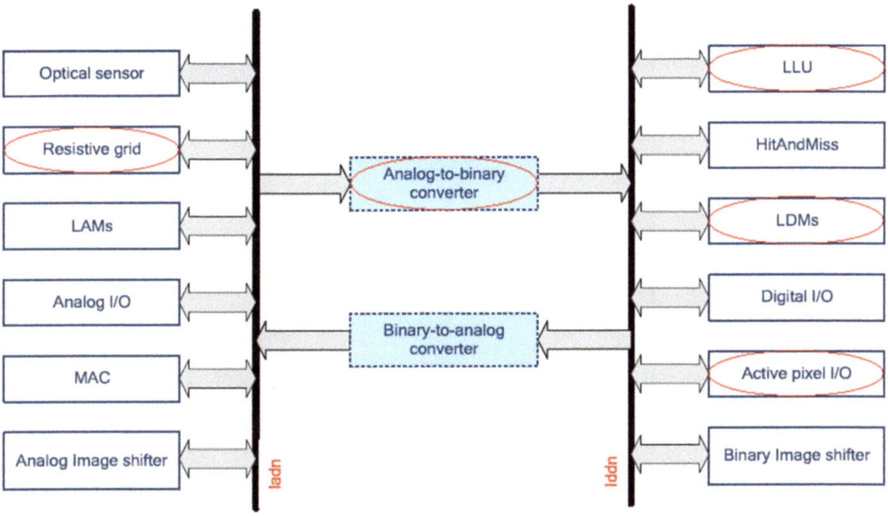


**Fig. 9.20** SIS Q-eye block diagram for object reconstruction

**Fig. 9.21** Intermediate steps in object reconstruction

The Active Points download is a very fast way of downloading binary sparse images from the SIS Q-Eye (a sparse image is an image with the majority of its pixels set to zero, and only a few containing ones), this functionality extracts very quickly the coordinates of the active pixels (pixels set to 1) and sends this information to the image memory, at the same time it counts the number of points that have been downloaded.

In this case, we use the active points functionality to check if the image is empty or not. At the same time we store the coordinates of the active points in a certain image memory. If the image is not empty we perform another iteration of the directional erosion.

In this way we are deleting first the far away objects, since those have the smaller displacement, and thinning the nearest objects with the larger displacements.

Finally, we remove all the points in the image, and the Active Points Download counts zero points, at this moment we go back to the last stored active points, that correspond to the last deleted object, the nearest object. And use these coordinates as seed to reconstruct the nearest object in the motion image. Figure 9.23 shows some sample images of this processing.

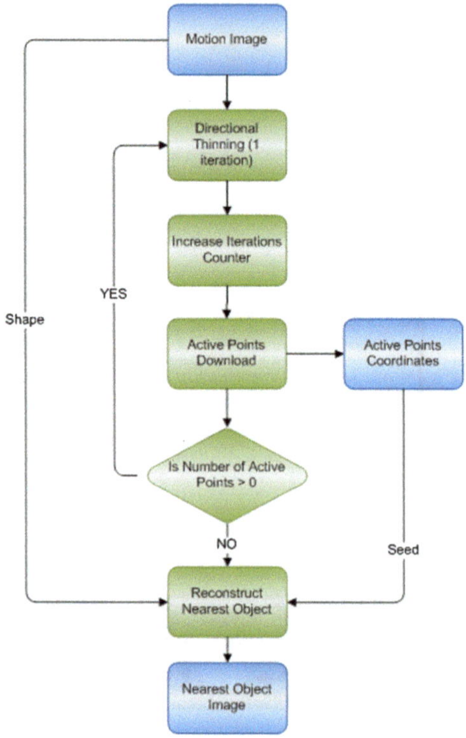


**Fig. 9.22** Flow diagram for nearest object detection

The implementation of this algorithm is also quite straightforward in the SIS Q-Eye thanks to the Hit and Miss block which allows a fast implementation of directional erosion, and the already explained Active Points block.

*Conclusions*

A new method has been developed to detect the nearest object to the camera by using motion parallax in a similar way to the insect brain. This algorithm is made up of three steps:

1. Normal Flow Extraction: using an easily implementable approach to the optical flow extraction we get the motion information contained in a sequence of images.
2. Binary Matching: temporal coherence is checked for moving objects to distinguish them from noise.
3. Nearest Object Detection: Far away objects are iteratively deleted from the motion image making use of the parallax.

All these steps have been implemented in the SIS Q-Eye achieving a very robust algorithm able to detect nearest objects at very high frame rates.

In particular, the implementation of the complete motion based segmentation algorithm takes between 2.4 and 3.2 ms, depending on the characteristics of the

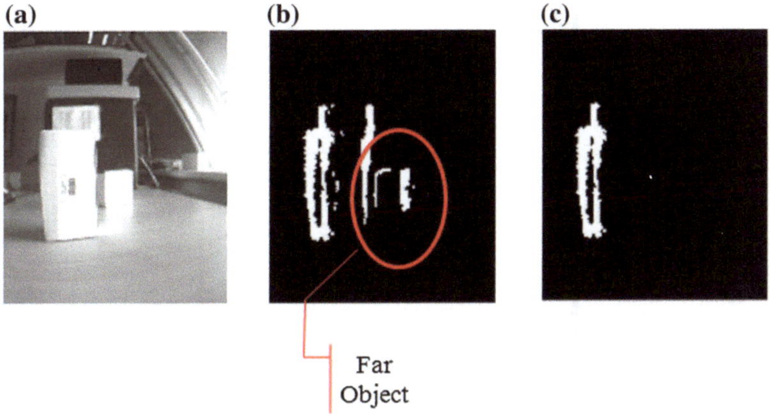


**Fig. 9.23** Sample image for nearest object detection

scene. This would allow extracting the nearest object at a frame rate of up to 416 frames per second.

### *Edge Based Segmentation*

The importance of edge information for early machine vision is usually motivated from the observation that under rather general assumptions about the image formation process, a discontinuity in image brightness can be assumed to correspond to a discontinuity in either depth, surface orientation, reflectance or illumination. In this respect, edges in the image domain constitute a strong link to physical properties of the world. A representation of image information in terms of edges is also compact in the sense that the two-dimensional image pattern is represented by a set of one-dimensional curves. Thus, applying an edge detection algorithm to an image may significantly reduce the amount of data to be processed and may therefore filter out information that may be regarded as less relevant, while preserving the important structural properties of an image [14]. For these reasons, edges have been used as main features in a large number of computer vision algorithms. A large number of approaches have been developed for detecting edges. The earliest schemes focused on the detection of points at which the gradient magnitude is high. In the ideal case, the result of applying an edge detector to an image may lead to a set of connected curves that indicate the boundaries of objects, the boundaries of surface markings as well as curves that correspond to discontinuities in surface orientation. If the edge detection step is successful, the subsequent task of interpreting the information contents in the original image may therefore be substantially simplified. Most edge detection methods can be grouped into two categories, search-based and zero-crossing based. The search-based methods detect edges by computing a measure of edge strength, usually a first-order derivative expression such as the gradient magnitude. The zero-crossing based methods search for zero crossings in a second-order

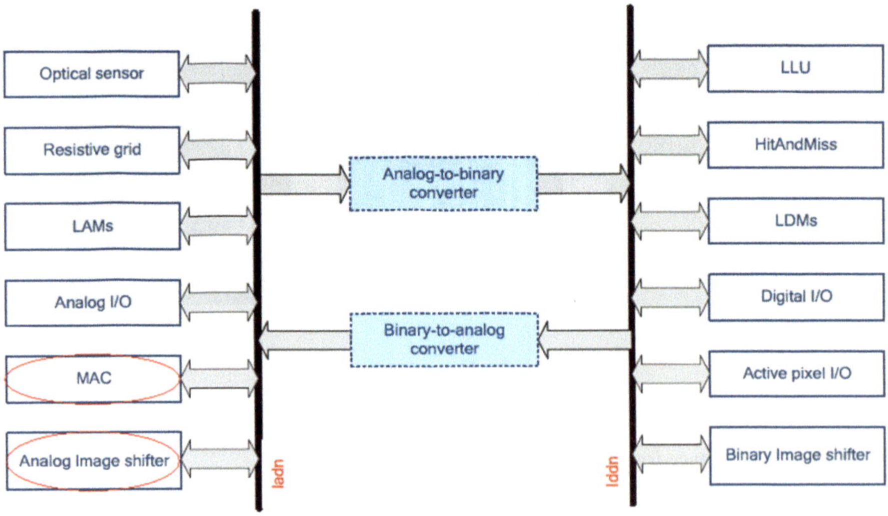


**Fig. 9.24** SIS Q-Eye blocks used for Laplacian by convolution

derivative expression computed from the image in order to find edges, usually the zero-crossings of the Laplacian or the zero-crossings of a non-linear differential expression. As a pre-processing step to edge detection, a smoothing stage, typically Gaussian smoothing, is almost always applied (see also noise reduction). Both categories, first and second derivative methods for extracting edges, can be efficiently implemented on SIS Q-Eye.

Three types of edge detection methods have been implemented in the Eye-RIS VSoC. The robustness and accuracy of the results for the current application has been analyzed, and the best method has been chosen. Besides, a new method has been developed to remove edges that are of no interest for the detection of the target landmarks.

*Second-derivative methods: Laplacian and Laplacian of Gaussian (LoG)*

In image processing, the discrete Laplacian is calculated as sum of differences over the nearest neighbors of the central pixel. Or what is the same, a convolution with the following kernels:

$$D_{xy}^2 = \begin{array}{|c|c|c|}\hline 0 & 1 & 0 \\ \hline 1 & -4 & 1 \\ \hline 0 & 1 & 0 \\ \hline\end{array} \text{ or } D_{xy}^2 = \begin{array}{|c|c|c|}\hline 1 & 1 & 1 \\ \hline 1 & -8 & 1 \\ \hline 1 & 1 & 1 \\ \hline\end{array}$$

depending on the desired connectivity, 4 or 8 respectively [15].

If we focus on the 8 connectivity filter, two implementations are possible using the processing blocks in the SIS Q-Eye:

1. The first one consists on implementing the convolution itself, using the blocks displayed in Fig. 9.24, the Analog Image Shifter is used to bring each of the 8

**Fig. 9.25** Laplace Filter implementation in the digital domain (*left*) versus Laplace Filter implementation using balanced convolution in the analog domain of SIS Q-Eye (*right*)

neighbors to the central pixel, and the MAC adds or subtracts the values and applies the gains, so that the Laplacian of a pixel p(x,y) is:

$L(p(x, y)) = p(x - 1, y - 1) + p(x, y - 1) + p(x + 1, y - 1) + p(x - 1, y) - 8\,p(x, y) + p(x + 1, y) + p(x - 1, y + 1) + p(x, y + 1) + p(x + 1, y + 1)$

However, this convolution can not be implemented like that, since the signal range would be exceeded after the second addition (as was explained in a preceding paragraph) and this information can not be recovered. A reordering of the factors must be applied to keep the image information within the signal range. The resulting Laplacian is as follows:

$L(p(x, y)) = p(x - 1, y - 1) - p(x, y) + p(x, y - 1) - p(x, y) + p(x + 1, y - 1) - p(x, y) + p(x - 1, y) - p(x, y) + p(x + 1, y) - p(x, y) + p(x - 1, y + 1) - p(x, y) + p(x, y + 1) - p(x, y) + p(x + 1, y + 1) - p(x, y)$

To implement this convolution fifteen MAC operations are required (additions and subtractions) and also eight shiftings to bring the 8 neighbors to the central pixel. But, although this operation may look quite complex and intensive, the obtained results are almost identical to a digital implementation (see Fig. 9.25) and the execution time amounts to only 27 $\mu$s.

2. In the second case we are decomposing the Laplacian kernel as follows:

$$\begin{array}{|c|c|c|}\hline 1 & 1 & 1 \\ \hline 1 & -8 & 1 \\ \hline 1 & 1 & 1 \\ \hline \end{array} = \begin{array}{|c|c|c|}\hline 1 & 1 & 1 \\ \hline 1 & 1 & 1 \\ \hline 1 & 1 & 1 \\ \hline \end{array} + \begin{array}{|c|c|c|}\hline 0 & 0 & 0 \\ \hline 0 & -9 & 0 \\ \hline 0 & 0 & 0 \\ \hline \end{array}$$

That is, this convolution can be expressed as a Low Pass Filter of the image to which an amplified version of the image itself is subtracted.

As can be seen in Fig. 9.26 the low pass filter can be implemented using just the resistive grid (which operation is very efficient in terms of speed and accuracy)

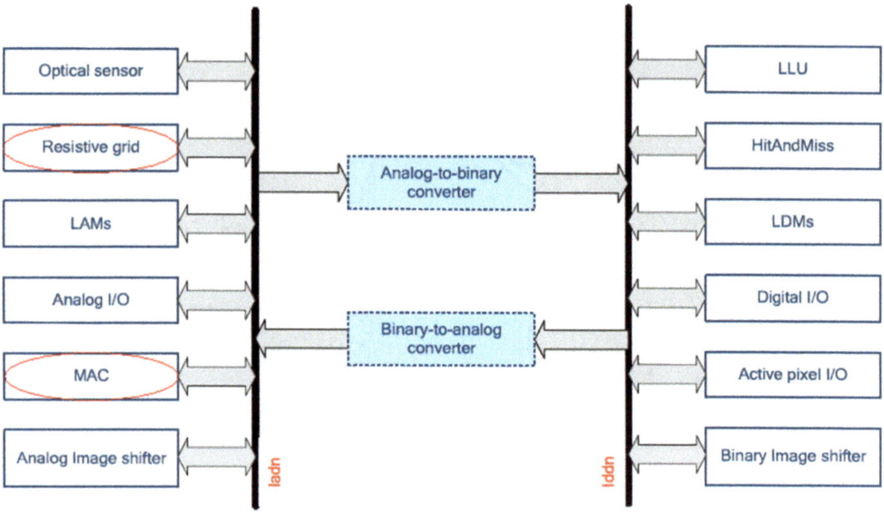


**Fig. 9.26** SIS Q-Eye Blocks used for Laplacian by decomposition

and several MAC operations. Thus, the number of operations can be reduced from 23 to only 5 (since the MAC can only multiply by 2, 1 or 1/2; 4 iterations are necessary to achieve the -9 factor). In this way we can reduce the execution time of this filter down to 6 μs, and theoretically the degradation due to the intensive operation in the analogue domain should also be reduced.
However, with this method we have no means of controlling the saturation of the signal level during the operation, as a consequence the results are not as good as expected. In Fig. 9.27 the results of applying this method are compared to a digital implementation. This method is discarded for implementing the Laplacian in the SIS Q-Eye.

Figure 9.28 shows the results of applying the Laplacian by convolution on a test image that contains samples of the different landmarks to detect. This image is thresholded to obtain a better appreciation of the quality of the extracted edges.

It can be noticed that the filtered image is a bit noisy, as is usual when a high pass filter is applied. To improve the noise a low pass filter is usually applied before the Laplacian filter, conforming what is called a Laplacian of Gaussian or LoG filter. In SIS Q-Eye the Resistive Grid block provides a very efficient Gaussian filter, so this LoG can also be easily implemented. Figure 9.29 shows the results of applying the LoG filter to the same image of Fig. 9.28. It can be seen how the results are less noisy, however some edges are lost due to the smoothing effect of the Gaussian filter.

*First-derivative method: Sobel filter*

The Sobel operator is very used within edge detection algorithms in image processing. Technically, it is a discrete differentiation operator, computing an approximation of the gradient of the image intensity function. At each point in the image, the result of the Sobel operator is either the corresponding gradient vector or the norm of this

**Fig. 9.27** Laplace Filter implementation in the digital domain (*right*) versus Laplace Filter implementation in the analog domain of SIS Q-Eye (*left*)

**Fig. 9.28** Results of applying Laplacian by convolution on a test image containing the different landmarks. This image is thresholded to better appreciate the quality of the results

**Fig. 9.29** Results of applying LoG on a test image containing the different landmarks. This image is thresholded to better appreciate the quality of the results

vector. The Sobel operator is based on convolving the image with a small, separable, and integer valued filter in horizontal and vertical direction, and is therefore relatively inexpensive in terms of computations.

Mathematically, the operator uses two $3 \times 3$ kernels which are convolved with the original image to calculate approximations of the derivatives—one for horizontal changes, and one for vertical. These kernels are [15]:

$$G_y = \begin{array}{|c|c|c|} \hline 1 & 2 & 1 \\ \hline 0 & 0 & 0 \\ \hline -1 & -2 & -1 \\ \hline \end{array} \text{ and } G_x = \begin{array}{|c|c|c|} \hline -1 & 0 & 1 \\ \hline -2 & 0 & 2 \\ \hline -1 & 0 & 1 \\ \hline \end{array}$$

At each point in the image, the resulting gradient approximations can be combined to give the gradient magnitude, using: $\sqrt{G_x^2 + G_y^2}$.

The square and square root can not be easily implemented using SIS Q-Eye, so we will use a different approach that will be explained later on this document.

Due to the anisotropic nature of the Sobel filter, it can not be optimized by using the Resistive Grid + MAC operation scheme (like in the Laplacian filter), since the Resistive Grid is an isotropic filter. The only possible implementation consists on a convolution. An appropriate reordering of the convolution terms is also needed in order to avoid saturations in the signal range.

Therefore the Sobel is converted from:

$$\begin{aligned} G_y = {} & p(x-1, y-1) + 2 * p(x, y-1) + p(x+1, y-1) - p(x-1, y+1) \\ & - 2 * p(x, y+1) - p(x+1, y+1) \end{aligned}$$

to

$$\begin{aligned} G_y = {} & p(x-1, y-1) - p(x-1, y+1) + p(x+1, y-1) - p(x+1, y+1) \\ & + p(x, y-1) - p(x, y+1) + p(x, y-1) - p(x, y+1) \end{aligned}$$

and analogously with the $G_x$ component.

To implement this convolution seven MAC operations are required (additions and subtractions) and also eight shiftings to bring the 8 neighbors to the central pixel. But, as happened with the Laplace implementation, the obtained results are almost identical to a digital implementation (see Fig. 9.30) and the execution time amounts to only 15 μs.

Figure 9.31 shows the results of applying the Sobel filter on a test image that contains samples of the different landmarks to detect. This image is thresholded to obtain a better appreciation of the quality of the extracted edges.

Figure 9.32 shows a composition of the Horizontal and Vertical Sobel components to obtain the complete edges of the image. It can be noticed that these edges are more complete and less noisy than the Laplacian o LoG approaches. Therefore, the Sobel filter will be the one used for the edge based segmentation.

*Edges Post-processing*

**Fig. 9.30** Sobel Filter implementation in the digital domain (*left*) versus Sobel Filter implementation in the analog domain of SIS Q-Eye (*right*)

Several issues are treated in this paragraph.

1. As was explained in the previous paragraph, the Sobel gradient magnitude is obtained by combining the Horizontal and Vertical components according to: $\sqrt{G_x^2 + G_y^2}$
   However, square and square root operations can not be easily implemented using SIS Q-Eye, so another approach has been developed to overcome this issue.
2. The resulting edges image, contains noise and many edges that do not belong to the objects of interest, but to the background. To release the post-processing tasks of dealing with many information of no interest, we have developed a new method for removing edges that are of no interest for the detection of the target landmarks.

The Sobel magnitude $\sqrt{G_x^2 + G_y^2}$ is typically approached by $|G| = |G_x| + |G_y|$. Then, usually a threshold is applied to get only the high frequency components. Since SIS Q-Eye codifies the sign of the Sobel components as greylevel, the sum of the absolute values can not be easily implemented. To emulate this operation we threshold separately the positive and negative components of the Horizontal and Vertical Sobel and apply an OR operation to sum the components. Finally the Horizontal and Vertical binarized components also undergo an OR operation to get the complete edges image. Figure 9.33 shows the flow diagram of these operations.

As can be seen, the resulting edges image has some noise, edges that don't belong to the landmarks, and in general, edges are more than one pixel wide. To release the post-processing algorithms from unwanted information, that would increase the computational effort that must be carried out in the digital micro-processor, we have applied new processing steps in the SIS Q-Eye.

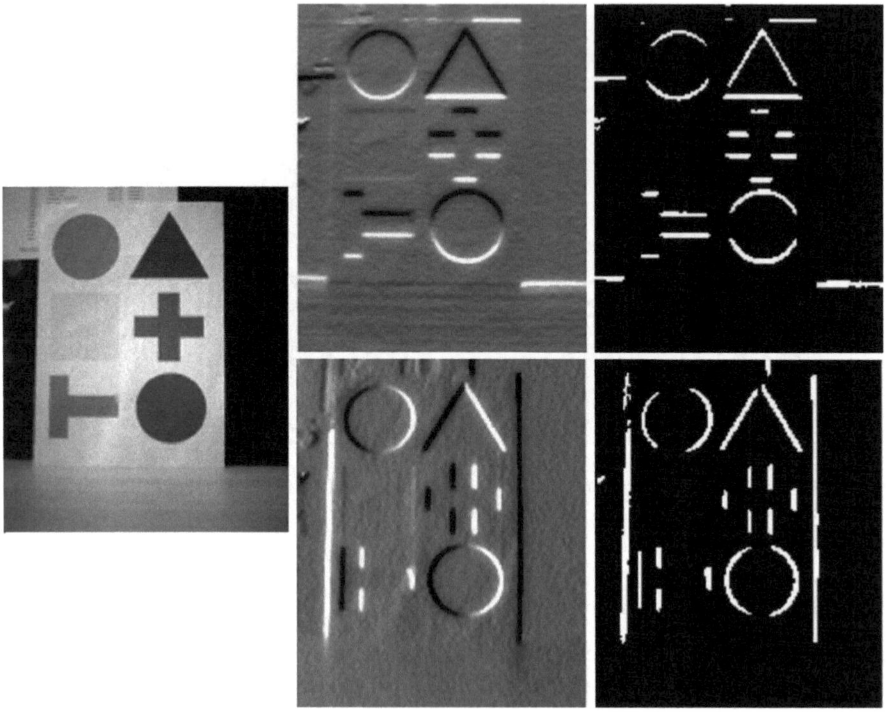

**Fig. 9.31** Results of applying Sobel filter by convolution on a test image containing the different landmarks. This image is thresholded to better appreciate the quality of the results

Figure 9.34 shows the flow diagram of the post-processing that is carried out on the edge images.

First of all, a skeletonization is applied so that edges are reduced to only one pixel wide. Skeletonization is carried out by successive iterations of the Hit And Miss and Logic Operations, so a very efficient execution can be achieved. In the sample images of the previous figures, 4 iterations of the Skeleton operation are carried out in 23 $\mu$s.

The number of necessary iterations of this operation is automatically obtained within the Skeleton function itself. The function checks that further iterations of the skeleton operation make no change in the resulting image, at this moment the function exits. This checking is achieved using SIS Q-Eye blocks also, therefore the execution is very efficient and does not introduce a significant overload in the execution time.

After the skeletonization, the blobs touching the borders of the image are erased. A blob is a Binary Large Object, or a set of white connected pixels, that usually (and ideally) represent one object of the real world. These objects are erased since, as they touch the image borders, we can not assess if they are complete or incomplete, and if their shape inside the image is representative or not of their real shape.

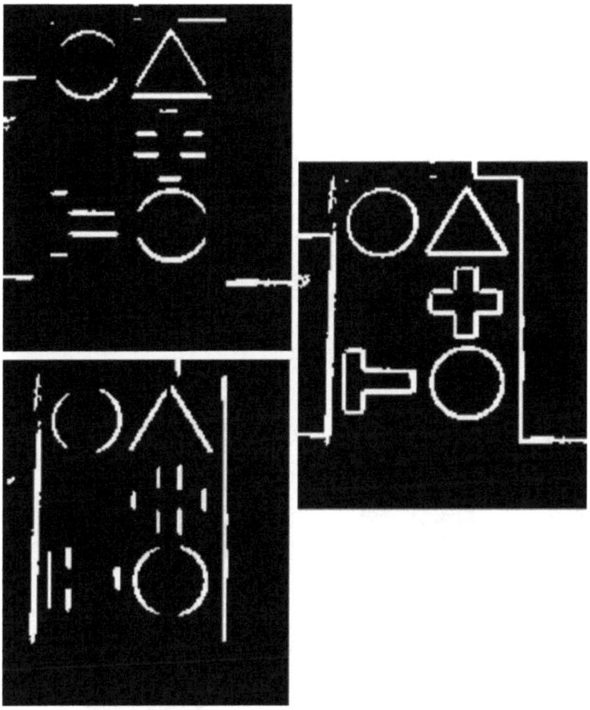

**Fig. 9.32** Edges image result of composing the horizontal and vertical Sobel components

To erase the border blobs we make use of two interesting capabilities of the SIS Q-Eye. The first one is the capability of quickly reconstructing blobs using the Flood Fill operation that was previously explained. The second one is the possibility of specifying the contour conditions for the different operations of the SIS Q-Eye. That is, the pixels outside the cell array can be treated as having:

- A fixed determined value. This value can be either greyscale (between 1 and 255) or binary (0 or 1).
- The same value as that of its nearest cell. What we call Replicate.

Figure 9.35 shows the results of applying a close operation with three different contour conditions, Black, White or Replicate, it must be noted how results differ.

Figure 9.36 shows the flow diagram of the Erase Border Blobs. First of all a black image with white borders is created by means of using the dilate operation with the contour conditions set to white. This image is multiplied by the source image to obtain white pixels in the points where the blobs are touching the borders of the image. This image is used as seed to perform a Flood Fill on the original image. The result is the blobs touching the borders of the image. This resulting image is subtracted from the original image so that only the blobs completely inside the image remain.

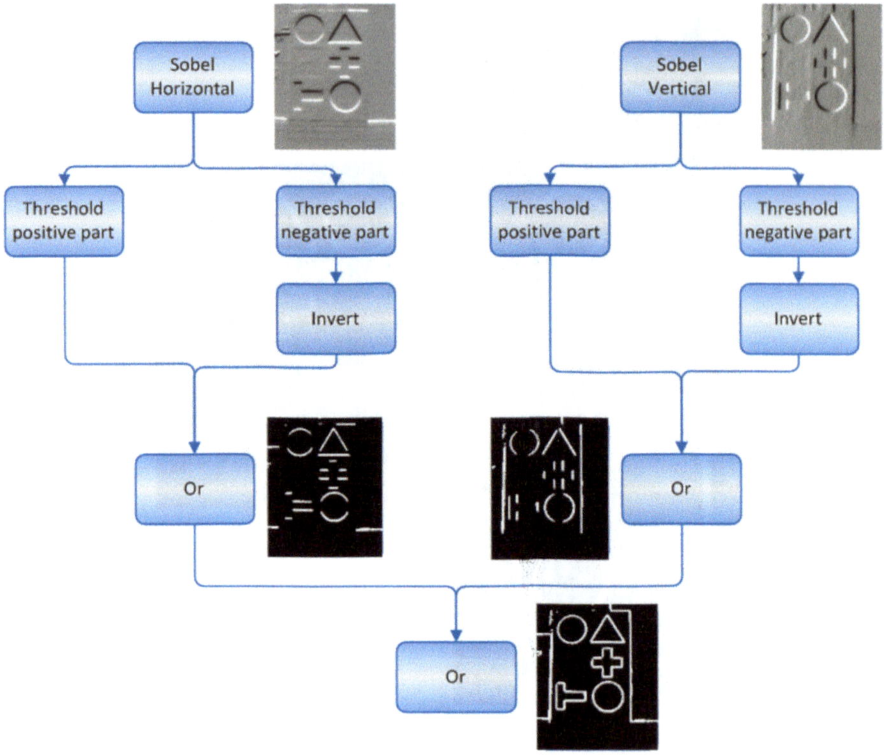


**Fig. 9.33** Flow diagram for sobel threshold and composition

This operation is carried out completely in the SIS Q-Eye using its processing blocks, so that the execution is very efficient, taking around 30 μs.

Finally, since the landmarks should be closed contours, all the open contours are removed. For this purpose the algorithm displayed in Fig. 9.37 has been developed.

A closed contour is a contour that doesn't have any end point, where an end point is a pixel that only has one or none active neighbors. Removing end points from an image iteratively, we would remove open contours and leave closed contours. This can be done by successively performing Hit And Miss and Logic Operations in what is called a Pruning Operation. However, for long open contours this operation can take around 2.5 ms, so we have developed a method to improve the efficiency of the pruning operation by using the Flood Fill operation.

As can be seen in Fig. 9.37, first of all a number of Pruning Operations is carried out in order to remove the outliers of the Skeletonization. Then, the End Points of the resulting image are obtained, and these points are used as seed to reconstruct the open contours by means of the Flood Fill operation. The reconstructed contours are subtracted from the original image so that only closed contours remain.

**Fig. 9.34** Flow diagram for edges post-processing

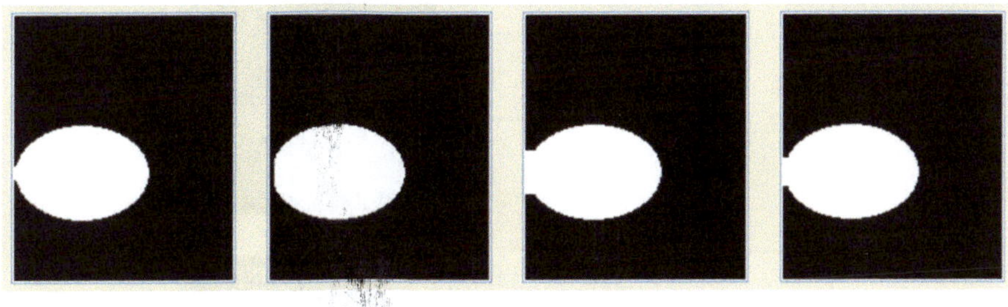

**Fig. 9.35** Example of the effects of different contour conditions for a close operation. From *left* to *right*: original image, result of 3 iterations of close with contour conditions set to *black*, result of the same operation with contour conditions set to *white*, result of the same operation with contour conditions set to *replicate*

This method has reduced the execution time to only 73 μs, which is almost 35 times faster.

### *Diffusion Based Segmentation*

As we have explained before, the diffusion of images is an operation that can be very efficiently implemented in the SIS Q-Eye thanks to the Resistive Grid Block. To take advantage of this efficiency we have developed a method for segmentation based on the Diffusion operation.

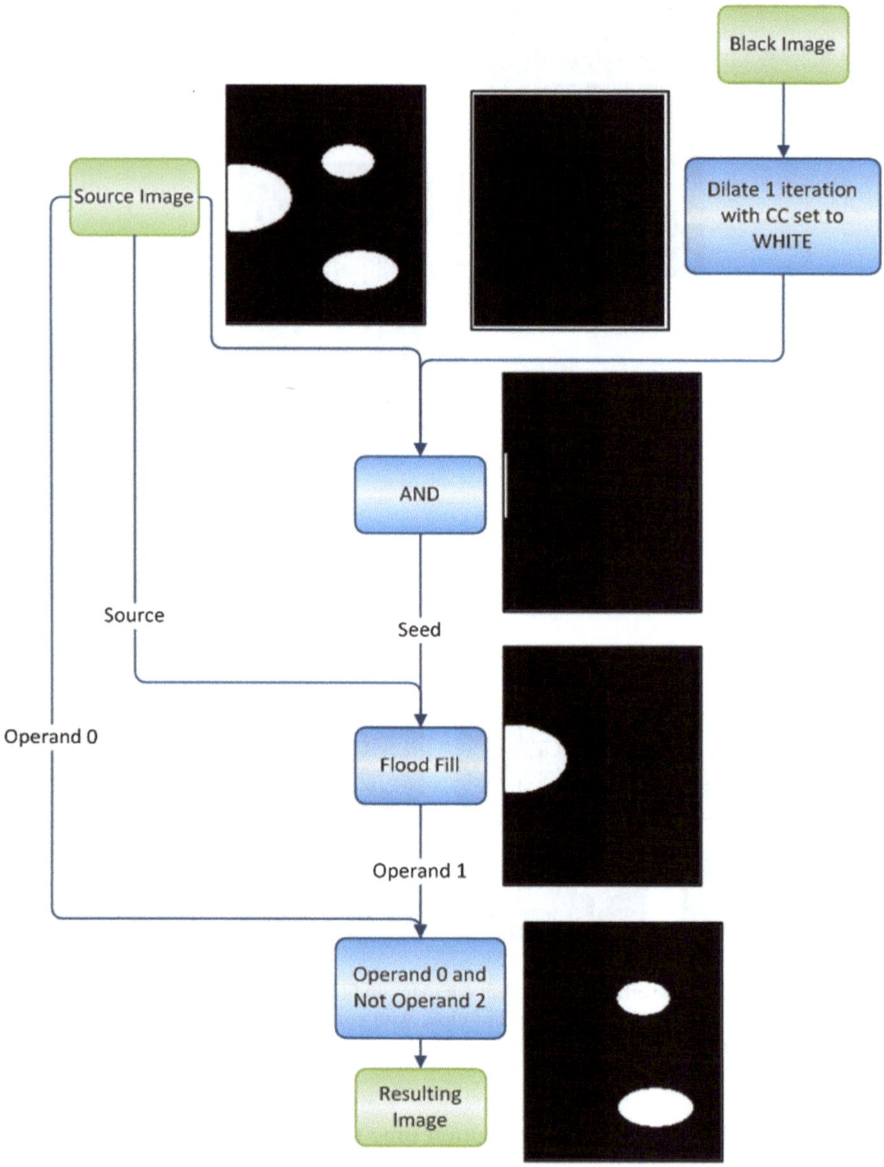


**Fig. 9.36** Flow diagram of the erase border blobs operation

As can be seen in Fig. 9.39, first the Resistive Grid is used to obtain a low pass filtered version of the original image. Then this image is used to threshold the original image using the MAC Block (see Fig. 9.38), using the following equation:

$$Destination(i, j) = \begin{cases} 1 \text{ if } origin(i, j) > LP(origin(i, j)) + offset \\ 0 \text{ in other case} \end{cases}$$

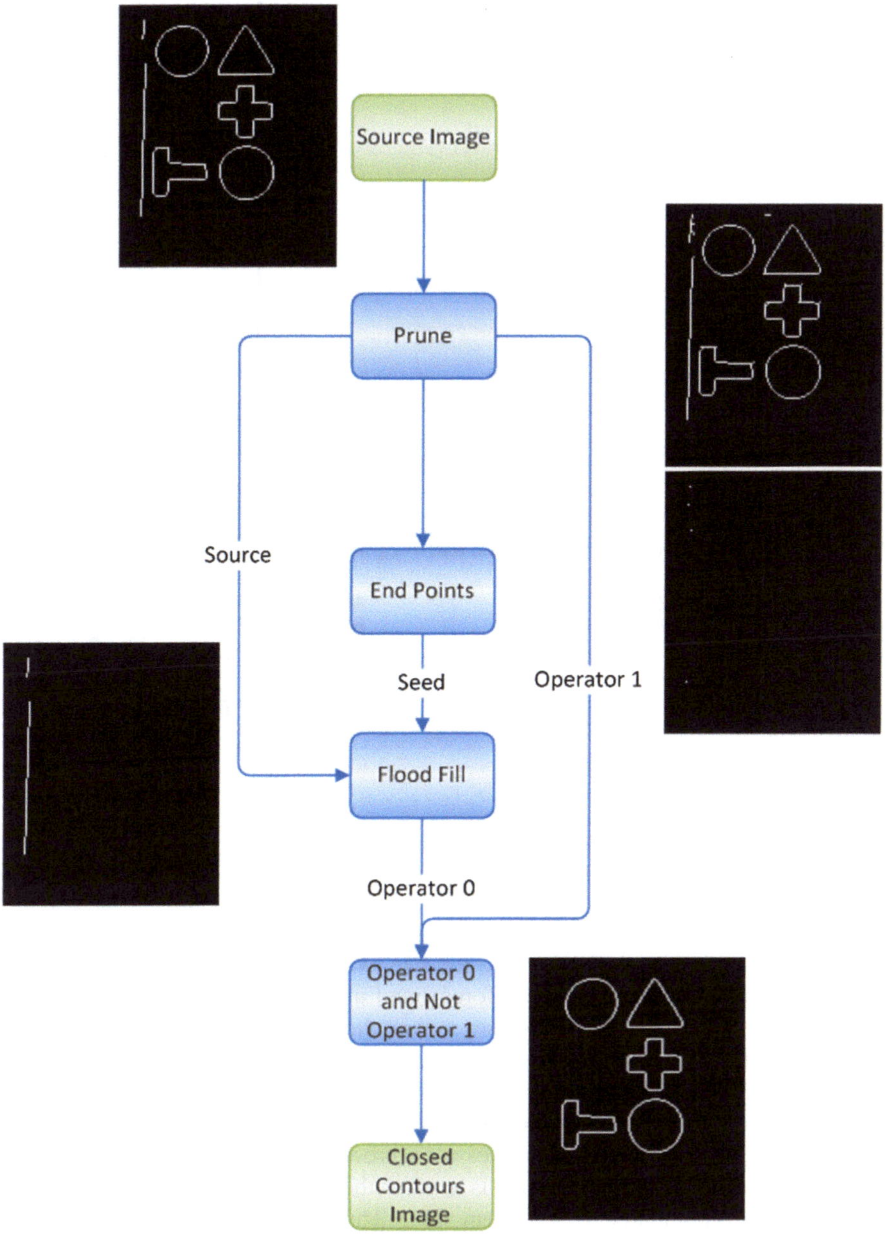


**Fig. 9.37** Flow diagram of the remove open contours operation

The threshold is calculated by statistically examining the intensity values of a local neighborhood around each pixel. In this case, the spatial low-pass filter of the local intensity distribution which is performed over the input image. The size of

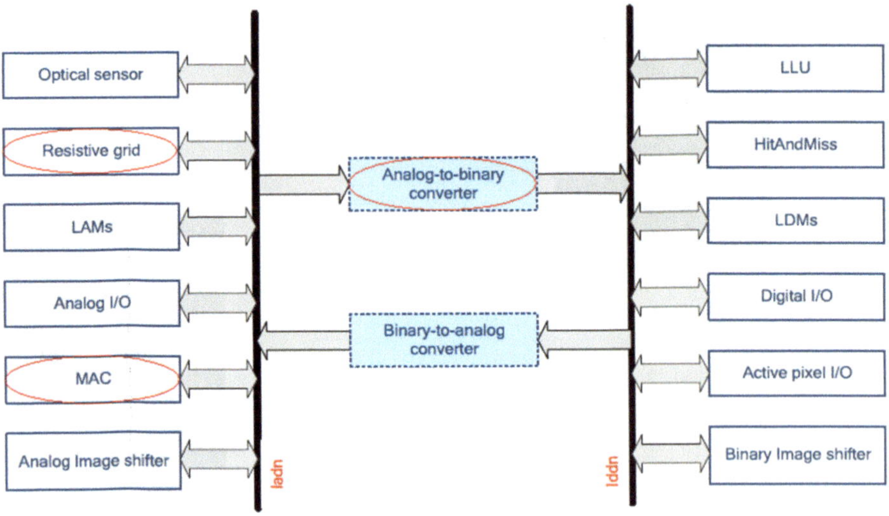


**Fig. 9.38** SIS Q-Eye blocks used in diffusion based segmentation

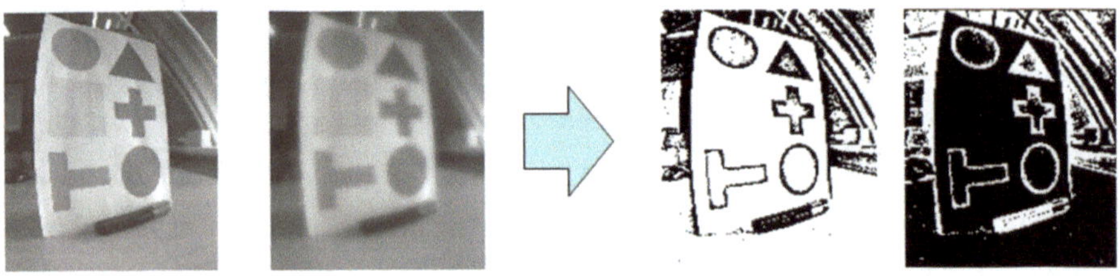

**Fig. 9.39** Images of diffusion based segmentation. From *left* to *right*: original image, filtered image, segmented image and inverted segmented image

the neighborhood (or, in other words, the bandwidth of the low-pass filter) has to be large enough to cover sufficient foreground and background pixels, otherwise a poor threshold is chosen. The offset makes the threshold more appropriate for neighborhoods where the range of intensity values is very small and their mean is close to the value of the center pixel. The offset contains signed information, which means negative values are represented by offset values between 1 and 127, an offset of 128 represents 0, and positive values are represented by offsets between 129 and 255. Note that the threshold operation is local, as the threshold image is not a constant and each pixel is compared to a different value. This operation is extremely efficient, since comparing an image to another is a unitary operation for the SIS Q-Eye and is performed in all its pixels simultaneously.

Adjusting the bandwidth of the filter to the size of the edges that must be obtained, we can obtain the binarized edges in only 4 $\mu$s (see Fig. 9.39).

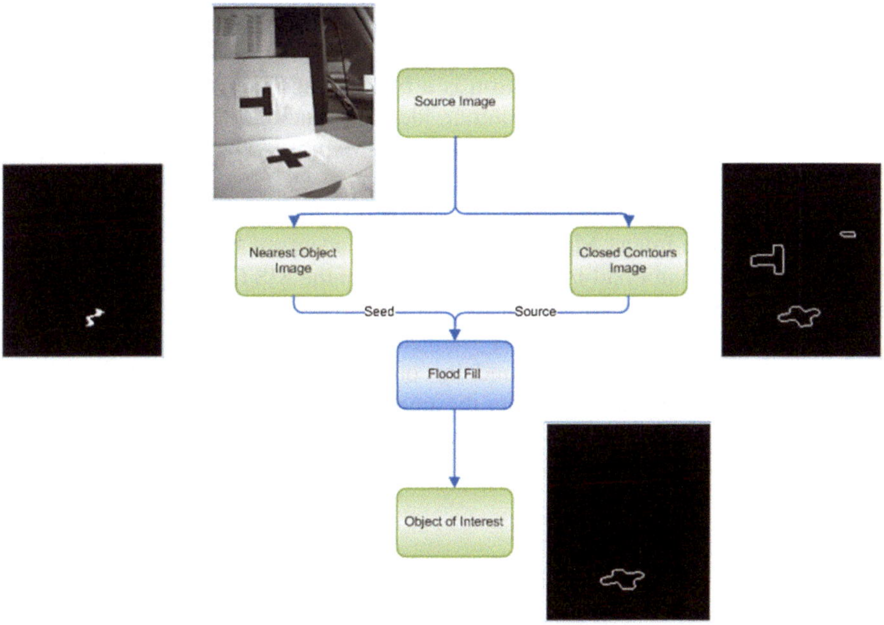


**Fig. 9.40** Flow diagram of the cues combination method

However, this approach has been discarded due to the difficulty of adjusting the bandwidth of the filter to cover the wide range of working distances that are necessary for this application.

### *Cues Combination*

To improve the robustness of the segmentation of the landmarks we make it rely on multiple cues. What makes something possible to segment in a natural environment is that it differs significantly in some property from its immediate surroundings. In a rich environment we can not know a priori what property that could be. Hence the system should in parallel process several different cues to be able to capitalize on the relevant ones [16].

In the present case the cues are motion and edges, which segmentation has been explained in detail in the previous paragraphs.

Our cue combination method can be summarized as extracting the nearest object showing a closed contour. This method has been chosen since it is known that the "insect" will move first to the object that is nearest, as it provides the most parallax. Besides, landmarks are geometric figures with a closed contour.

As can be seen in Fig. 9.40, the algorithm is based on reconstructing the detected closed contours using as seed the nearest object, by means of the Flood Fill operation that we have explained before.

This operation takes 150 μs, including the load of the motion and edge images to the SIS Q-Eye.

The result of the cues combination is a black image with a few active (white) points, corresponding to a closed contour, ideally a landmark.

This information is downloaded to the Image Memory to be further processed by the NIOS II microprocessor.

Although the binary image containing the active points could have been directly downloaded to the Image Memory to be further accessed by the NIOS II microprocessor. The Active Pixel I/O Block contained in the SIS Q-Eye, provides a more efficient solution.

In this Block, the image is quickly scanned from left to right and from top to bottom, and when an active pixel is found, its coordinates are downloaded to the Image Memory. Therefore, for sparse images the amount of information to download is very small, and the execution very efficient. Additionally, during the post-processing, the microprocessor doesn't have to search the image for the active points, but it can access these points directly.

### *Implementation of the Segmentation Algorithms in SIS Q-EYE*

In the described pre-processing algorithms, we have reduced the huge amount of information contained in a greyscale image, most of which is of no use, to just a few active points corresponding to the object of interest. Managing this quantity of information has been done very efficiently thanks to the parallel processing architecture of the SIS Q-Eye.

We have made use of almost all the processing blocks included in SIS Q-Eye. Specially those that provide very efficient functionalities, like the Resistive Grid, have been used very intensively, and in very innovative ways (e.g. Flood Fill functionality).

To have an idea of the size of the algorithms implemented a few numbers are introduced next:

Program Memory The program memory for SIS Q-Eye in Eye-RIS VSoC is 4K Instructions. The size of the SIS Q-Eye application explained in the previous sections is 3979 Instructions, so the 97 % of the program memory is used.

Vectors Memory SIS Q-Eye is programmed by applying vectors through some specific registers. The maximum number of different vectors that can be applied is 256. The developed application generates 216 vectors, so 84 % of the Vectors Memory is used.

Despite the size of the processing, and the amount of data to manage, the complete preprocessing algorithms implemented in SIS Q-Eye are carried out in an average of 3.8 ms, which could lead to a frame rate of 260 frames per second (although as we will see, the post-processing carried out in the NIOS II will reduce this frame rate).

The information that is left to the digital processor for post-processing is of small size and high reliability, so that not very complex algorithms are necessary to extract the final results. This kind of algorithms are the most suitable for the existing digital microprocessor, since it is not very powerful. On the other hand, more powerful

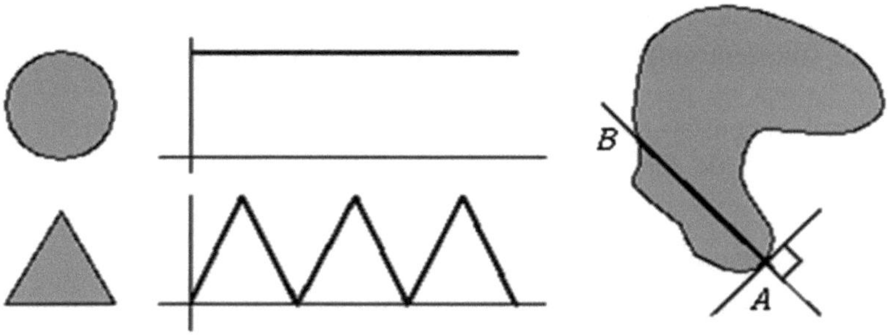


**Fig. 9.41** Different methods of signature extraction

microprocessor couldn't have been implemented in chip so efficiently in terms of silicon area and power consumption, as the existing one.

### Signature Extraction

With the signature extraction we consider finished the pre-processing, as the implementation in the Eye-RIS VSoC regards.

Now the post-processing starts; in this processing the input is information of interest about an image, in this case the contour of the object of interest, and the output will be decisions about the scene, or useful information about the same, in this case the existence or not of a landmark to be approached by the robot, and several features of the same. This processing must be done within the NIOS II digital microprocessor, since this processor is not very fast and powerful the amount of data to be managed by the processor and/or the complexity of the implemented algorithms should be reduced in order to keep high frame rates. Thanks to the robustness of the pre-processing algorithms running on the SIS Q-Eye we can develop now not very complex post-processing algorithms, and still achieve very robust and fast results.

The first step in post-processing consists of extracting the signature of the segmented contour.

A signature is a 1-D functional representation of a boundary which basic purpose is to reduce the boundary representation to a 1-D function, which generally is easier to manage than the original 2-D boundary [17].

It only makes sense to consider using signatures when the boundaries do not have self-intersections, nor deep, narrow concavities or thin, long protrusions. In these cases it can be guaranteed that the vector extending from its origin to the boundary intersects the boundary only once, thus yielding a single-valued function of increasing angle [15].

A signature can be generated in different ways:

- One of the simplest and most usual is to plot the distance from an interior point (typically the centroid) to the boundary as a function of the angle. See Fig. 9.41 left.
- Other method consists on determining for each point of the boundary, its distance to the opposite side. See Fig. 9.41 right.

### *Signature Implementation*

For our implementation we have made use of the first approach, which obtains distances from the points in the boundary to the centroid, with some modifications to improve execution efficiency.

First of all we calculate the centroid as the average of the respective coordinates of the points of the boundary. Thanks to the Active Pixel I/O operation that closed the pre-processing, now the microprocessor doesn't have to search the image for the active points, but these points are directly accessible:

$$x_c = \frac{\sum\limits_{contourpixels} x}{\text{Number of contour pixels}}$$

$$y_c = \frac{\sum\limits_{contourpixels} y}{\text{Number of contour pixels}}$$

Then the angle and distance relative to the centroid is calculated for each point of the contour. Typically these values are obtained as:

$\Delta x = x - x_c$
$\Delta y = y - y_c$
$\rho = \sqrt{\Delta x^2 + \Delta y^2}$
$\theta = atan \frac{\Delta y}{\Delta x}$

For efficiency issues, we are using the square of $\rho$, so that we don't have to use the square root function that is very time consuming. In order to further save execution time, we don't use the atan() C function, but we use a Taylor Series approach with three terms:

$atan(x) = \sum_{n=0}^{2} \frac{(-1)^n}{2n+1} x^{2n+1}$ for $|x| < 1$

Finally, we are subsampling the signature into 6 degrees slots. That is, points of the contour with $\theta = 0, 1, 2, ..., 5$ are converted to the same point in the signature. The module of this point is the average of the modules of the included points. This way the resulting signature has fixed length equal to 60 (Fig. 9.42).

It must be remarked that thanks to the simplicity of the shape of the landmarks to be classified, these approaches can be made without loss of accuracy in the results.

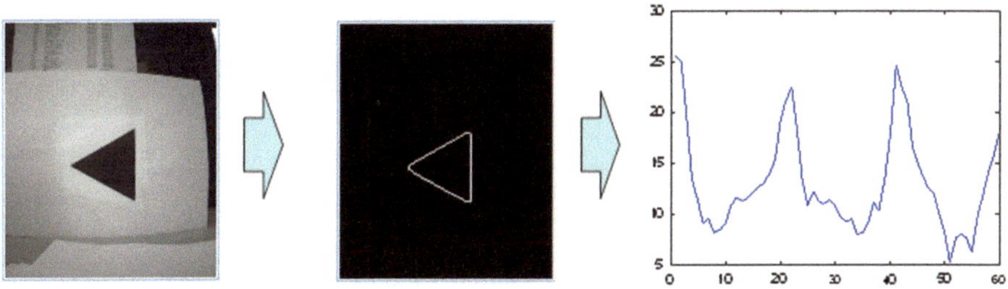


**Fig. 9.42** Example of signature extraction. From *left* to *right*: original image, extracted contour, signature of the triangle reduced to 60 bins

### *Signature Normalization*

Signatures described in the previous paragraph are invariant to translation, since they are referenced to the centroid of the shape, but they do depend on rotation and scaling.

Normalization with respect to rotation can be achieved by finding a way to select the same starting point to generate the signature, regardless of the shape's orientation. Different methods for rotation normalization can be used:

- The most commonly used method consists on selecting the starting point as the point farthest from the origin of the vector, in our case the centroid.
- Another way is to select a point on the major eigen axis. This method requires more computation but is more robust because the direction of the eigen axes is determined by using all contour points.

For the shapes to be recognized within the scope of this project, the first method will suffice. We could resort to the second method if some issues with the recognition rates would arise, but as will be seen in the Results paragraph this has not been necessary.

Based on the assumptions of uniformity in scaling with respect to both axes and that sampling is taken at equal intervals of $\theta$, changes in size of a shape result in changes in the amplitude values of the corresponding signatures. The basic idea under scale normalization is to remove dependency on size while preserving the fundamental shape of the waveforms. Also different methods can be devised:

- The most commonly used method consists on scaling all functions so that they always span the same range of values, typically [0,1].
- A more robust, but also more computational intensive, approach is to divide each sample by the variance of the signatures, assuming that the variance is not zero, or so small that it creates computational difficulties. Use of variance yields a variable scaling factor that is inversely proportional to changes in size.

Also for the scope of this project, the first method provides really good results, so that there is no need of resorting to the more complex method.

To improve the execution time of this method, our normalization is in the range [0,255] so that float point variables are avoided. The NIOS II microprocessor doesn't

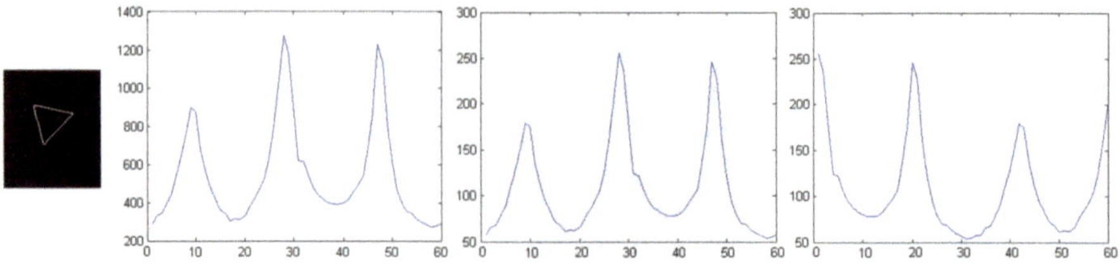

**Fig. 9.43** Example of signature normalization. From *left* to *right*: original image, signature in 60 bins, signature scale normalized, signature scale and orientation normalized

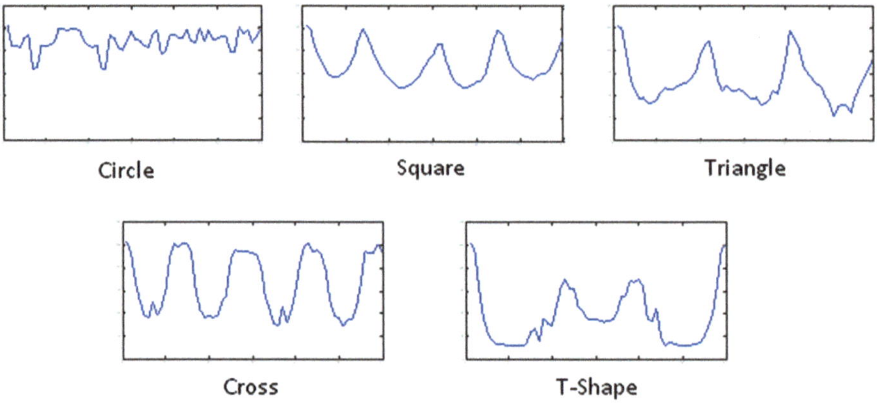


**Fig. 9.44** Normalized signatures of the different existing landmarks

include a Float Point Unit, and calculation using float point variables must be software emulated, so their execution is slower. Avoiding the use of float point variables we achieve important speed ups in the execution.

In Fig. 9.43 the signature for a triangular shape is displayed. As can be seen, the original signature expands between 200 and 1400, after the first normalization the range is between 1 and 256, and after the second normalization the signature starts at the farthest point from the centroid.

### *Results*

The extraction and normalization of the signature from the segmented shapes has been implemented in the NIOS II digital microprocessor using the explained optimizations to achieve short execution times. The extraction of the signature takes only 1–2 ms, depending on the size of the landmark. The complete normalization process (orientation and scaling) can be run in only 85–90 μs.

In Fig. 9.44 the normalized signatures of the different existing landmarks is displayed.

## Classifier

Having reduced the representation of an object to its normalized signature, the problem of classifying different structures can be solved by defining proper distance measures on the signatures with respect to a model [18]. More specifically, consider a landmark with signature representation $S_l$ and the set of models of the different landmarks $S_{m_i}$ with i representing the different landmark shapes.

Different distances can be defined:

- Minkowski Distance: or $L_p$ norm is defined as:

$$d_M(S_l, S_{m_i}) = \left(\sum_{s \in S} |S_l(s) - S_{m_i}(s)|^p\right)^{\frac{1}{p}}$$

  where typical values for p are 1 (Manhattan distance), 2 (Euclidean distance) or infinite (uniform norm or Chebyshev distance).
- Jaccard Distance: this measure is commonly used for measuring similarity between fuzzy sets. Its underlying idea is to compare the size of the intersection of the two sets with the size of their union. In terms of a distance, it can be defined as follows:

$$d_J(S_l, S_{m_i}) = 1 - \frac{\sum_{s \in S} min(S_l(s), S_{m_i}(s))}{\sum_{s \in S} max(S_l(s), S_{m_i}(s))}$$

  The main advantage of the Jaccard distance is that the result is always in the interval [0,1], therefore defining a threshold is much easier. For this reason we have chosen this measure for implementation.

### *Classifier Implementation*

To create the landmark models, several acquisitions of each type of shape have been made offline. Their signatures have been extracted and averaged to conform the model. These models are stored in variables that will be used in each iteration of the visual algorithm to perform the classification.

Then, for each segmented landmark, it is compared to each model using the Jaccard distance. The model giving the minimum distance is considered to be the real shape of the landmark. This minimum must be below a determined threshold, otherwise it is considered that the detected contour doesn't correspond to any valid landmark shape.

The implementation of this classifier in the NIOS II microprocessor, with the habitual optimizations takes: 65 μs for the extraction of each Jaccard distance, which amounts to 350 μs for the extraction of the Jaccard distances to each of the five possible landmark shapes, and search for the minimum value.

Figures 9.45 and 9.46 show the results of detecting and classifying two different landmarks. The shape of the different landmarks is codified as follows:

UNKNOWN 1

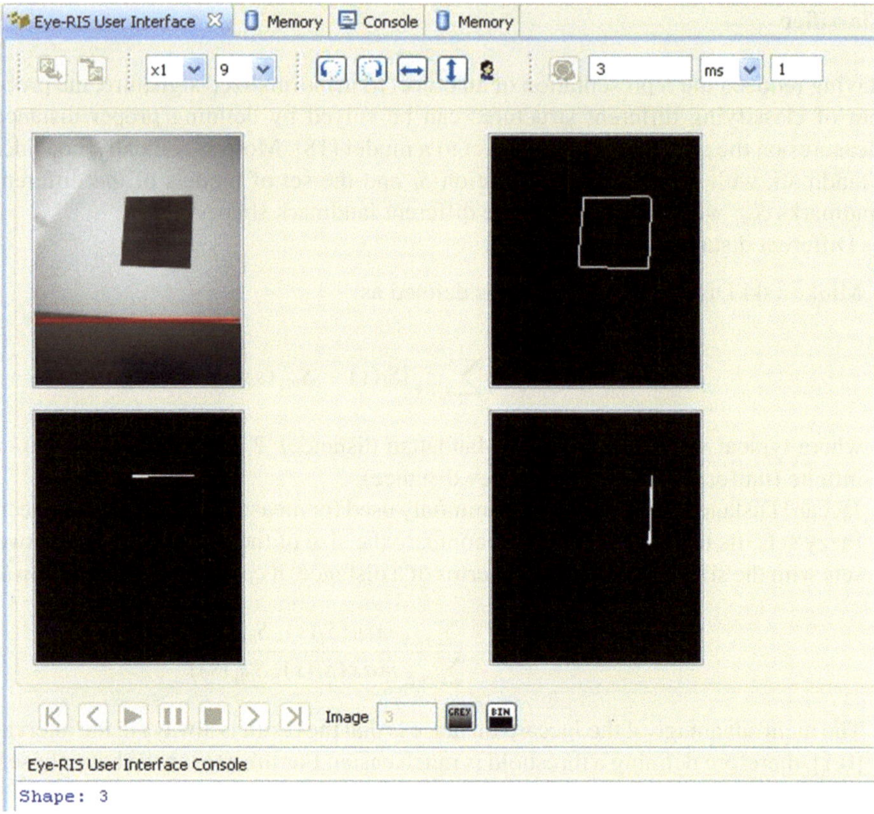


**Fig. 9.45** Image with square landmark correctly segmented and classified

CIRCLE 2
SQUARE 3
TRIANGLE 4
CROSS 5
TSHAPE 6

where UNKNOWN means that no correct landmark shape has been recognized.

Globally, the processing of an acquired image to segment and classify the present landmark is carried out within Eye-RIS VSoC in around 5.6 ms. This would allow a frame rate of about 180 frames per second. It must be noticed that since pre-processing and post-processing are done in different processors, SIS Q-Eye and NIOS II respectively, it would be possible to parallelize execution in both processors reduce further the processing time, however due to the nature of the application we consider the achieved frame rate is more than satisfactory.

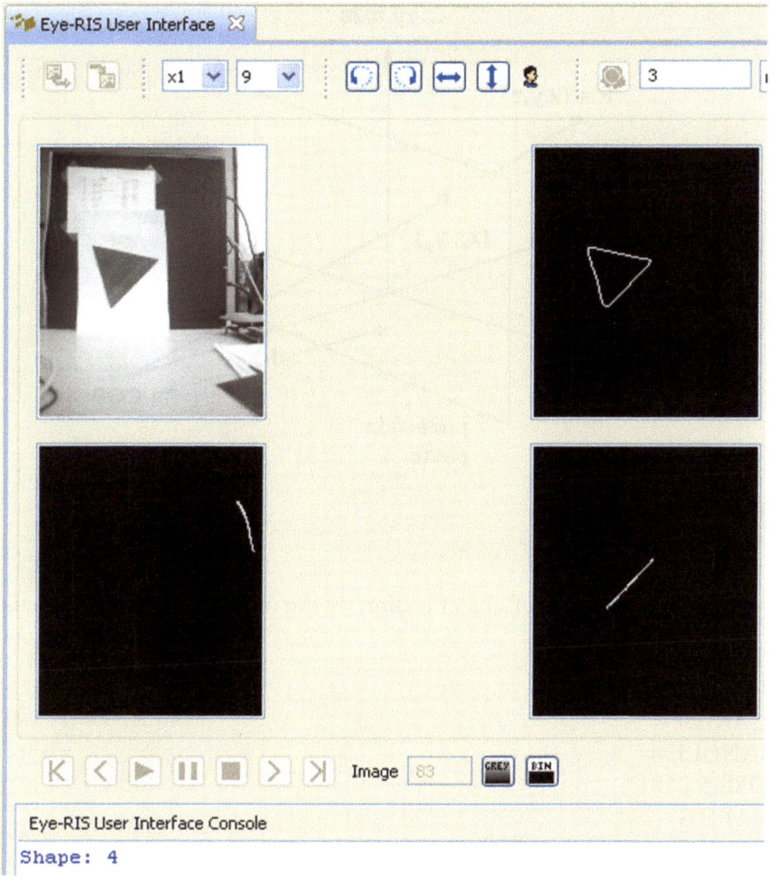


**Fig. 9.46** Image with triangular landmark correctly segmented and classified

## Extraction of Features for the Insect Brain

A number of features has been obtained from the detected landmarks that are needed by the Insect Brain. The extraction of these features makes use of both, operations in the SIS Q-Eye and operations in the NIOS II microprocessor. The distribution of the execution has been made according to the rules previously explained for the development of the main visual routines. Moreover, the extra processing needed for the extraction of features for the insect brain, is intermingled with the main code in order to optimize the use of resources and minimize the execution time.

Next we describe the insect brain features and their implementation:

*Movement* The direction of the main motion is provided. This value is directly obtained in the motion based segmentation algorithm, therefore no extra processing is necessary.

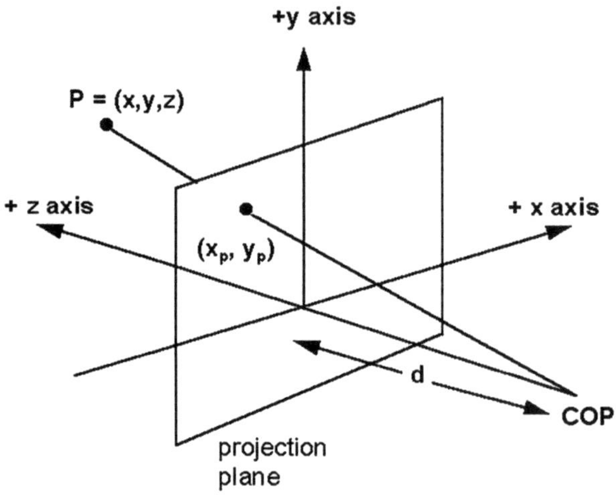


**Fig. 9.47** Projective geometry

*Type of object:* The type of object is directly the output of the classifier, that is:

UNKNOWN 1
CIRCLE 2
SQUARE 3
TRIANGLE 4
CROSS 5
TSHAPE 6

so no extra processing is needed. See Figs. 9.45 and 9.46.

*Distance:* For the distance calculation we make use of the projective geometry. See Fig. 9.47 According to this, let $(x_p, y_p)$ be the image coordinates of a point and (x, y, z) the 3D coordinates (Real World coordinates) of the same point P. A direct application of Thales theorem shows that:

$x = \frac{zx_p}{d} y = \frac{zy_p}{d}$

Being d the focal distance of the camera. Expanding these equations from coordinates to distances we can see that:

$ObjectWidth = \frac{zImageWidth}{d} ObjectHeight = \frac{zImageHeight}{d}$

So, knowing the size of the object in the image, and the real size of the object, the remaining unknown is the distance to the object, or the z coordinate:

$z = \frac{ObjectSized}{ImageSize}$

The algorithm developed to extract this feature has two main parts:

1. Extracting the object size. This step is performed in the SIS Q-Eye just after finishing the segmentation of the object of interest. We take as input the contour

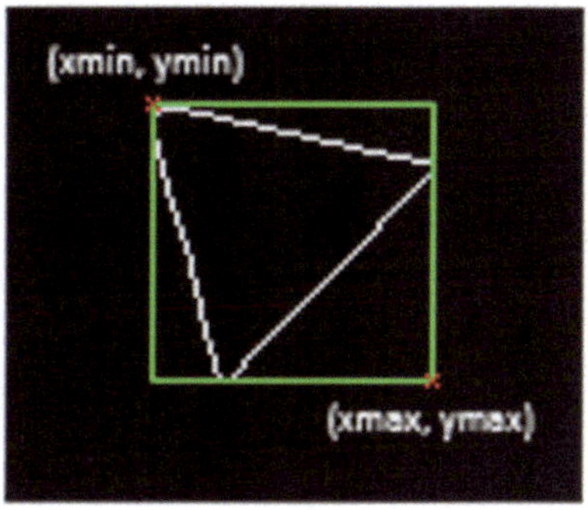


**Fig. 9.48** Bounding box of a landmark contour

image, and use a new feature of the Active Pixel I/O block. This block is capable of providing the coordinates of the upper left and bottom right corners of the bounding box of an object.

In Fig. 9.48 the bounding box of a landmark, and its extreme coordinates are displayed. This functionality takes only 4 μs in the SIS Q-Eye. The coordinates obtained from the Bounding Box function give us the size of the object in the image as:

$ImageSize = ((x_{max} - x_{min}) + (y_{max} - y_{min}))/2$

2. Extracting the distance to the object. This operation is carried out in the NIOS II microprocessor, since the SIS Q-Eye cannot carry out products or divisions. Since the focal distance of the optics and the real size of the object are known, we directly apply the equation: $z = \frac{ObjectSized}{ImageSize}$ This operation takes only 13–15 μs.

*Perimeter:* the perimeter is obtained as the number of pixels of the segmented contour. This number is directly given by the Active Pixels I/O block as we download the contour to the Image Memory, therefore, no extra processing is needed.

*Area:* To obtain the area of the landmark from its segmented contour, the algorithm displayed in Fig. 9.49 has been developed. This algorithm is fully performed using the SIS Q-Eye. In this algorithm, first of all the contour image is inverted using the Logic Operation block, this gives an image with two blobs (or sets of connected pixels), one is the object of interest itself, and the other is the background. Using the Erase Border Blobs functionality (that was previously explained) we can remove the background blob very efficiently. The number of pixels of the remaining object, or area, can be also very efficiently obtained using the Active Pixels I/O functionality. The execution of this algorithm takes around 60–90 μs depending on the size of the object.

*Centroid:* The centroid was obtained as first step for the signature extraction, so no extra processing is needed.

*Horizon position:* Most horizon detection algorithms rely on the existence of two main areas in an image that are separated by a quasi-horizontal line, and that present different features regarding mainly their color [19–21]. For our horizon

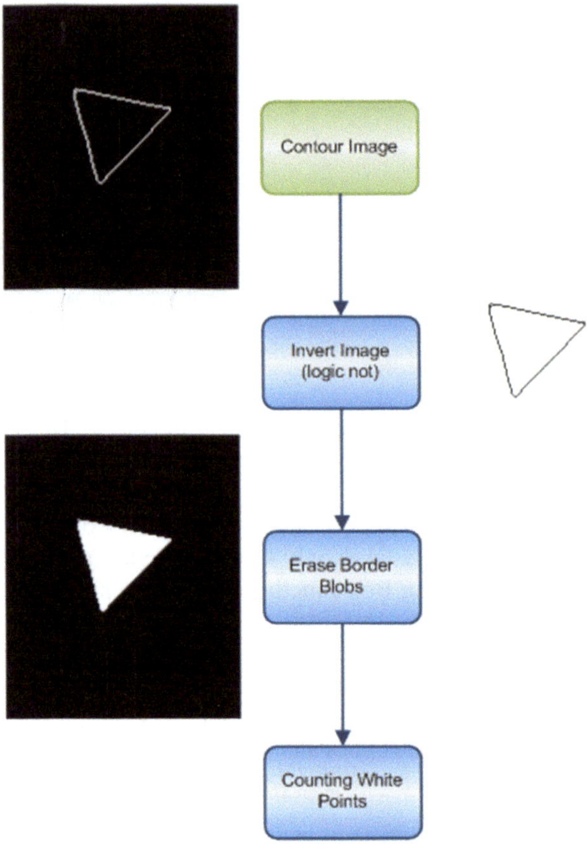


**Fig. 9.49** Flow diagram of the area from contour algorithm

position algorithm we must take into account that no real horizon is seen, so we have to make some assumptions as to what we are going to detect:

- Our horizon will be the main horizontal or quasi-horizontal line appearing in the image.
- The "sky" will present brighter grey-levels than the "earth". This is true for all "real horizon" detection algorithms and also for the Tribot arena [22, 23]. This condition will be easily reverted in our algorithm, but must be a priori specified.
- The horizon will not change abruptly form one image to the next.

Based on these conditions the algorithm displayed in Fig. 9.50 has been developed.

1. First of all the Horizontal Sobel of the image is obtained.
2. A threshold is applied to obtain the negative parts of the Sobel, that is, the parts where a bright to dark (from top to bottom) transition happens.

3. The resulting points are downloaded using the Active Pixel I/O block. Until here the algorithm is carried out in the SIS Q-Eye as is depicted in Fig. 9.50. Next steps are implemented in the NIOS II microprocessor.
4. The Active Points are analyzed to create a vertical projection histogram. A vertical projection histogram counts the number of active pixels in each row of the image. This histogram allows us to easily detect the existence of dominant horizontal lines in the image.
5. The search for the horizon in the projection histogram starts.
6. If the horizon was detected in a previous image, a small interval of the projection histogram around the previous location is searched for the maximum.
7. If a maximum is found the algorithm finishes.
8. If a maximum is not found, or there was no previous horizon detected, then the complete projection histogram is searched for the maximum.

It must be pointed out, that this algorithm follows the optimal scheme for an Eye-RIS application. First of all the image is analyzed in the SIS Q-Eye, and the amount of information is reduced to only information of interest. Then, this information is downloaded to the NIOS II, where the information is further reduced, and analyzed with medium complexity algorithms to extract the final results. Thanks to this optimized implementation, the horizon detection can be achieved in only 330–400 μs.

*Height with respect to horizon:* Once obtained the horizon, this feature is calculated as the difference in the y coordinate between the centroid of the landmark and the horizon.
See Fig. 9.51.

*Alpha (angle between landmark and center of image):* This feature is obtained from the centroid of the landmark as:

$$\alpha = \arctan\left(\frac{LandmarkCentroid_y - ImageCenter_y}{LandmarkCentroid_x - ImageCenter_x}\right)$$

*Vertical Distributedness:* The vertical distributedness is the ration between the height and the width of the landmark. These two values are extracted from the Bounding Box measures.

*Horizontal Distributedness:* Is the inverse of the vertical distributedness.

*Bounding Box orientation:* The bounding box obtained by the SIS Q-Eye is aligned with the image axis, but for this feature we need to extract the minimum bounding box containing the landmark, which may not be aligned with the image axis.
In Fig. 9.52 the difference between the aligned bounding box (in green) and the bounding box provided by the SIS Q-Eye (in red) can be appreciated.
The second order moments $M_{02}$, $M_{20}$, $M_{11}$, known as the moments of inertia can be used to determine orientation. In general, the orientation of an image describes how the image lies in the field of view, or the directions of the principal axes. In terms of moments, the orientations of the principal axes, $\theta$, are given by:

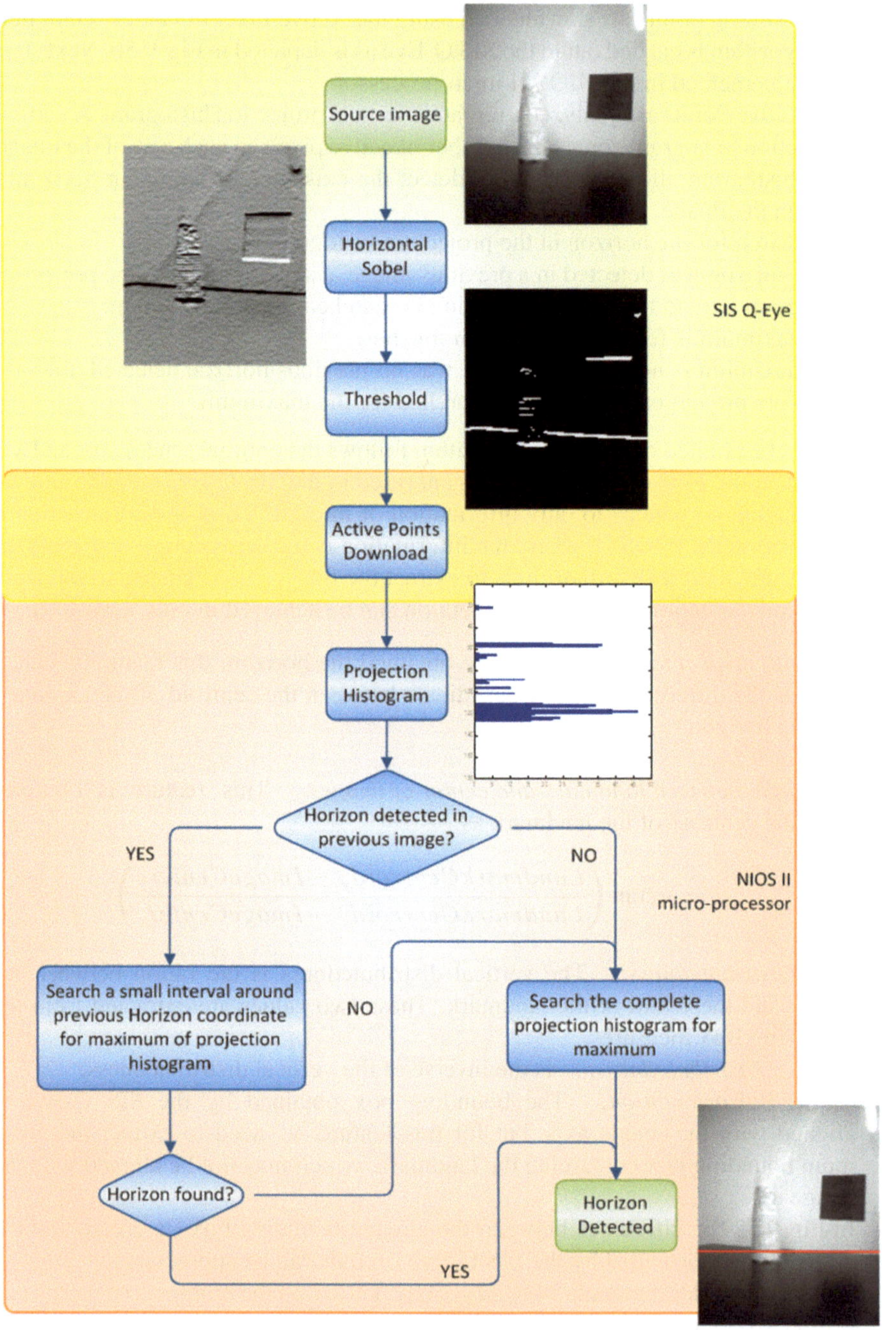


**Fig. 9.50** Flow diagram of the horizon detection algorithm

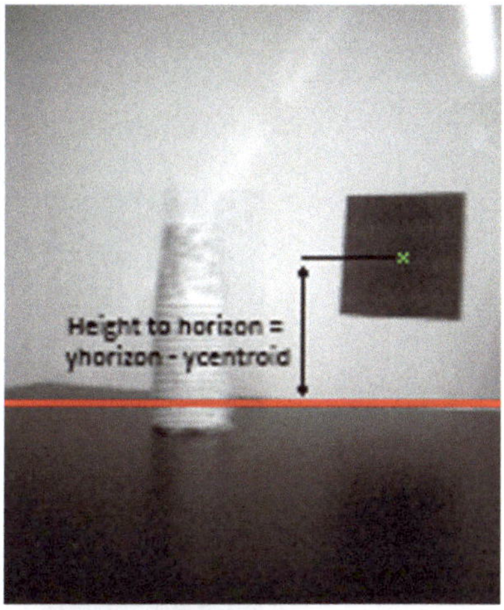


**Fig. 9.51** Height with respect to horizon feature

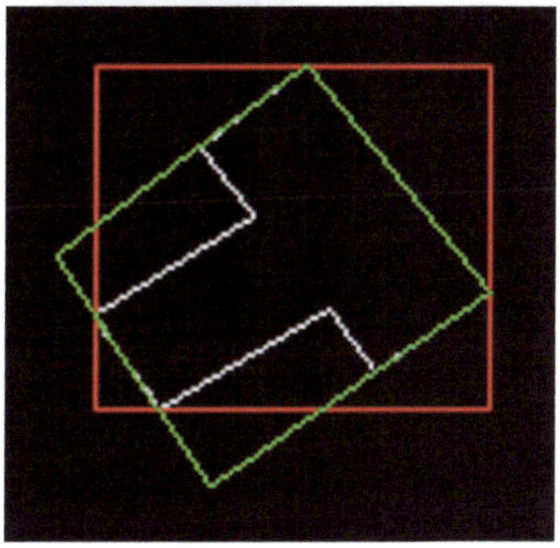

**Fig. 9.52** Aligned bounding box (*Green*) versus SIS Q-Eye bounding box (*Red*)

$\theta = \frac{1}{2}\arctan\left(\frac{2\mu_{11}}{\mu_{20}-\mu_{02}}\right)$

where the moments of order pq are:

$m_{pq} = \sum_x \sum_y x^p y^q f(x, y)$

and the central moments of order pq:

$\mu_{pq} = \sum_x \sum_y (x - \bar{x})^p (y - \bar{y})^q f(x, y)$

where:

$\bar{x} = \frac{m_{10}}{m_{00}} \bar{y} = \frac{m_{01}}{m_{00}}$

The calculation of $\theta$ is performed in the NIOS II microprocessor from the contour coordinates takes around 350–700 μs depending on the size of the shape.

*Grey value of landmark:* For getting the grey value of the landmark, the algorithm depicted in Fig. 9.53 has been implemented. In this algorithm we take advantage of the capabilities of SIS Q-Eye of performing masked operations. We use the blob image of the landmark that was obtained to get its area as a mask on the original

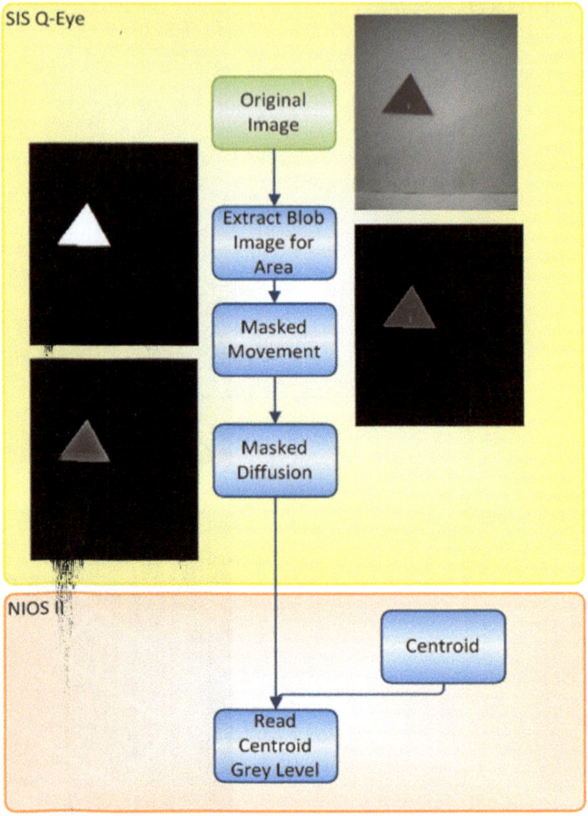


**Fig. 9.53** Flow diagram for grey value extraction

image. Then we apply a masked diffusion, so that only the unmasked pixels are diffused. Doing this we get an average of the grey-level within the object, where noise and defects are removed. Then, reading the grey level of one of the pixels within the shape, we have the average grey-level of the object. In this case we use the centroid point, since it had been already obtained.

*Grey value of background (around landmark).* An algorithm, very similar to the previous one, has been developed for extracting the grey value of the background. See Fig. 9.54. In this case the mask is obtained as a thick outer border of the shape, and the point whose grey value is read is outside the shape. This algorithm takes only 16,8 μs.

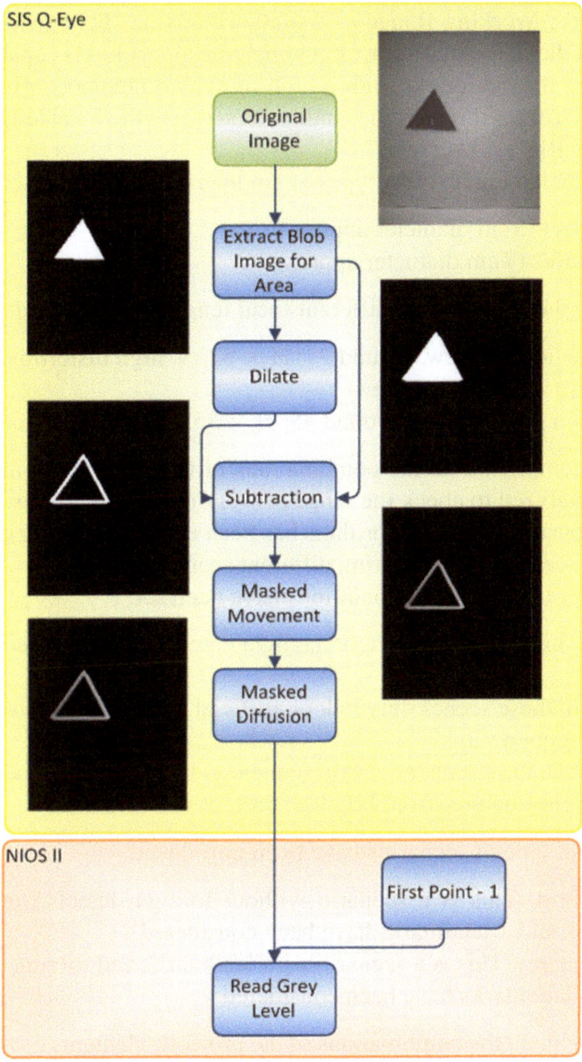


**Fig. 9.54** Flow diagram for grey value of background extraction

## Algorithms Testing and Refinement

### *Test Plan*

A test plan for the algorithms developed has been specified. These tests have been devised to check the robustness and functionalities of the algorithms developed for the Eye-RIS v2.1 platform.

First of all the Working Range of the algorithms has been tested. By working range we mean the range of distances at which the algorithm is capable of detecting and recognizing the landmarks with a high rate. Two elements affect the working range of the algorithms directly: the landmarks size and the focal length of the optics used in the Eye-RIS v2.1 system.

Regarding the landmarks, two types of landmark will be considered:

- Big landmarks (15 cm diameter approx.)
- Small landmarks (7 cm diameter approx.)

Considering the optics, three different focal lengths will be used:

- 3.5 mm: wide field of view, around 71° H × 82° V, high distortion.
- 4.5 mm: not so wide field of view
- 6 mm: medium field of view, around 45° H × 55° V, low distortion.

Sequences comprising all the combinations of the previous elements have been captured and analyzed to check the range of work of the algorithms.

Then the robustness of the algorithms has been evaluated setting different illuminations for the scene and considering different scenarios.

Three different illumination conditions have been set:

- Bright scene: in these scenes bright daylight enters the room where the sequences are captured.
- Dark scene: in these scenes only indoor artificial light is present during the acquisition of the sequences.
- Variable illumination scene: in these sequences the blinds are opening or closing so that the scene changes from bright to dark or viceversa.

Also different type of scenarios have been introduced:

- Simple scenario. This is a scenario without foreign objects (only landmarks). Scenes with 1 and 2 landmarks have been considered.
- Complex scenario. This is a scenario with landmarks and foreign objects. Scenes with 1 and 2 landmarks have been considered.

Sequences with all the combinations of the previous elements have been captured and analyzed to check the robustness of the algorithms.

Also different measurements have been carried out in the described sequences. The measures are made on the most relevant features:

- Movement: the percentage of successful movement estimation versus the distance to the landmark.
- Type of object: the percentage of successful object classifications versus the distance to the landmark.
- Distance: Precision and resolution of the estimated distances to the landmarks.

All the remaining features:

- Perimeter.

- Area.
- Height.
- Width.
- Centroid.
- Horizon position.
- Height with respect to horizon.
- Alpha (angle between landmark and center of image).
- Vertical Distributedness.
- Horizontal Distributedness.
- Bounding Box orientation.
- Grey value of landmark.
- Gray value of background (around landmark).

have been tested qualitatively in the analyzed sequences, but not quantitatively as the first three features.

Finally, the algorithms have been quantitatively tested against different sequences acquired at the Tribot Arena [22, 23].

*Final demo setup*

After testing the Visual Algorithms using two different lens, the following conclusions can be stated.

The proposal of using 6 mm focal length lens involves a limitation on the maximum recognition distance. This maximum achievable distance is 150 cm for the Cross shape with a size of 24 cm, the maximum size achievable by the LCD monitors used in the demo.

A new solution using 12 mm focal length lens was tested. Using a lens with a focal length double the previous one will double the maximum achievable recognition distance. The results of these tests show that the maximum distance exceeded the size of the arena, approximately 250 cm in its widest side. Nevertheless the use of big landmarks, 21 cm in this case to increase robustness, together with the use of 12 mm lens increases the minimum recognition distance. In this case no landmark was completely seen on the monitor at 50 cm distance, being necessary to increase the distance to the landmark in order to acquire a complete shape in the image.

The use of the 6 mm lens is finally discarded, as the only way of achieving greater recognition distances with that lens would be increasing the size of the landmarks, which cannot be done due to technical reasons.

The recommended lens is therefore the 12 mm focal length lens. This lens involves a limitation on the minimum recognition distance. As the final objective has been set in recognizing a landmark at distances shorter than 50 cm, a suitable solution was needed. The final solution is based in the use of landmarks of different sizes:

- Big sized landmarks will be used for long distances.
- Small sized landmark will be used for short distances.

In the tests 21 and 12 cm landmarks were used. Several tests show these landmarks ensured a wide range of work, covering from close distances up to the limits of the arena. The main controller will be responsible for changing the size of the landmarks on the monitors, as well as the Real Height value on the visual application, at the same time. If not satisfied this condition, the distance estimations will offer erroneous values. Basic real-situation tests show that the solution works properly for the initially defined objectives.

## 9.3 Conclusions

A complete set of visual routines has been developed for segmenting, classifying and describing a set of landmarks. This algorithms have proven completely robust to illumination changes, and very rugged against different environments.

The main innovation within this work is not on the algorithms themselves, since most of them are well-known widely-used vision algorithms, but on the selection and implementations of these techniques on the particular architecture of the Eye-RIS Vision Systems.

Thanks to the existence of the processing circuitry at the pixel level, a new complex algorithm for image acquisition has been developed that adapts both to temporal and spatial variations almost without extra processing time. This algorithm has permitted the remaining processing to be completely independent from the illumination conditions as has been shown in the testing.

After that, a number of pre-processing techniques have been developed to segment the objects of interest. It must be noticed that the complexity of these algorithms would have required an important amount of computation in conventional systems, however thanks to the implementation in the SIS Q-Eye the global processing time for these algorithms is really reduced. A new motion parallax based segmentation algorithm has been devised to extract the location of the nearest object. Also the technique of edge segmentation, though not original itself, includes a refinement post-processing whose efficiency is due to the particularities of the SIS Q-Eye. During the development of these techniques a special use of some processing blocks of the SIS Q-Eye has been developed. This use, although not the original for these blocks, has provided us with some very efficient functions that have been widely used thereafter. To mention just two, the use of the resistive grid in the digital domain to reconstruct objects, and the use of the contour conditions (originally meant to be used in convolutions and hit and miss operations) to create images with predefined borders.

The combination of the different pre-processing algorithms developed, has resulted in a very robust segmentation of the objects of interest, and in a considerable reduction of the amount of data to be further processed. As a consequence, the post-processing algorithms developed for the NIOS II didn't need to be very complex or computationally intensive to achieve the expected results. Being the NIOS II not a very powerful microprocessor, the developed algorithms have been also optimized for their implementation, achieving thus a quite high frame rate.

The extraction of features for the insect brain has also followed the philosophy of including most of the intensive processing within the SIS Q-Eye, and re-using the available information in order to obtain very short processing times.

We have to point out how also the communication ports present in the Eye-RIS VSoC have been taken advantage of for sending information to the Master System, the robot.

Finally, the Eye-RIS VSoC has been integrated in the rover robot platform using the communication ports as interface. The tests performed on the complete platform have proved the successful integration of the vision system. A feasible scheme for the demo has been defined, and the preliminary tests have shown the correct performance of the VSoC within the complete application.

## References


1. P. Arena, S. De Fiore, L. Patané, P.S. Termini, R. Strauss. Visual learning in drosophila: application on a roving robot and comparisons, in *5th SPIE's International Symposium on Microtechnologies*, pp. 1–12
2. P. Arena, S. De Fiore, L. Fortuna, L. Nicolosi, L. Patané, G. Vagliasindi, Visual homing: experimental results on an autonomous robot, in *18th European Conference on Circuit Theory and Design (ECCTD 07).*
3. P. Arena, L. Patané, P.S. Termini, Decision making processes in the fruit fly: a computational model, in *Frontiers in Artificial Intelligence and Applications - Proceedings of the 21st Italian Workshop on Neural Nets*, vol. 234
4. P. Arena, M. Cosentino, L. Patané, A. Vitanza, Sparkrs4cs: a software/hardware framework for cognitive architectures, in *5th SPIE's International Symposium on Microtechnologies*, pp. 1–12
5. P. Arena, L. Patané, P.S. Termini, A. Vitanza, R. Strauss, Software/hardware issues in modelling insect brain architecture, in *4th International Conference on Intelligent Robotics and Applications (ICIRA)*
6. S. Battiato, A. Castorina, M. Mancuso, High dynamic range imaging for digital still camera: an overview. J. Electron. Imaging **12/3**, 459–469 (2003)
7. A. el Gamal. High dynamic range image sensors. *Tutorial ISSCC'02*, 2002
8. M. Mronz, B. Poeck, K. Neuser, T. Triphan, R. Strauss, Analysis of a spatial orientation memory in drosophila. Nature **453**, 1244–1247 (2008)
9. P. Arena, S. De Fiore, L. Patané, L. Alba, R. Strauss, Drosophila-inspired visual orientation model on the eye-ris platform: experiments on a roving robot, in *Proceedings of the SPIE Microthecnologies Conference*, Prague
10. A.K. Mishra, Y. Aloimonos, Active segmentation. Int. J. Humanoid Rob. **6/3**, 361–386 (2009)
11. L. Alba, P. Arena, S. De Fiore, L. Patane, R. Strauss, G. Vagliasindi. Implementation of a drosophila-inspired model on the eye-ris platform, in *Proceedings of the IEEE International Conference of CNNA*, Berkley
12. W.T. Freeman, D. Anderson, P. Beardsley, C. Dodge, H. Kagez, K. Kyumaz, Y. Miyakez, M. Roth, K. Tanakaz, C. Weissman, W. Yerazunis, Computer vision for interactive computer graphics. IEEE Comput. Graphics Appl. 42–53 (1998)
13. D. Proffitt, T. Banton, Perceived depth is enhanced with parallax scanning. University of Virginia - Cognitive Science Department, March, Technical report, 1999
14. T. Lindeberg, Edge detection and ridge detection with automatic scale selection. Technical report, Computational Vision and Active Perception Laboratory (CVAP) Department of Numerical Analysis and Computing Science KTH (Royal Institute of Technology), August, 1998

15. R.C. Gonzalez, R.E. Woods, *Digital Image Processing* (Addison-Wesley Publishing Company Inc., Reading, 1992)
16. P. Nordlund, Figure-ground segmentation using multiple cues. Technical report, Computational Vision and Active Perception Laboratory (CVAP) Department of Numerical Analysis and Computing Science KTH (Royal Institute of Technology), May , 1998
17. E. Alegre, R. Alaiz-Rodrguez, J. Barreiro, J. Ruiza, Use of contour signatures and classification methods to optimize the tool life in metal machining. Est. J. Eng. **15–1**, 312 (2009)
18. T. Fober, H. Hullermeier, Similarity measures for protein structures based on fuzzy histogram comparison, in *Proceedings World Congress on Computational Intelligente* (2010)
19. D. Dusha, W. Boles, R. Walker, Fixed-wing attitude estimation using computer vision based horizon detection, in *Proceedings 12th Australian International Aerospace Congress*, pp. 1–19 (2007)
20. B. Zafarifar, H. Weda, P.H.N. de With, Horizon detection based on sky-color and edge features, in *Visual Communications and Image Processing (VCIP)*, vol. 682220, pp. 1–9 (2008)
21. T. Cornall, G. Egan, Calculating attitude from horizon vision, in *Proceedings AIAC-11 Eleventh Australian International Aerospace Congress*, pp. 1–8 (2011)
22. P. Arena, S. De Fiore, L. Patané, M. Pollino, C. Ventura, Insect inspired unsupervised learning for tactic and phobic behavior enhancement in a hybrid robot, in *WCCI 2010 IEEE World Congress on, Computational Intelligence*, pp. 2417–2424
23. P. Arena, L. Patané, M. Pollino, C. Ventura, Tribot: a hybrid robot for cognitive algorithm implementation, in *18th IEEE Workshop on Nonlinear Dynamics of Electronic Systems (NDES 2010)*, pp. 1–6

# Part IV
# Scenarios and Experiments

# Chapter 10
# Experimental Scenarios

**P. Arena and L. Patanè**

**Abstract** This Chapter describes a number of different scenarios and related experiments envisaged according to the various design principles for cognitive architectures. The primary approach taken into consideration was the biologically driven one, so the scenarios have been prepared at the aim to demonstrate how efficiently the insect brain computational model built succeeds in reproducing and enhancing the insect capabilities already addressed in Neurobiology.

## 10.1 Introduction

When a Cognitive architecture is designed and developed, one of the most important aspects to be considered is the definition of evaluation criteria and the creation of scenarios where the system performance can be evaluated.

One of the most relevant attempts in this direction, can be found in the Darpa Project named BICA (Biologically Inspired Cognitive Architectures) that began in 2005 with the intent to create the next generation of Cognitive Architecture models of human artificial intelligence. The second phase of BICA would have involved the actual construction of new intelligent agents that live and behave in a virtual environment; however, this phase was canceled because it was seen as being too ambitious. Among the activities of the BICA, a core set of functions typical of a two-year-old human child was depicted with particular attention to cognitive, perceptual, and motor skills. These include behavioral tasks related to search, navigation, manipula-

P. Arena · L. Patanè (✉)
Department of Electrical, Electronic and Computer Science Engineering,
University of Catania, 95125 Catania, Italy
e-mail: parena@dieei.unict.it

L. Patanè
e-mail: lpatane@dieei.unict.it

P. Arena and L. Patanè (eds.), *Spatial Temporal Patterns for Action-Oriented Perception in Roving Robots II*, Cognitive Systems Monographs 21,
DOI: 10.1007/978-3-319-02362-5_10, © Springer International Publishing Switzerland 2014

tion, memory, language, and three pathways to procedural knowledge: instruction, demonstration, and reinforcement/exploration [1].

Another interesting attempt to define a roadmap for selecting a cognitive architecture for an unmanned vehicle was done in [2]. The idea is to identify criteria for evaluation and common experiments.

From these experiences it is evident how several difficulties can be encountered when testing scenarios for cognitive architectures have to be designed. To facilitate the problem, we considered the formalization of scenarios like an integration process that involves several different elements that take part in the definition process. Due to the complexity of the problem, four different design principles have been considered [3].

**Bottom–Up** A bottom-up process can be followed starting from the information about the basic behaviours of the robotic platforms taken into account. Each structure due to its mechanical peculiarities allows the definition of different basic behaviours that can be used in a scenario to solve a given task (e.g. the hybrid robot Tribot thanks to the frontal manipulator is able to move objects just pushing or carry them).

**Top–down** The top–down approach is based on the formalization of scenarios that should be of interest for the robotic field. The idea is to include real life scenarios where the cognitive architecture can be evaluated considering the impact in solving open problems of interest for people producing an impact in the society (e.g. Rescue robots helping in post-disaster scenarios).

**Biologically-driven** The validation of the insect brain model passes through tests and experiments that can show the similarity between the robot behaviour and the Drosophila. Starting from the experiments reported in literature, the elementary blocks that constitute the insect-inspired cognitive architecture have been evaluated and compared with the biological data. The biologically-driven scenario definition is important to confirm the hypotheses done during the cognitive architecture design phase.

**Robot oriented** Finally the definition of scenarios can be considered following a robot-driven approach. Starting from the capabilities of the insect brain architecture, it is possible to extend the structure to solve tasks that are far beyond the drosophila capabilities. The relevant blocks of the architecture can be improved, in a simple way, expanding the memory available or duplicating elementary blocks. Following this design flow, the complexity of the scenarios and the insect brain computational model can grow up in parallel on the basis of the real fly's abilities to create a more powerful agent. This is in line with the biological evidence that, for example bees brain can be considered as a grown model of the fly brain, where essentially the same areas can be identified (MB, CX, etc). This enables the enhanced capabilities of the bees (eg. labyrinth solving, social behaviour emergence, etc) with respect to the fly.

The prepared scenarios will include situations of increasing complexity that will show the acquisition of cognitive capabilities (starting from the basic behaviors but going beyond these skills) like multisensory integration, contradictory cues resolution, efficient information storage and retrieval. Those capabilities will be tested starting from classical exploration and foraging tasks, going up to tasks involving motivation, based on reward functions.

The neural structures responsible for these behaviors have been very recently enriched with new substructures inspired by very recent discoveries in the insect counterparts. This gave the opportunity to design and simulate new networks that, via processes based on reaction-diffusion dynamics, are able to lead to the emergence of complex behaviors like attention and expectation, which can be considered the basis of a cognitive behavior, i.e. the capability of planning ahead.

On the basis of the experimental data available, the primary approach considered was the biologically driven one, so the scenarios have been prepared accordingly. Following these considerations the various scenarios that were prepared aimed at demonstrating the emergence of the same environmentally mediated behaviours met in insects, and, when possible, also the effect of computationally induced alterations of the same blocks in showing the analogous effect measured in mutants will be demonstrated. At this aim, since all the fly experiments were performed in well prepared arenas, mostly endowed with landmarks consisting in shapes projected using LEDs, a similar setup was prepared in order to allow an easy comparison with the biological data.

The experiments will be tailored to the sensory-motor loop available in each robotic structure. The scenarios envisaged will have to emphasize the model capabilities, using the robots as test beds [4].

## 10.2 Experiments and Scenarios

The proposed experiments were tailored to the testing of different blocks, part of the insect brain, in scenarios directly related to the biological experiments. The question that we want to answer is "*Does our robot behave how the insect do?*" (Biologically-driven scenarios) and "*Is it simple to go further beyond the fly capabilities just extending the basic neural structure models?*" (Robot-oriented scenarios).

To answer to the first question it is important to develop experimental set-up similar to the experiments performed on the flies. Another interesting aspect consists into investigate if the proposed insect brain architecture is able to show the same behaviours as those found in the mutant flies simply modifying either the blocks or the learning mechanisms that are inhibited in the mutant while perfectly work in the wild type flies.

In the following sections an overview of the proposed experiments is given whereas the discussion on the results is reported in the next chapter. The robotic platform used for the experiments are both roving robots and hybrid ones like the Tribot [5].

To evaluate the insect brain architecture several experiments have been identified. Each experiment is oriented to a specific insect capability as reported in Table 10.1.

**Table 10.1** Proposed experiments and related parts of the Insect brain involved

| Experiment | Brain structures involved |
|---|---|
| Visual orientation | PB, FB (CX) |
| Extended detour paradigm | EB (CX) |
| Visual learning | FB (CX) |
| Adaptive termination | MBs |
| Decision making | MBs & CX |

*PB* Protocerebral bridge, *FB* Fan-shaped Body, *EB* Ellipsoid Body, *CX* Central Complex, *MBs* Mushroom Bodies

For each set-up the key aspects have been identified, in fact it is reported a brief description of the biological experiments carried out with the fly, recalling what introduced previously, but here addressing the involved blocks used and the processed sensory information. Moreover the behaviour of mutant flies is also taken into consideration underlying the corresponding modification in the insect brain model, the data for comparison between the real fly and the robot are also well focalized. All these elements have been used to define the robot simulations and experiments and to think about "how to go further" extending the insect capabilities in a simple way, slightly modifying the model.

### *10.2.1 Visual Orientation*

**Description:** Walking fruit flies are attracted by nearby objects. They estimate the distance to these objects by the parallax motion of their images on the retina. The process in flies is not only selective to image motion created by the self-motion of the fly but also sensitive to object motion. Added visual motion (e.g. oscillations) makes objects more attractive than their stationary counterparts. Front-to-back motion, the natural parallax motion on the eyes of a forward-translating fly, is preferred. A group of several more distant objects can be more attractive than one close object. Objects, that are most attractive in the fronto-lateral eye-field, act as deterrent in the rear visual field.

**Involved Blocks:** Visual orientation is completely controlled by the Central Complex and in particular the basic behaviours, orienting towards nearest blobs, are coordinated by the PB and the FB through descending pathways that reach the ventral lobes for motor control.

**Mutants:** When the central part of the PB is ablated, the fly is no longer able to move towards a target. The elimination of specific sectors in the model could be enough to replicate the mutant behaviour.

**Arena:** The testing environments, like for the fly experiments, consist in a squared arena with black stripes placed on the walls in different configurations. The robot navigates in the environment attracted by these black lines that create a high contrast with the white walls.

**Sensory information:** The robot is equipped with distance sensors to avoid collisions with the walls and with the Eye-Ris visual processor where the control architecture is implemented. The camera includes a fish-eye lens useful for mimicking the 320° visual field present in flies.

**Data for comparison:** Statistical data acquired from biological experiments can be compared with data acquired from the robotic campaign. In particular the variables of interest are: robot trajectory, density probability of visits in the arena and gaze distribution.

**Experiment:** The experimental campaign was performed both on a roving robot and on the Tribot where the presence of whegs can replicate more realistic conditions.

**Extension of the model:** The model can be improved considering additional sectors in the PB description.

### *10.2.2 Extended Detour Paradigm*

Short-term orientation memory is an important element used by the fly when targeting an object. If the object disappears the ellipsoid body, a part of the Central Complex, is responsible for tracking the target, creating an afferent virtual copy of the object directly mapped in the Protocerebral Bridge. The presence of this memory capability has been demonstrated using the detour paradigm as discussed in the following.

**Description:** Neuser et al. (2008) showed that Drosophila *melanogaster* possesses a short-term spatial memory during locomotion [6]. Flies can remember the position of an object for several seconds after it has been removed from their environment. The memory enables them to continue the approach. The detour consists into temporarily attract the fly away from the direction towards the chosen target, putting a distractor in the arena for a few seconds while the target is switched off. When the distractor is eliminated, the fly is able to aim for the former target. In most cases wild-type flies turn towards this new target if it is presented for more than 500 ms. Experimental data strongly suggest that flies stored the relative position of the first target in a spatial orientation memory (i.e. EB) for at least 4 s. This flexible goal-driven orientation requires that the target position is stored, especially in case the target moves out of sight. The capability to retain, recall and integrate such positional information into guiding behavior has been summarized under the term of spatial working memory. Neurophysiological studies in primates, which possess a comparable memory, indicate that sustained activity of neurons encodes the sensory information even though the object is no longer present. "Extended" means that the experiment includes also the exploration behaviour. During the exploration phase, the fly is not subject to relevant external stimuli and navigate into the arena satisfying one of its internal drive (i.e. Curiosity). The exploration procedure is based on the increase of the mean free path algorithm: the fly increases gradually the step length, maintaining the turning angle constant.

**Involved Blocks:** Mainly the Central Complex (PB, FB and EB), the memory location is in the EB that through a inner and a outer ring realize a conversion between idiotethic and allothetic reference systems.

**Mutants:** The study of Neuser et al. [6] also shows that the GABAergic ring neurons of the ellipsoid body in the central brain are necessary and their plasticity is sufficient for a functional spatial orientation memory in flies. What if the EB model does not correctly work?

*Case 1:* Ablation of EB: No visual short term memory. The robot can reach only visible objects.

*Case 2:* Defective EB: Altering the time constants or the parameters in the EB model, the mutant-robot will show a defective short-term visual memory.

**Arena:**

- Experiments in an empty arena with two fading objects (black stripes on two adjacent walls).
- Different behaviors depending on the persistence time of the two objects.
- Introduction of obstacle avoidance.

**Sensory information:** Visual flow from a panoramic camera; object identification through a frontal camera, distance and contact sensors for collision avoidance.

**Data for comparison:** Relation between the robot trajectory and the object positions, different behaviours occur depending on the time spent to reach a target and on the time spent from its last detection.

**Experiment:** This experiment is inspired by the Neuser et al. [6] experiment on Detour Paradigm with fruit flies. The robot is placed into an arena, at the beginning of the experiment objects are not visible. The robot should activate the Exploration behaviour. When the first object is presented, the PB and FB model are stimulated and the robot should switch the behaviour to an Object Approach. When the first object fades out, a second object is presented. The robot tries to approach this last object and, while it is moving, it updates the last remembered position of the first one through the EB model. Each step, the EB model will present to the PB model an alternative (low weighted) input, relative to the ghost-object. When also the second object has been faded out, the robot will reach the updated remembered position of the first object depending on the time spent in reaching the former and the latter object.

**Extension of the model:** It is possible to improve the fly capabilities related to visual orientation, in view of robot oriented scenarios. Simple changes in the time constant can guarantees an increase in the memory depth and the replication of the EB block is a simple way to track more than one object concurrently.

### *10.2.3 Visual Learning*

Visual learning is an important aspect of fly life. In the previous scenarios the orientation mechanism guides the insect behaviour and the objects are indistinguishable.

However flies are able to extract visual clues from objects (e.g. color, center of gravity position and others), that can be used to learn to associate a meaning to specific features (i.e. a reward or a punishment). The proposed scenario shows how the fly treats visual inputs and learns through classical and operant conditioning the proper behaviour depending on the object visual clues.

**Description:** In an interesting paper Liu et al. [7] trained stationary flying flies to avoid flying towards specific visual objects appearing on a surrounding white cylinder. If flies approached such an object a heat beam was automatically activated punishing the on-going behaviour. Wild-type flies effectively learn to avoid those objects but this is not the case for the learning mutant rutabaga defective in the cyclic AMP dependent pathway for plasticity.

The dangerous and the harmless objects shown on the cylinder were characterized by distinct visual features: their height over ground, their angular orientation or their size. In a partial rescue strategy for localizing the memory in the fly brain, small areas of the brain were "repaired" by expressing the wild-type form of the rutabaga gene in otherwise rutabaga-defective flies. When exclusively plasticity was restored in the first layer of the fan-shaped body the flies were able to recognize objects distinct in their centers of gravity. Differences in the angular orientation of the objects with respect to the visual field are not relevant. Likewise, if layer five of the fan-shaped body was repaired the flies could remember the orientation of the dangerous objects but not their height. The presumed object features stored in layers 2, 3, 4, and 6 are not yet identified, but wild-type flies, in general, can recognize visual objects also by their color, size or vertical distributedness [8].

**Involved Blocks:** Mainly the Central Complex is involved in the visual process (PB and FB) but also other structures like the Behaviour Selection Network (BSN) that can be considered like a distributed function of the insect brain.

**Arena:** Four objects with common features are placed on four different walls of the arena.

**Sensory information:** Visual flow is acquired from a camera with a frontal view of the scene.

**Data for comparison:** Robot trajectories and object positions, trends of the variables inside the involved neural blocks subject to learning.

**Experiment:** This experiment is inspired by the Liu et al. (2006) experiment on fruit flies [9]. The robot is placed into an arena where some objects are present. These objects are different but they also have common features. We assume to consider dangerous one of the possible features (for example, the green color). Every time the robot tries to approach a green object, it will be punished. Every time the robot detects an object, it switches the behaviour in order to approach the selected object. The FB will work as a classifier increasing or decreasing the weights relative to each feature of the detected objects. In this way the robot learns to avoid dangerous objects and to recognize the dangerous features.

**Mutants:** What if the the Fan-shaped Body is defective? The robot cannot be subject to visual conditioning. The trajectories of the robot will be uniformly distributed near the neutral and dangerous objects that in any case represent attractive points in the arena.

**Extension of the model:** The experiment can include also attractive objects and the classifier can distinguish objects depending on affordances: good for eat, good for drink, and so on.

Moreover other visual features can be used instead of those ones related to the biological case, in order to be consistent with the experiment needs. Of course, additional features will be equivalent to enlarging the number of layers in the fly's FB.

### *10.2.4 Adaptive Termination*

When a behaviour wins the competition and is chosen as the most suitable for the current context, the fly prolongs its duration until either the reference point (e.g. the distance from the chosen target) is reached or another behaviour wins the competition. To avoid continuous flickering from one choice to another a bias is added to the winner creating a kind of hysteretic function. Experiments demonstrate that MB-defective flies have difficulties to give up an ongoing behavioral activity in an adaptive manner. This is mostly due to the absence of an inhibitory effect associated to the MB as discussed below.

**Description:** wild-type flies, unable to fly, that were attracted by an object presented beyond a water-filled moat, stopped walking at the edge of the moat and turned into a new direction. The behavior of MB-defective flies was very different: they kept walking toward the attractive but inaccessible object, sometimes for minutes. However, they were able to turn immediately, when the attractive object was removed. The sensing of the water (with gustatory receptors in the front legs) and the ability to turn were unimpaired. The hypothesis, to explain this behaviour is that the MBs might be involved in the weighing of behavioral options in the light of incoming sensory information. Other examples of this kind of behaviour are present in the fly, for instance, if a small spot of sugar was deposited on the walking surface (not smelling, not visible) wild-type flies mostly stopped and ate when they accidentally encountered the spot. Sugar receptors on the front legs allow flies to sense the sugar. In the same situation MB defective flies overrun the sugar. According to the above hypothesis the ongoing behavior "walking" had a high weight and the highly adaptive alternative option "eat sugar" could not overwrite the walking. Other examples in which the MB-defective flies went on with their previously chosen behaviour has been identified. One example is given by gap crossing behavior. Once the decision to climb is taken by the fly at a gap of surmountable width and the fly has thus started climbing, the gap has been broadened in the aftermath. Wild-type flies terminated the ongoing approach significantly earlier than the MB defective flies.

**Involved Blocks:** The relevant blocks involved in the adaptive termination process are a sub-part of the MB named MB(2), the BSN and the CX for visual targeting.

**Arena:** In the arena one object is introduced and considered as target while other objects limit the robot movement toward the aimed target. Sometimes the obstacles makes the target unreachable by the robot.

**Sensory information:** Visual input, contact and proximity sensors, low level target sensors (i.e. color sensors used to detect strips on the ground used as sugar traces).

**Data for comparison:** Robot trajectories, time spent in trying to reach the unreachable object, object-to-robot distance.

**Experiment:** This experiment is not referred to a precise experiment with fruit flies, but to a category of experiments that lead to the MBs for the implementation of the adaptive termination and the evaluation of the ongoing behaviour in the insect brain. The robot is placed into an arena where some attractive objects are also present. Some obstacles can make unreachable the objects. When an object has been detected, the robot should switch the behaviour from Exploration to Object Approach. The MB(2) model should periodically and progressively inhibit the ongoing behaviour. In this way, Exploration is the behaviour useful to escape from local minima.

**Mutants:** What if MB(2) is defective? The robot will be "trapped" in the ongoing behaviour. However, to make the experiment more realistic, a spontaneous decay could be also introduced in the BSN, in order to allow the robot to escape from local minima like in the real fly but after a long time. In this case, the time spent by the robot following an unreachable object is a key variable used to evaluate the system performance.

**Extension of the model:** The model can be extended thinking to the possibility to have adaptive weights in the BSN. In this way the behavioural hierarchies can be adapted in time depending on the past experiences.

### *10.2.5 Decision Making*

A decision making process has been identified in the flies that have been trained to avoid objects with specific visual features. In presence of a conflict the fly has to decide which features are the most relevant to make a choice. The decision making strategies is guided by a pre-wired hierarchical categorization of the features that, for instance, leads the fly to give more importance to color with respect to shape.

**Description:** Decision making is an interesting aspect of the insect brain. The number of experiments related to this process are limited due to the complexity of the set-up that needs to exclude a multitude of possible concurrent variables to perform a correct analysis of the results. However an interesting experiment was performed by Tang and Guo (2001) [9], they trained stationary flying flies, using a heat beam, to avoid a visual pattern in a choice situation where two times two pairs of objects are presented. One pair of objects was green and had an upright-T shape and the other pair was blue and had an inverted T-shape. In the test situation the colors were switched between objects. This measure created a conflicting situation for the flies. Is the predictor of heat the form or the color of the previously punished object? The wild-type flies invariably took their decisions to avoid the previously punished objects based on the color of the objects. Next, flies were trained again on objects with saturated colors but during the test the colors were not only switched but also faded out to variable degrees. Below a certain saturation point of the colors the wild-type flies suddenly and consistently select the shape instead of the color of the objects to be avoided. There was always a clear decision in all flies of the population. MB-less flies had no clear decision point at which they would switch from color to shape. Rather their population average started with a purely color dependent decision at full saturation and then gradually shift towards shape as the color fade. Therefore MBs are involved in decision making processes.

**Involved Blocks:** The part of the insect brain mainly involved in this kind of experiments are the PB and FB as part of the CX for the visual recognition and classification and part of the MBs named MB(3) as concern the decision process.

**Arena:** The robot is free to move in the arena where two objects different in shape and color are fixed to the walls. After learning the color of the two objects is switched and got fainter.

**Experiment:** This experiment is inspired by the Liu and Heisenberg (1999) [10], experiment on fruit flies. The robot is placed into an arena in which two different objects are present in two copies placed in opposite sides of the arena. The robot is punished when it tries to approach one type of objects. In a second phase one of the characteristic is exchanged (e.g. the color) between the objects. I n this case a dilemma arise and the system have to choose which is the most dangerous object. A simulation campaign was performed at different levels of color saturation.

**Sensory information:** Relevant sensory information are acquired from a frontal camera, while the punishment signal consists of either a sound pattern acquired though microphones or colored patches on the ground acquired with low level color sensors.

**Data for comparison:** Robot trajectory and object position, color fading parameters and robot gaze direction as a function of the color fading degree.

**Mutants:** What if the MBs is ablated? The decision making process is affected and a clear choice is not evident also in the case of saturated colors, in this case statistical analysis of the data shows a different distribution of the choice.

**Extension of the model:** The proposed model can be easily extend including more complex objects with several features involved in the recognition and classification process. Moreover the decision process can be guided considering multisensory modalities concurrently involved in the process.

## 10.3 Conclusions

A series of experimental setups useful to evaluate the capabilities of the proposed insect brain model are illustrated. Particular attention was devoted to the comparison with the real fly experiments and the matching between the mutant defective abilities and the robot when specific parts of the brain-like structure are compromised. For each experiment details about the arena, the sensory information needed, the mutant behaviour and the potential extensions of the insect capabilities are provided. The obtained results are described in the Chap. 11.

# References


1. B.S. Minnery, S.T. Mueller, M. Jones, The bica cognitive decathlon: a test suite for biologically-inspired cognitive agents, in *Internal Report* (2007)
2. L.N. Long, S.D. Hanford, Evaluating cognitive architectures for intelligent and autonomous unmanned vehicles, in *AAAI Workshop* (2007)
3. P. Arena, L. Patané, P.S. Termini, A. Vitanza, and R. Strauss, Software/hardware issues in modelling insect brain architecture, in *4th International Conference on Intelligent Robotics and Applications (ICIRA)* (2011)
4. P. Arena, M. Cosentino, L. Patané, A. Vitanza. Sparkrs4cs: a software/hardware framework for cognitive architectures, in *5th SPIE's International Symposium on Microtechnologies*, pp. 1–12 (2011)
5. P. Arena, L. Patané, M. Pollino, C. Ventura. Tribot: a hybrid robot for cognitive algorithm implementation. in *18th IEEE Workshop on Nonlinear Dynamics of Electronic Systems (NDES 2010)*, pp. 1–6 (2010)
6. K. Neuser, T. Triphan, M. Mronz, B. Poeck, R. Strauss, Analysis of a spatial orientation memory in drosophila. Nature **453**, 1244–1247 (2008)
7. G. Liu, H. Seiler, A. Wen, T. Zars, K. Ito, R. Wolf, M. Heisenberg, L. Liu, Distinct memory traces for two visual features in the drosophila brain. Nature **439**, 551–556 (2006)
8. R. Ernst, M. Heisenberg, The memory template in drosophila pattern vision at the flight simulator. Vision Res. **39**, 3920–3933 (1999)
9. S. Tang, A. Guo, Choice behavior of drosophila facing contradictory visual cues. Science **294**, 1543–1547 (1999)
10. L. Liu, R. Wolf, R. Ernst and M. Heisenberg, Context generalization in Drosophila visual learning requires the mushroom bodies, Nature **400**, 753–756 (1999)

# Chapter 11
# Robotic Experiments and Comparisons

**P. Arena, S. De Fiore, L. Patanè, P. S. Termini and A. Vitanza**

**Abstract** This Chapter will present the final results obtained in testing the insect brain computational model. A series of experiments have been assessed in order to evaluate the capabilities of the proposed control architecture, in comparison with the behaviour shown by real flies while performing the same tasks.

## 11.1 Introduction

The development of an insect brain model is a complex task that can be handled and then evaluated considering several different prospectives. An important issue related to the final aim of the proposed research activity consists into trying to unraveling the neural mechanisms and assemblies that are involved in insect cognitive behaviors. The research was principally oriented towards the fly *Drosophila melanogaster* and the neural blocks identified and modeled, even if common to most of the insect species, are tuned in the fly direction. Recently a virtual fly brain project has been activated to create neural structures trying to mapping biologically plausible network architecture. Other results on the insect brain reconstruction focused on brain-wide wiring networks are reported in [1]. Our objectives, that are partially shared with [1], are however mainly directed toward the formalization of an architecture, looking at the neural assemblies rather at a functional level, trying to formalize the fly brain from the dynamical system prospective considering the reaction-diffusion paradigm.

P. Arena · S. De Fiore · L. Patanè (✉) · P. S. Termini · A. Vitanza
Department of Electrical, Electronic and Computer Science Engineering,
University of Catania, 95125 Catania, Italy
e-mail: parena@diees.unict.it

L. Patanè
e-mail: lpatane@dieei.unict.it

P. Arena and L. Patanè (eds.), *Spatial Temporal Patterns for Action-Oriented Perception in Roving Robots II*, Cognitive Systems Monographs 21,
DOI: 10.1007/978-3-319-02362-5_11, 

For these reasons the direction followed in the architecture evaluation process, consisted into comparing the results obtained by neurobiologists in fly experiments with the robot behaviour when performing the same tasks [2, 3].

Biological experiments that involves flies are mainly based on behavioural analysis, this approach motivates the use of statistical analysis of the insect behaviours in order to obtain the relevant key parameters filtering out noise and the fact that only few variables are considered in each experiment discarding the other intrinsic internal variables. These results have been replicated on the robot obtaining a good match between fly and robot behaviours. Moreover another interesting aspect useful for the model evaluation consists in the analysis of the mutant fly behaviour. Several experiments concerning flies can be replicated using, instead of Wild type exemplars, genetically modified lines where either specific blocks of the brain or basic learning mechanisms are altered (e.g ablating neural assembly in the brain or using gene expressed toxins that can modify specific learning cascades making the fly deficient in specific behaviours).

Experimental results obtained with mutant flies can be obtained also by using the insect brain model following the same alteration that neurobiologists perform on real insects. This is an added value that can be used to further justify the feasibility of the developed structure.

## 11.2 Simulations and Robot Experiments

The proposed insect brain model structure has been developed on the basis of the neurobiological data acquired through specific experiments together with the information already available in literature. When the process was not supported by biological evidences for Drosophila, other insect species have been taken into account (e.g. locust and bee). Finally, when needed, working hypotheses were used to finalize parts of the block model, trying to obtain from the robotic experiments results to be used for design new fly experiments for confirmation.

A block diagram of the final implementation of the insect brain architecture is reported in Fig. 11.1. The discussion about the interactions among the implemented blocks is given in the following sections in relation to the specific experiments taken into consideration to compare the real fly with its robotic counterpart.

The experiments were performed both in simulation and on real robots. The developed SW/HW framework, described in details in Chap. 6, includes a 2D simulation environment useful for roving robots (see Fig. 11.2a) and a 3D dynamic simulator based on ODE, where legged and hybrid robots can interact with objects in a virtual arena (see Fig. 11.2b). Different robotic platforms used during the experiments are: the roving robot equipped with an on-board PC (Fig. 11.2c), two different prototypes of the hybrid robot Tribot (see Fig. 11.2d) [4, 5] and a six-legged robot called Minihex (see Fig. 11.2e) [6].

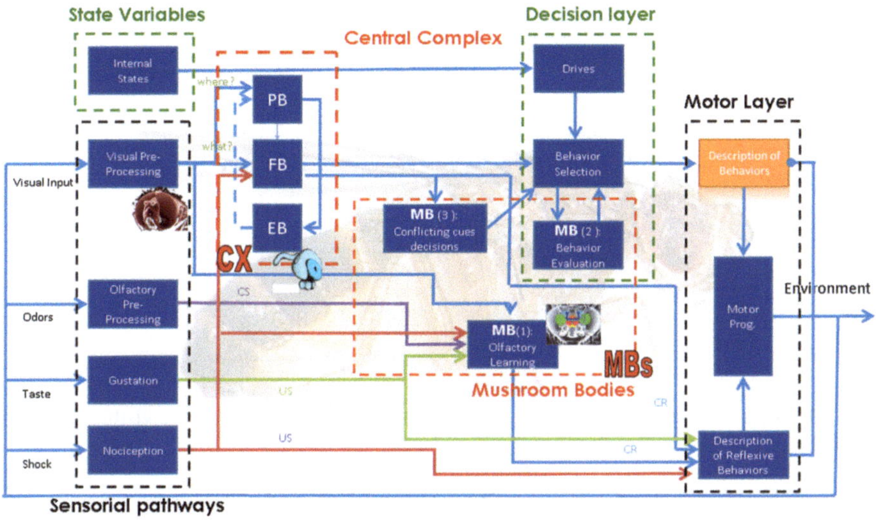


**Fig. 11.1** Block diagram of the insect brain model

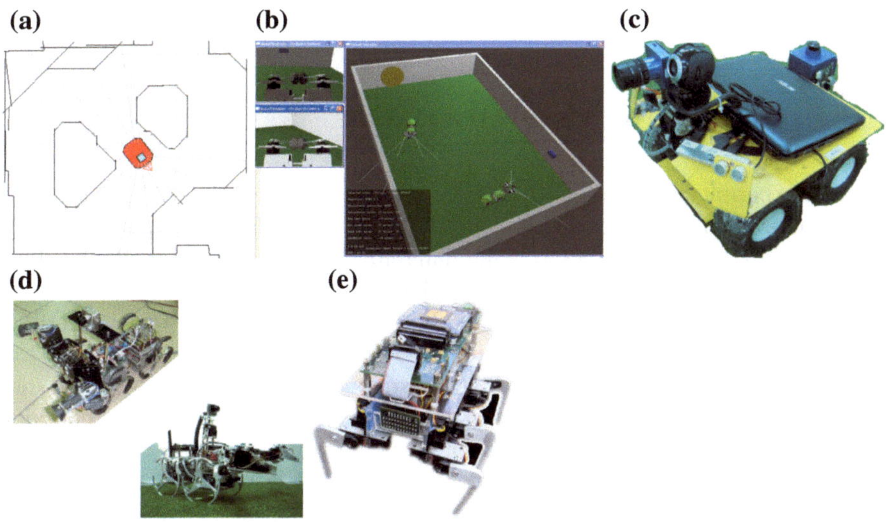


**Fig. 11.2** The experiments have been performed both in simulation, using a 2D simulation environment **a** and a 3D dynamical simulator **b** and by using robotic prototypes: **c** a rover, **d** hybrid robots (Tribot I and II) **e** and a hexapod (MiniHex)

The framework developed to test the insect brain [7, 8], includes a graphical interface that allows to identify the different active blocks while the robot is performing the experiment as shown in Fig. 11.3.

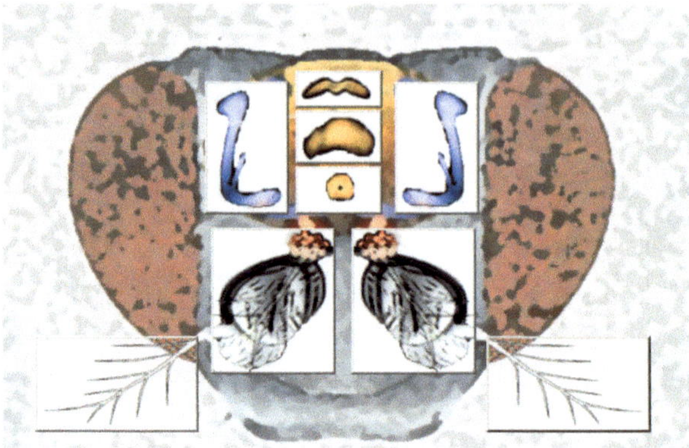

**Fig. 11.3** Graphical interface where the active elements of the insect brain are visible during the experiments

The following sections describe the effectiveness of the proposed structures and the capabilities of the insect brain-inspired system in comparison with the real fly experiments.

## 11.3 Experiments on Visual Orientation

The first experiment carried out to validate the insect brain architecture, consists in the visual orientation task as already introduced in the previous chapter.

Figure 11.4 shows the block scheme diagram of part of the CX including functionalities related to the PB and FB involved in the visual orientation task. For visual orientation the validation was performed both in simulation by using the 2D Pioneer simulator and the 3D dynamic simulation environment and by means of real prototypes, in particular the Pioneer P3AT, a small rover and the Tribot were tested.

In Fig. 11.5 the front-end of the dynamic simulation environment is reported; it is interesting to notice that the presence of elementary motion detectors. The system is sensible to motion and the sector integrators through a winner takes all strategy allows the identification of the direction where the most appealing object (e.g. nearest one) is present.

The robotic experiments were firstly performed using a roving robot equipped with distance sensors for collision avoidance and with the Eye-Ris 1.3 visual system where the whole visual orientation structure has been implemented.

The implementation on a smart visual processor guarantees high speed performances: using the Eye-Ris a controlling cycle needs 129 ms for the sensor acquisition, 32.5 ms for the motor commands, 26.8 ms for the signal processing and 16,6 ms for communication routines. Each control step is then performed in about 200 ms.

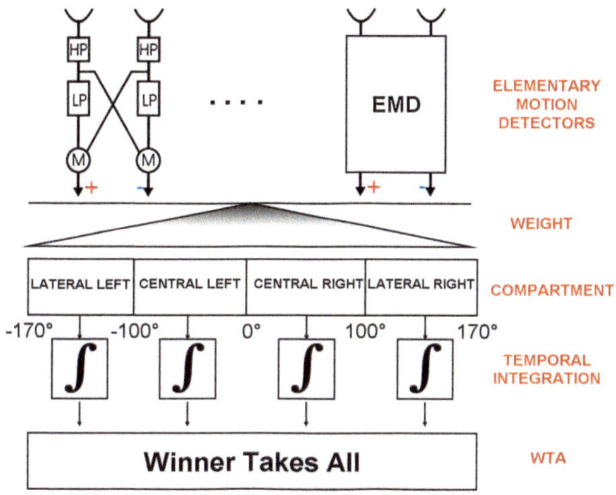


**Fig. 11.4** Minimal model of the orientation behavior towards visual objects (*HP* high pass filter, *LP* low pass filter, *M* multiplication stage)

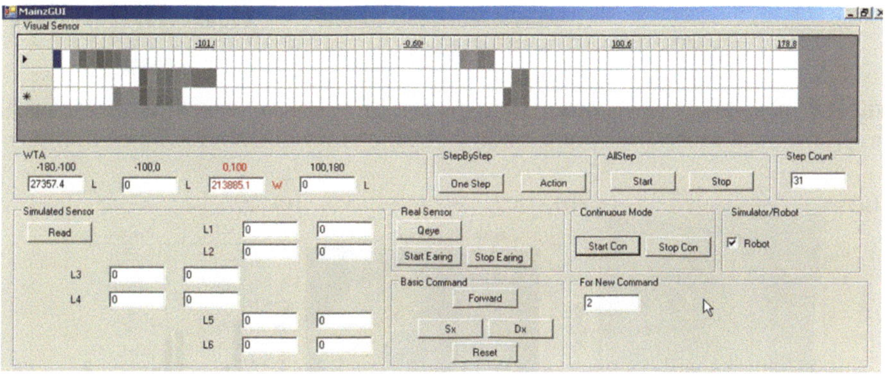


**Fig. 11.5** Simulation interface, the three rows indicates the time evolution of the information collected by the 72 sensible points distributed in a ring that covers 320° of the visual field. When a sector wins the competition the robot heading changes accordingly and the previous updated memory locations are reset to start a new competition

To make a comparison with fly experiments, the fly behaviour has been considered and relevant data, as reported in [9], have been taken into consideration for comparison.

To show the matching between the real fly and a robot controlled through the insect brain architecture, the same experiment has been reproduced. The results are reported in Fig. 11.6. The squared arena is $2 \times 2\ m^2$ and ten different runs have been recorded for 5 minutes each.

Depending on the configuration of the attractive landmarks in the arena, the behaviour of the fly changes and the robot behaviour changes accordingly as shown in

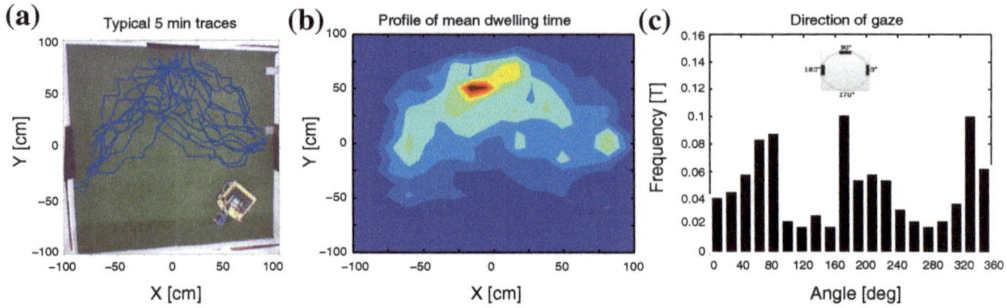


**Fig. 11.6** Behaviour of a roving robot involved in a visual orientation task with 3 landmarks. **a** robot trajectory, **b** density probability of visits in the arena, **c** gaze distribution

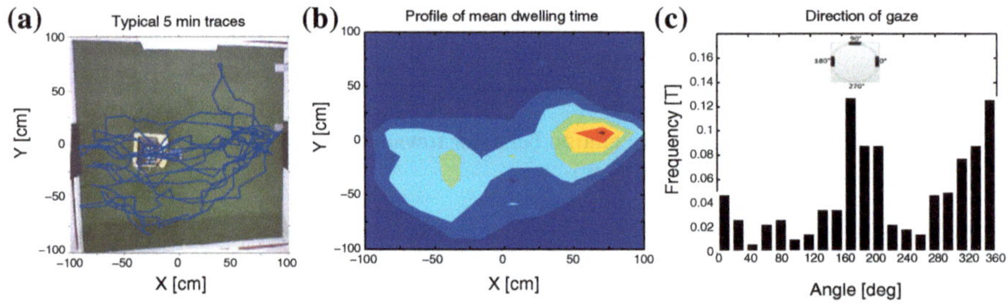


**Fig. 11.7** Behaviour of a roving robot involved in a visual orientation task with 2 landmarks located in opposite walls. **a** robot trajectory, **b** density probability of visits in the arena, **c** gaze distribution

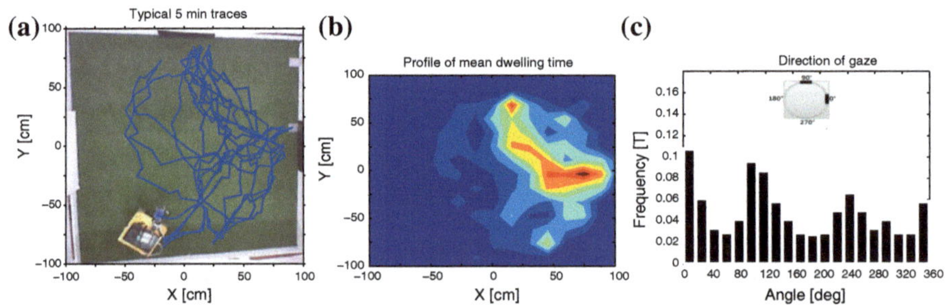


**Fig. 11.8** Behaviour of a roving robot involved in a visual orientation task with 2 landmarks located on two adjacent walls. **a** robot trajectory, **b** density probability of visits in the arena, **c** gaze distribution

Fig. 11.7 where two objects are placed on the opposite walls and in Fig. 11.8 where the objects are located in two adjacent walls.

Finally the behaviour with a single landmark was also investigated and the results are reported in Fig. 11.9.

Details about the orientation model and the experiments are reported in [10, 11].

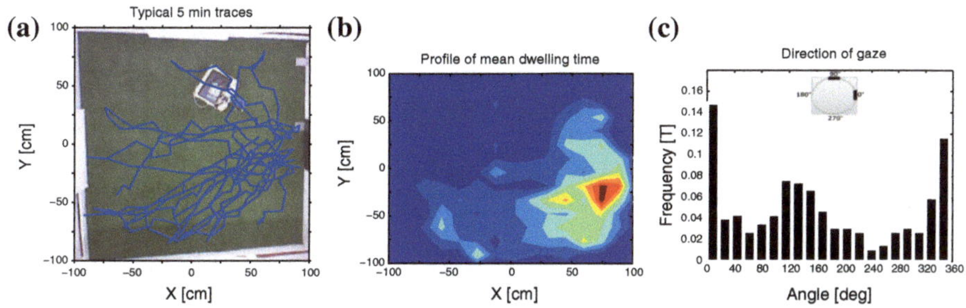


**Fig. 11.9** Behaviour of a roving robot involved in a visual orientation task with only 1 landmark. **a** robot trajectory, **b** density probability of visits in the arena, **c** gaze distribution

## 11.4 Extended Detour Experiment

In order to navigate with path integration the animal needs a way to determine its orientation relative to a fixed direction and some means to compute distances. Using this information it can compute an estimation of its current position which is equivalent to vectorially adding all distances traveled. The "Extended Detour Paradigm" is an extension of the visual orientation experiment. In this experiment the visual flow processed through the CX is elaborated also by the Ellipsoid Body where information about the position of interesting objects can be memorized behaving like a short term spatial memory. The robot, that initially performs an exploration strategy in an open arena, is suddenly attracted by the appearance of a pair of strips in two opposite walls of the arena. In our paradigm, the stripes disappear when the robot crosses the invisible midline of the circular walking platform, and a new target appear laterally at a 90° angle to the robot.

Exploration is the behaviour that leads flies to find new sources. In the model the exploration is thought as a default-behavior: if the fly has not strong needs, it explores the environment and searches objects or places that could be useful in the future. The searching strategy of the fly Drosophila melanogaster has been investigated by [12]. Following this analysis, it has been decided to implement the exploration using the "increase of the mean free path" algorithm: the fly increases gradually the step length, maintaining the turning angle constant. However this algorithm is only an approximation of the global trajectory of the fly.

We can distinguish two kinds of flies [13]:

Rover flies: Long steps lengths, wide turning angle.
Sitting flies: Short steps lengths, tight turning angle.

Rover flies are statistically about the 70 % of the population; this homogenous distribution is respected all around the world and resembles mechanisms of individual specialization that are common in social insects like bees and ants. Justifications of this characteristic in flies are under studies.

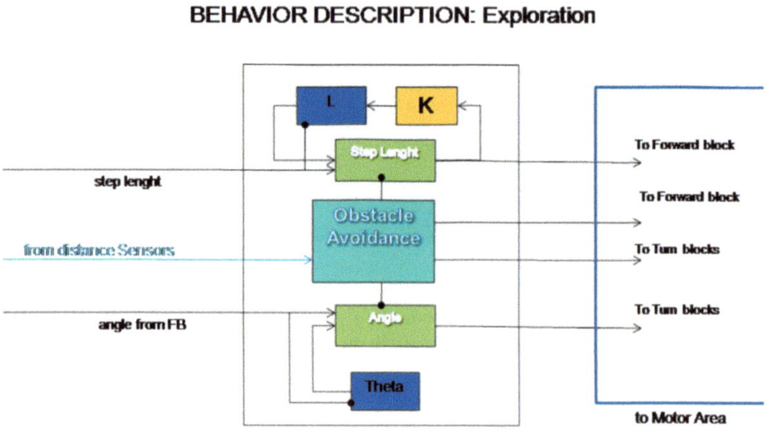


**Fig. 11.10** Exploration behaviour. The robot, during exploration, proceeds with a turning action using the theta angle parameter and a forward movement proportional to L with a gain k that increase step by step. The value for the parameters depends on the fact that we are considering a sitting or a roving fly. In presence of an object the PB control the step length and the turning angle and the obstacle avoidance mechanism, to avoid dangerous situations can inhibit the other blocks taking the lead of the robot behaviour

In the architecture the implementation of exploration needs to be integrated with an obstacle avoidance system, in order to avoid undesired collisions. In this case, if an obstacle is too close, the robot will suspend the realization of the increase of mean free path algorithm and send to the motor area the right behavior in order to avoid that obstacle. This is considered as a reflex induced reaction and a block diagram useful to describe the exploration strategy together with the role of the obstacle avoidance block is shown in Fig. 11.10.

The ellipsoid body in flies has an indispensable role in visual short term memory and orientation. The EB is the main actor of the detour experiment. Neurobiological experiments demonstrate the role of EB in solving problems of spatial orientation including memory capabilities. In the proposed insect brain model the EB has been considered as a neural structure linked to the PB and FB. The EB works as a spatial memory, the spatial position of interesting landmarks can be tracked and when the external stimulus is lost (e.g. because obscured by an obstacle), the EB creates a virtual copy projecting it in the corresponding position of the PB. In this way the architecture continues to work performing a visual orientation task following a 'ghost' landmark created by the EB. The landmark is chosen by the EB depending on the time spent in trying to reach it. The EB works as a fading memory so if a distractor is presented for more than a few seconds, the information stored by the EB are no more reliable.

In the following experiments, the EB has been modeled with the Wittman and Schwegler neural integration system [14], based on the sinusoidal array [15], a peculiar representation of spatial vectors with neurons. Before we can elaborate on how the computations can be performed by a neural network we have to decide which model we will use to describe the neurons. Various models of biological neurons have been suggested with various levels of abstraction. One of the most popular models is the average firing rate model [16]. It abstracts from the details of spike generation and describes the average number of spikes generated by a neuron in a fixed time interval.

In the model the compass information is encoded by a peak of activity on a dedicated group of neurons. The position of the peak corresponds to the compass direction. The neurons are closed in a ring to avoid a discontinuity in the encoding of angles. This representation must be transformed into a sinusoidal array representation. In order to use a neural network as a short-term memory we need some kind of feedback connections $w_{mem}$.

The complete model can be described by the following equation:

$$sa(i,t+1) = f[\sum_{j=0}^{N-1}(w_{i,j}^{mem} sa(j,t) + w_{(i,j)}^{input} input(j,t) + b_0)] \tag{11.1}$$

$$w_{i,j}^{mem} = 2cos[2\pi(\frac{i}{N} - \frac{j}{N})] \tag{11.2}$$

$$w_{i,j}^{input} = -k_0(v)cos[2\pi(\frac{i}{N^{input}} - \frac{j}{N})] \tag{11.3}$$

$$k_0(v) = \frac{v}{v_0} \tag{11.4}$$

$$f(x) = \begin{cases} 0 \ if & x < 0 \\ x \ \ if & 0 < x < 2b_0 \\ 2b_0 \ if & 2b_0 < x \end{cases} \tag{11.5}$$

where $sa$ is the neuron activity of the sinusoidal array, $w_{mem}$ are the connections between neurons of the sinusoidal array, $w_{input}$ the connection between the compass and the sinusoidal array and $k_0$ depends on the velocity of the robot.

The assessment of the model has been initially evaluated in the dynamical simulation environment using the Tribot robot. The robot behaviour is shown in Fig. 11.11 where a series of trials were reported. The robot initially attracted by the landmark on the right wall (i.e the first part of the trajectory until the arena midline is reached) is then distracted by a new landmark that appears in the upper wall after the disappearing of the first one. After a few steps the distractor is eliminated and the robot, using the copy projected by the EB to the PB, is able to change its heading trying to reach the no longer visible first landmark. The EB works as a fading memory and therefore the choice depends on the energy/time spent in following the different landmarks.

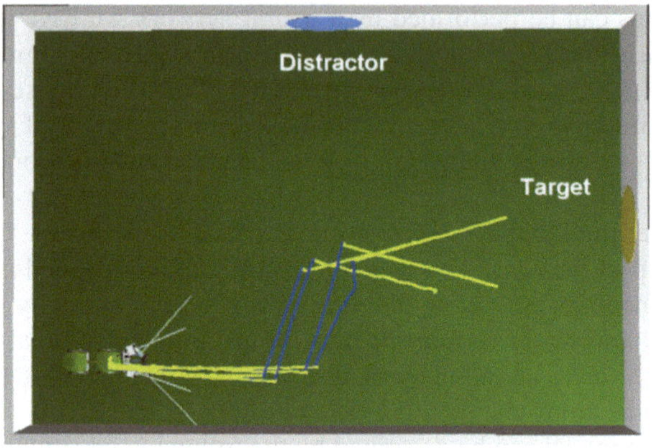


**Fig. 11.11** Behaviour of the simulated Tribot robot during the detour experiment. The trajectories followed by the robot during different trials are reported. It can be notice that during the first and second part of the trajectory the robot is attracted by a visible landmark, whereas the last part is guided by the EB that creates a virtual copy of the landmark most reliable (e.g. depending on the time spent in trying to reach it)

The same set-up has been also replicated in the robot arena and a series of experiments were performed with the Tribot robot. A typical trajectory obtained during the experiment is reported in Fig. 11.12 where the different phases of the experiment are emphasized. During the first part of the experiment the acquired visual data on the target position are used to update the EB status and then the system is able to estimate the position of the missing target for a few steps, using the EB fading memory.

The approaching behaviour toward either a visible or an estimated target, is elicited in the robot by the drives that guide the system (see for more detail Sect. 11.6). The introduction of the Behaviour Selection Network (BSN) allows to handle the results of the neurobiological experiments that statistically indicate the number of times the fly follows the virtual target instead of switching to an exploration phase. Figure 11.13 shows an example of a trajectory obtained during the detour experiment performed with the Rover in which the EB-driven targeting behaviour is stopped by the BSN eliciting the exploration behaviour.

This takes place at step 10 (Fig. 11.13) and is due to the presence of noise which is considered as an additional input to the BSN. This effect makes robot experiments more realistic and next to the biological counterpart, in which the behavior emerges statistically. The time evolution of the BSN output neurons is shown in Fig. 11.14 where the competition between the two behaviours is evident.

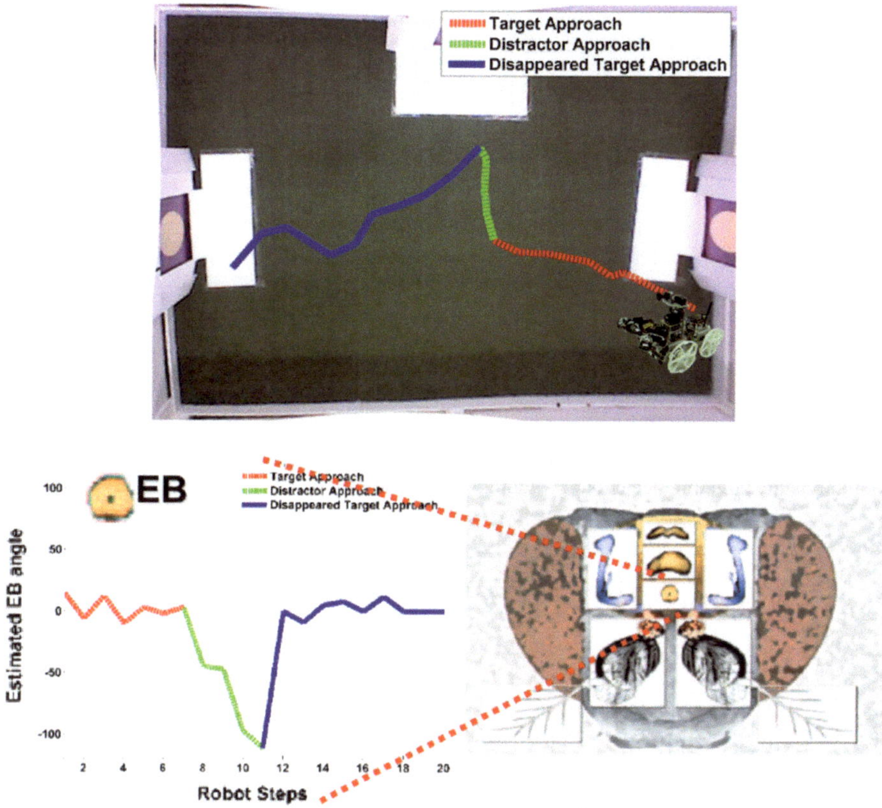


**Fig. 11.12** Trajectory acquired for the detour experiments carried out on the Tribot robot. The trajectory is divided into three parts that correspond to: a first approaching phase towards the visible target, a short time window while a distractor attracts the robot attention and a final part where the robot through the information coming from the EB is able to retrieve the position of the no longer visible target continuing the interrupted approaching behaviour. In the bottom part the role of the EB in tracking the target position is shown, this information is transferred to the PB creating a virtual copy of the missing target to recover the approaching behaviour

## 11.5 Visual Learning

The measuring of performance in animal cognitive tasks has been usually focused on accuracy as a fundamental parameter, independently of decision time. However, in more recent researches, it has become clear that tradeoffs exist between decision speed and accuracy in many relevant tasks, for instance, prey and predator detection and identification and spatial exploration strategies [17].

Insects, as previously discussed, are able to navigate in environments through visual orientation strategies that are useful for moving towards the nearest objects in a complex scene. This basic process represents a first decision level that in a parallel

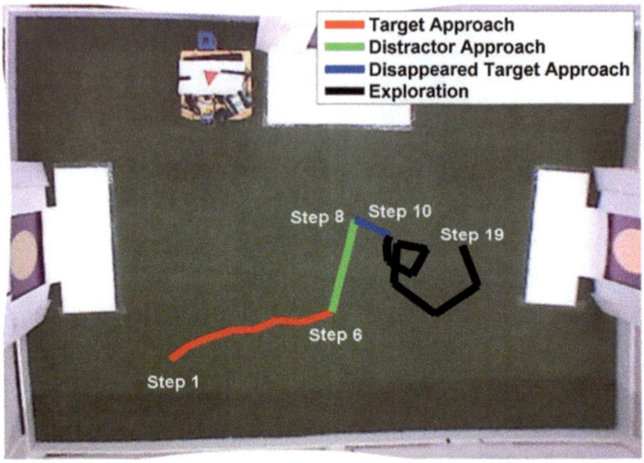


**Fig. 11.13** Robot trajectory during the detour experiment with the introduction of the BSN that can inhibits the targeting behaviour switching to exploration

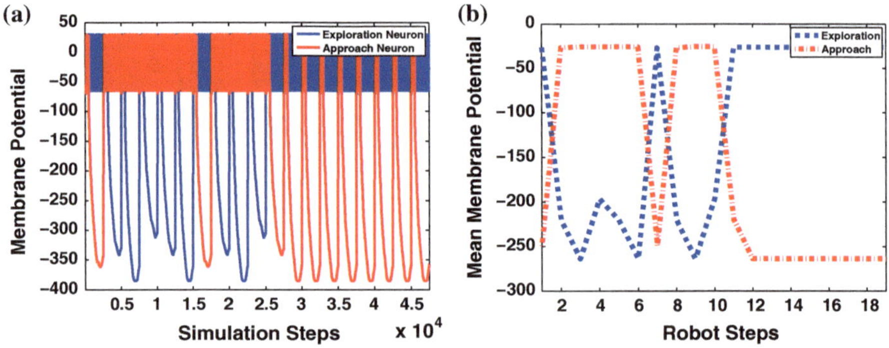


**Fig. 11.14** Evolution of the BSN output neurons **a** and mean membrane potential **b**. The approaching behaviour towards the virtually projected visual target form the EB is finally inhibited and the exploration takes the lead. During a robot step the BSN network evolves for 2,500 simulation steps

way can elaborate the visual scene (i.e. high decision speed) to determine the insect actions. On the top of this control mechanism a quite accurate decision process is present. When the targeted object is placed in front of the fly, the visual features are extracted through the FB and compared with the stored information about the same or similar objects. Experiments carried out on visual learning show the fly capability to associate either attractive or repulsive behaviours to specific objects or even single features if a training stage is performed using a reward or punishment-based procedure.

The proposed model inside the insect brain to demonstrate these cognitive processes, is mainly related to the CX block. In particular the previously mentioned sub-blocks, the PB, FB and EB, are able to support this new process. The objects placed just in front of the robot are segmented and all the relevant features

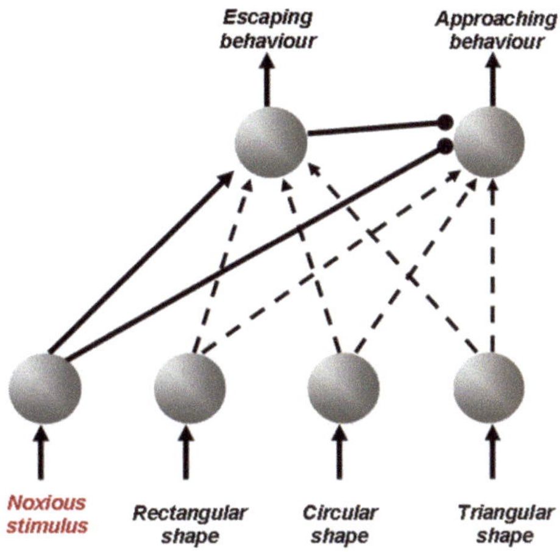


**Fig. 11.15** Elementary block of the FB structure. Unconditioned stimuli (e.g. an electric shock) guide the learning of a spiking network that associates a meaning to visual features extracted in the FB. The example refers to shapes, solid lines represent fixed synapses whereas dashed lines indicate plastic connections subject to learning through STDP

are extracted. The visual preprocessing phase, that mainly involves the optic lobe, has been performed both using standard libraries developed inside the framework and by means of the Eye-Ris visual system as discussed in the related document. Concerning the Fan-shaped body, the relay station role performed during the visual orientation, is now integrated and augmented with storing capability together with learning mechanisms. The developed network is a spiking structure as shown in Fig. 11.15 where unconditioned stimuli (i.e. reward and punishment) are used to elicit an STDP learning [18, 19].

The first layer of the network receives in input the visual pre-processed signals and is connected to a second layer with two neurons associated to the two main behaviours that can be elicited at this level: escaping and approaching. In particular, analyzing the synaptic connections, we can underly that: the shock neuron is connected through an excitatory synapsis to the escaping neuron and inhibits the approaching neuron. This is justified because in presence of a dangerous condition the robot escapes without taking into account the other sensory signals. The depicted connections are fixed (not subject to learning) because represent the basic level of apriori knowledge present in the system for safety reasons.

The other sensing neurons that receive information about the visual features of the object, are connected to both the second layer neurons through excitatory synapses subject to learning. The initial values have a bias toward the approaching behaviour giving to the robot an initial preference (i.e. curiosity) to be attracted from the objects in the scene. However the escaping behaviour, when elicited, is able to inhibit the object approaching. Similar structures can be replicated for different features in a modular way. Each subnetwork is devoted to handle specific visual features that concur to the behaviour selection process through a third layer as shown in Fig. 11.16. The third layer transfers the information processed trough the CX to the BSN where the merging with the other sensory modalities occurs.

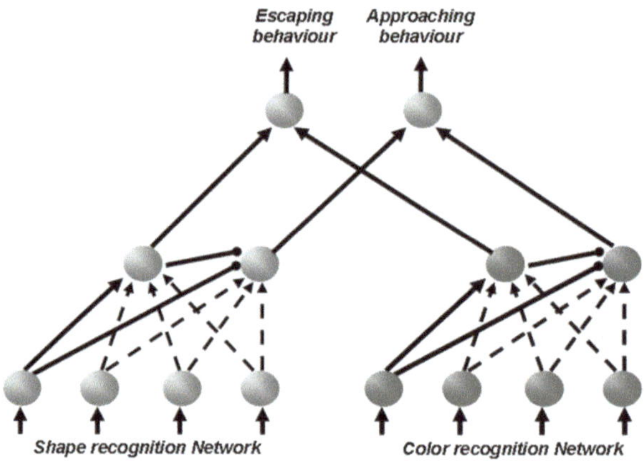


**Fig. 11.16** Multiple elementary blocks of the FB structure can be connected each other through a second layer where visual features are combined

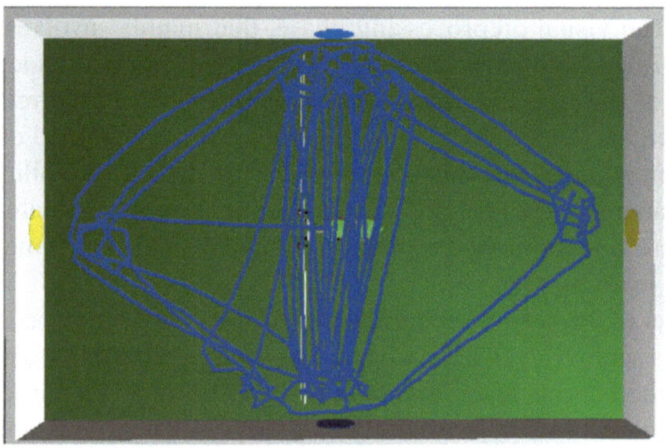

**Fig. 11.17** Visual learning experiment. The robot trajectories show the emerging behaviour when the robot is rewarded for reaching the targets placed on the center of the walls. The patrolling between the two targets on the farther walls does not emerge due to the preference in reaching the nearest target among the visible ones

In Fig. 11.17 the robot behaviour when four attractive objects are placed in the arena is shown. In this simulation the robot is able to extract the object features for learning. In this simulation the final behaviour depends on the fact that the robot has been rewarded when reaching all the targets.

In contrast, if during the simulation a pair of targets is considered dangerous the final behaviour is depicted in Fig. 11.18 where the robot safely reach the left and right targets while after a series of punishments in the downside one, learn to avoid both the down and upper side targets that have the same visual features.

In the following preliminary tests on the robot are reported which are currently under evaluation for further optimization.

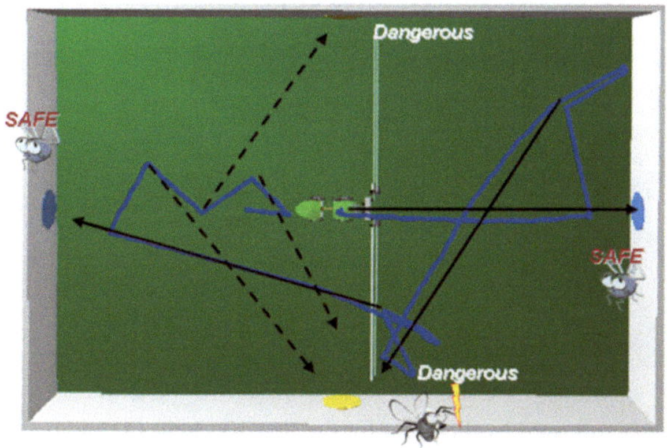


**Fig. 11.18** Visual learning experiment. The robot safely reaches the left and right targets (see the *solid arrows*) whereas is punished when the downside object is reached. The learning process allows the robot to avoid the objects and visual features associated to dangerous events, in this case the down and upper side ones (see *dashed arrows*)

The same set-up has been replicated in the arena for the robot experiments. Two yellow circles have been projected in the screens in the left and right side walls whereas a blue circle is shown in the upside of the arena. The Rover, exploring the arena, is attracted by these targets and is punished when the blue circle is reached. the trajectory obtained during the experiment are reported in Fig. 11.19. The robot is initially attracted by all the landmarks (Fig. 11.19a) and then avoid the blue circle (upper wall) due to the punishment signals (Fig. 11.19b). The changes in the robot behaviour are also evident in Fig. 11.20 where the gaze direction for the first and second part of the experiment is reported.

The level of punishment associated to each feature is shown in Fig. 11.21 where the escaping value for each object is also shown. The presence of a forgetting factor gives the possibility to modify the system behaviour in time: in the reported results the robot after several steps spent avoiding the blue circle, tries (at step 240 in Fig. 11.21) to approach the object previously considered dangerous but also in this case it is punished again and the associated punishment value is increased again. The basic spiking network can be extended in a modular way to include also other sensory modalities [20].

## 11.6 Adaptive Termination of Behaviours

Besides the CX and MBs blocks, to model the distributed functionalities that lead to the emergence of a behaviour in flies, a block called Behaviour Selection Network (BSN) has been considered. As also reported in the last progress report, it is till nowadays not clearly known where this structures is located in the insect brain; it

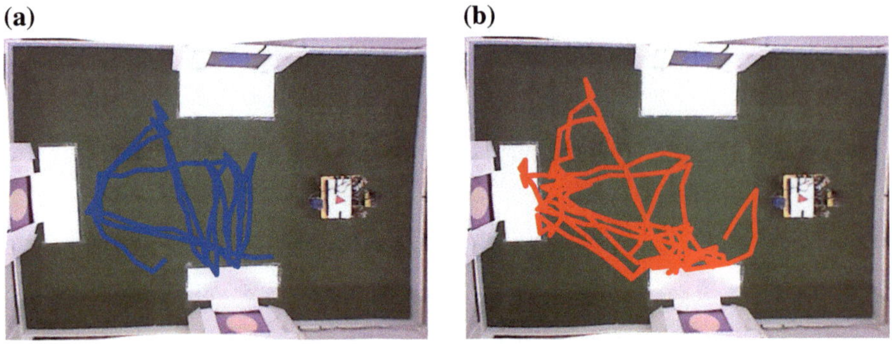


**Fig. 11.19** Trajectory followed by the robot for the first 100 steps (**a**) and for other 150 steps (**b**). The robot initially attracted by the blue landmark, after several punishments learns to avoid it

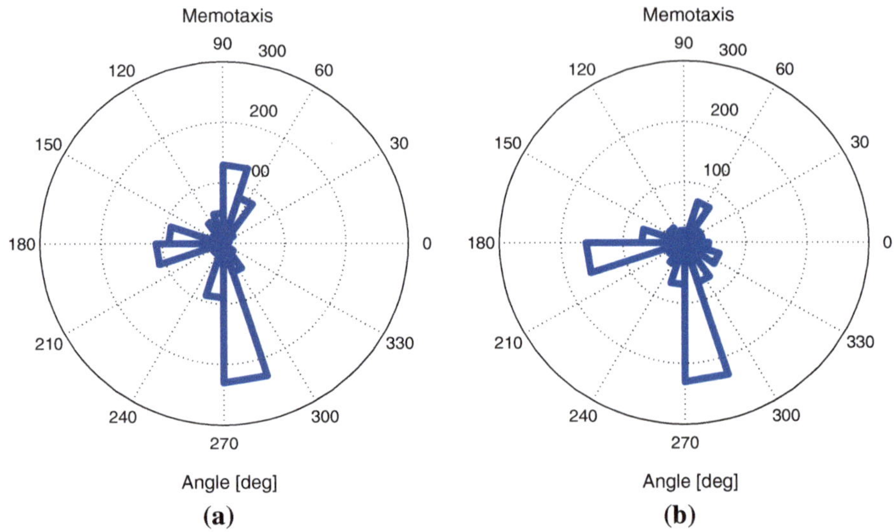


**Fig. 11.20** Gaze direction for the first 100 steps (**a**) and for other 150 (**b**)

is considered as a distributed functionality of the brain, that here was modeled as a block able to explain the behavioral experiments in flies.

The BSN takes information about the robot drives that depend on the system internal states and on the external context. The network is organized in a winner-takes-all topology and guarantees the emergence of a single behaviour on the basis of the robot needs.

A Scheme of the proposed block is shown in Fig. 11.22 where four possible behaviours have been taken into consideration: no activity, homing, explore, approach a target. The switching among the different behaviours is determined not only by the drives time evolution but also by the presence of an evaluation block that from the

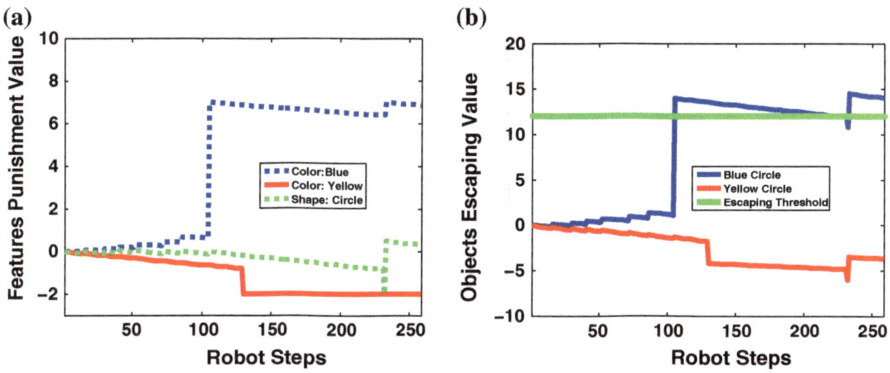


**Fig. 11.21** Temporal evolution of the punishment value associated to the visual features (**a**) and the escaping value associated to the two detected objects: Blue circle and yellow circle (**b**). After 100 steps the robot is able to discriminate the blue circle as dangerous. The presence of a forgetting factor leads to another punishment event after 200 steps

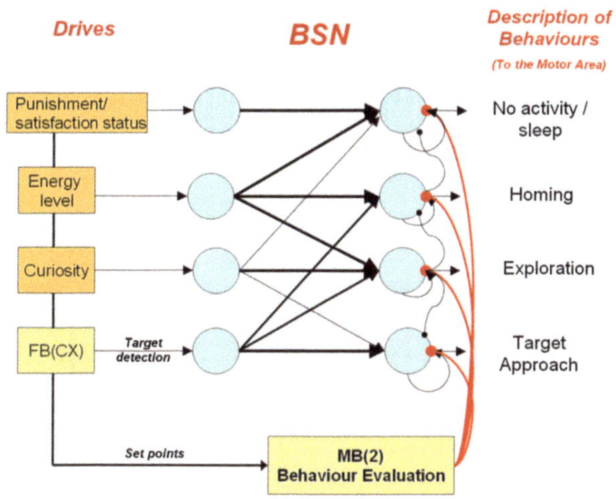


**Fig. 11.22** Scheme of the behaviour selection network. Four different behaviour can emerge depending on the robot drives, the drives-behaviours connection are fixed and the synaptic weights is shown using different line thickness. The MB(2) block evaluates the ongoing behaviour and inhibits the BSN output neuron accordingly

neurobiological experiments has been localized in the MBs. From the fly experiments the role of the two main centers in the behaviour establishment has been well underlined: CX is mainly responsible for the initiation of the behaviour, in fact CX-defective mutant flies show a reduction in terms of motor activity; whereas MBs are mainly involved in the termination of the ongoing activities (inhibitory effect). The neural structure devoted to the adaptive termination of the behaviours, named MB(2)

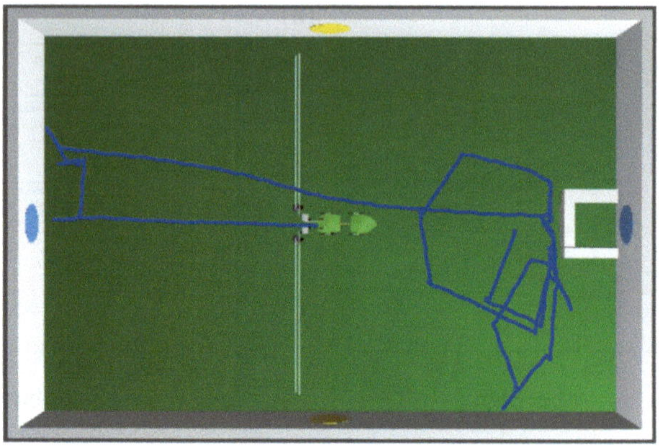

Fig. 11.23 Robot trajectory followed during the adaptive termination simulation. The robot is attracted by the blue target but among the two target available in the arena one is unreachable

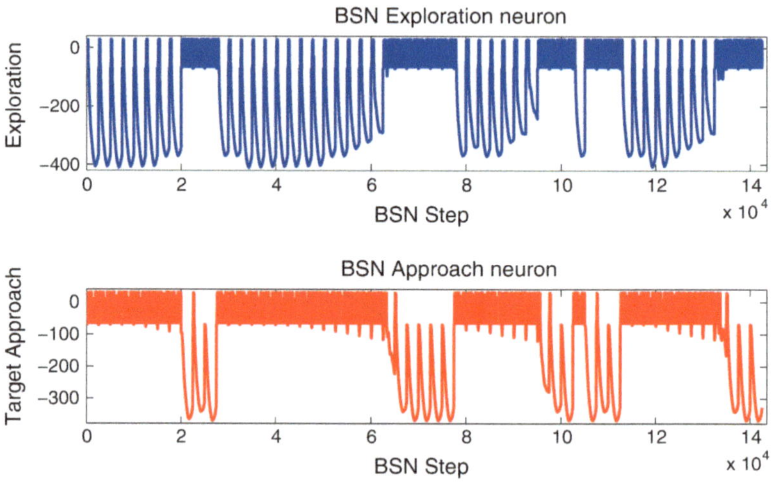


Fig. 11.24 Time evolution of the BSN output neurons considered in the experiment: exploration and target approaching. The winner-takes-all configuration allows that each step only one behaviour is elicited. For the BSN 2,500 steps correspond to 1 robot step

acts directly on the BSN output neurons through an inhibitory current that depends on the capability to reach the given set-point in a defined time windows.

Following the neurobiological experiments, a specific simulation set-up has been defined as shown in Fig. 11.23 where the arena used for the experiment is shown. The Tribot robot is attracted by two blue circles placed in opposite walls and one of the targets can not be reached due to the presence of a insurmountable barrier (the white framework surrounding one of the targets in Fig. 11.23). The trajectory followed by the robot is also shown in Fig. 11.23, the exploration and target approaching output neurons of the BSN are reported in Fig. 11.24.

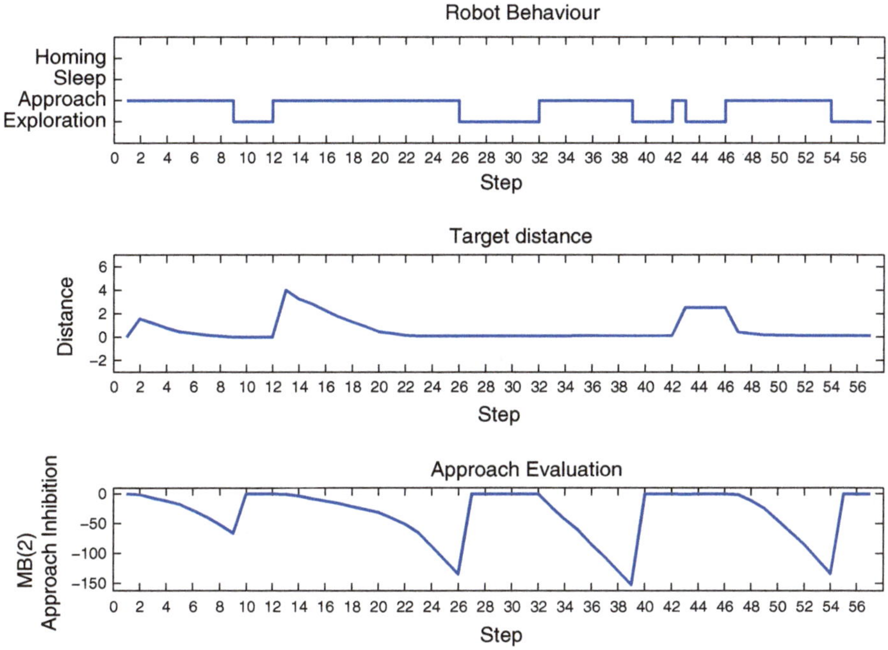


**Fig. 11.25** Time evolution of important variables during the simulation in Fig. 11.23. From the top: switching between the two competitive behaviours, estimated distance from the detected target, inhibition current from the MB(2) to the target approaching associated output neuron. The MB(2) evaluate the performance of the on-going behaviour, in the case of the target approaching through the robot-target distance trend. The robot is therefore forced to change behaviour when too much time is spent for a target approach behaviour without obtaining any significant improvement in reaching the set point (i.e. zero robot-target distance). It is to be underlined that the distance values are considered only in presence of a visible target. In fact, during the exploration behavior the "distance" measure is set to zero, since during exploration the target is still not detected

The Tribot is initially attracted by the reachable target, that is retrieved and then abandoned to explore the rest of the environment (see Fig. 11.25 step 9). When the second target is detected (step 12) the approaching behaviour is elicited, the robot-target distance decreases until the barrier is reached (step 20). The robot persists trying to satisfy the drives in approaching the target but the MB(2) block, evaluating the absence of improvements in terms of robot-target distance, increases the inhibition level until at step 26, the exploration behaviour wins the competition and the robot avoids to be locked on the unreachable target.

The actual MB(2) block has the role of evaluating the ongoing behavior in the specific case of unreachable targets; in fact, as it can be seen from the bottom part of Fig. 11.25, the inhibitory effect is a result of the combination of one basic and another inhibitory component that onsets in case of unreachable targets. It means that the model realized, even in the absence of MB(2), is in any case able to lead the agent to give up an ongoing behavior, but only in front of very long lasting deadlocks, as it can appreciated below Fig. 11.26.

When the MB(2) block is disconnected from the network, as shown also by MB-defective flies, the adaptive termination is not performed and the fly dedicates long time unsuccessfully attempting to reach the target (see Fig. 11.27).

The BSN signals obtained during a similar experiment performed with the Tribot II are reported in Fig. 11.27 where the switching between exploration and approaching behaviour is influenced by the evaluation effect introduced by the MB(2) block.

## 11.7 MB and Decision Making

In insects and in particular in flies, MBs is a key neural assembly for decision making. Visual information are mainly processed through the CX but interesting experiments on the fly show that even if no direct connections have been till now discovered between CX and MBs, the latter is responsible for decision making processes also in relation to visual inputs. The role of MB for generalization with foreground/background separation and in presence of contradictory cues has been demonstrated in different experiments.

The proposed model integrated inside the insect brain framework able to deal with this problem is shown in Fig. 11.28. The visual preprocessing block is able to extract the relevant features from the segmented objects present in the scene, the FB associates a meaning to the visual data depending on the rewarding or punishing signals coming from the environment. Finally the acquired information compete at the behaviour selection block in choosing the final robot behaviour. The role of MB in this control chain consists in modulating the information coming from the FB depending on the value of specific visual features that in the proposed experiments, in accordance with experiments in the fly, correspond to the level of color saturation. The part of the MB block devoted to this process, that we called MB3, is important to solve contradictory situations, as will be shown, but it is not essential for the visual orientation, targeting and learning.

As depicted in Fig. 11.28, direct connections between the two main centers (MBs and CX) are not present but an indirect interaction exists at the level of BSN that, as already discussed, is not a real physical structure of the brain but a block where all the distributed functionalities related to the final behaviour choice have been glued using the limited information known about this process.

The role of MB in handling visual information is up to now limited to two specific circumstances: in presence of contradictory cues and for foreground/background separation allowing a generalization process. In the following the model proposed to deal with the decision making process in presence of contradictory cues is reported and the structure has been modeled on the basis of the experimental results reported in [21].

The so-called MB3 block is depicted in Fig. 11.29. The network works as a gateway that modulates the information that from the FB are directed to the BSN depend-

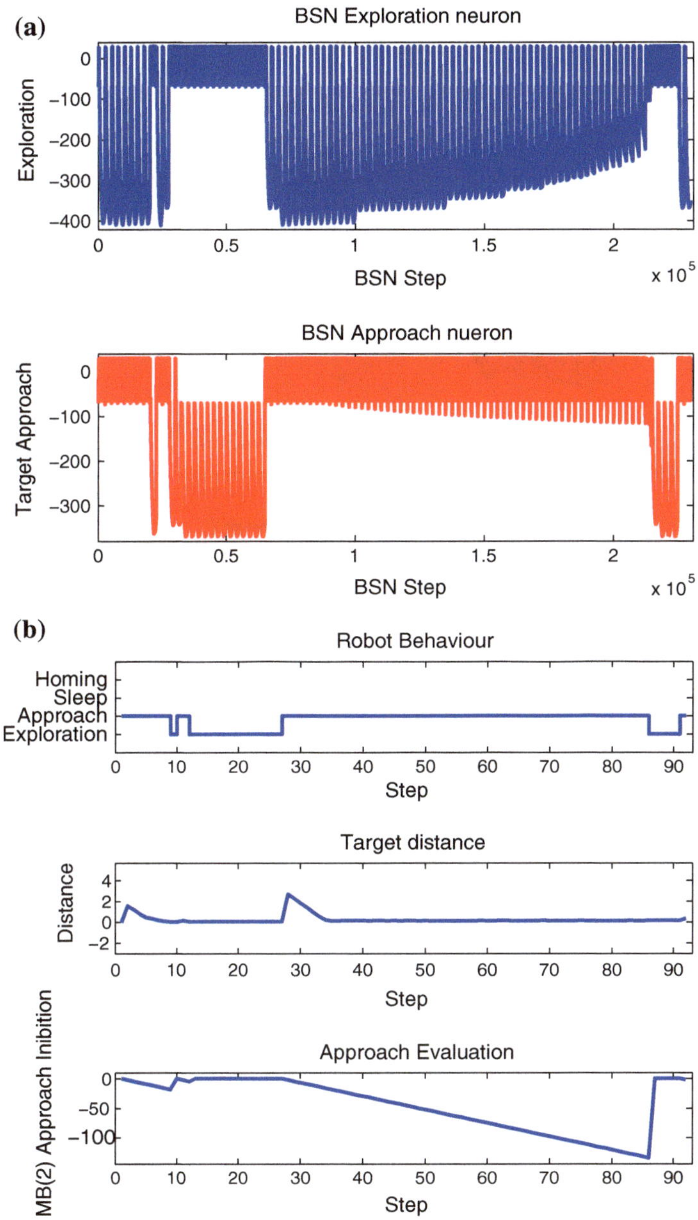


**Fig. 11.26** Behaviour of the robot when the MB(2) block is disconnected. The robot, after retrieving the reachable target (step 1–12), persists in approaching the target for long time without succeeding in doing it (step 27–85)

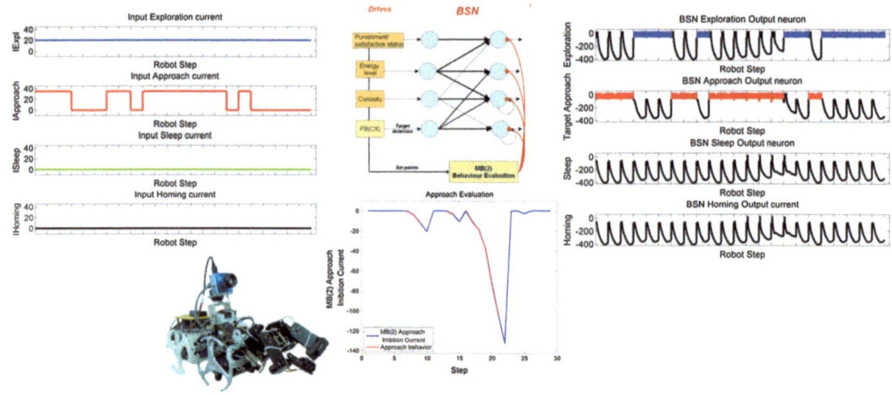


**Fig. 11.27** Input and output signals of the BSN block when the Tribot robot is performing an experiment of visual targeting in presence of obstacles

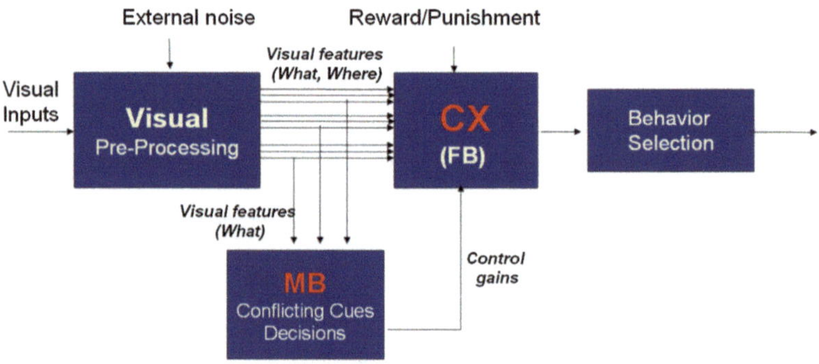


**Fig. 11.28** Block diagram of the elements of the insect brain involved in the decision making process in presence of contradictory cues in the processed visual field

ing on specific input signals. The key information processed at this level is the color saturation that allows to sharply decide the most convenient behaviour to be performed. The network consists of two neurons that receive an input current related to the color saturation value of the segmented object.

One of the neurons is sensible for high level and the other for low level of saturation. The output of these neurons inhibit the output of the two subnetworks of the FB (i.e. shapes and colors). These connections guarantee that when the color saturation level is high, the color network will lead with respect to the others and viceversa.

The MB3 network receives inputs from the FB, in particular the escaping and approaching signals are used to understand if a dilemma is present, due to the presence of contradictory cues. When both the neurons are active, the MB3 block inhibits one of the output of the FB subnetworks solving the dilemma depending on the level of color saturation, otherwise the MB3 is silent and does not interact with the process. Fig. 11.29 shows the whole network including FB, MB3 and part of the BSN.

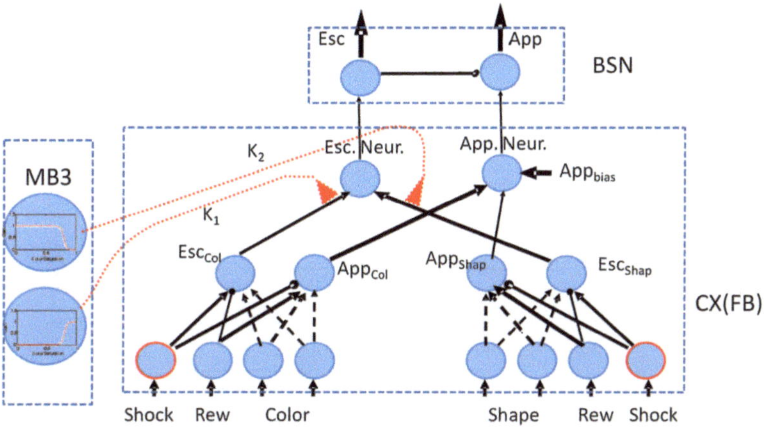


**Fig. 11.29** Interaction between the FB, MB3 and BSN blocks, the presence of MB3 is important for the final behaviour selection as part of the decision making process

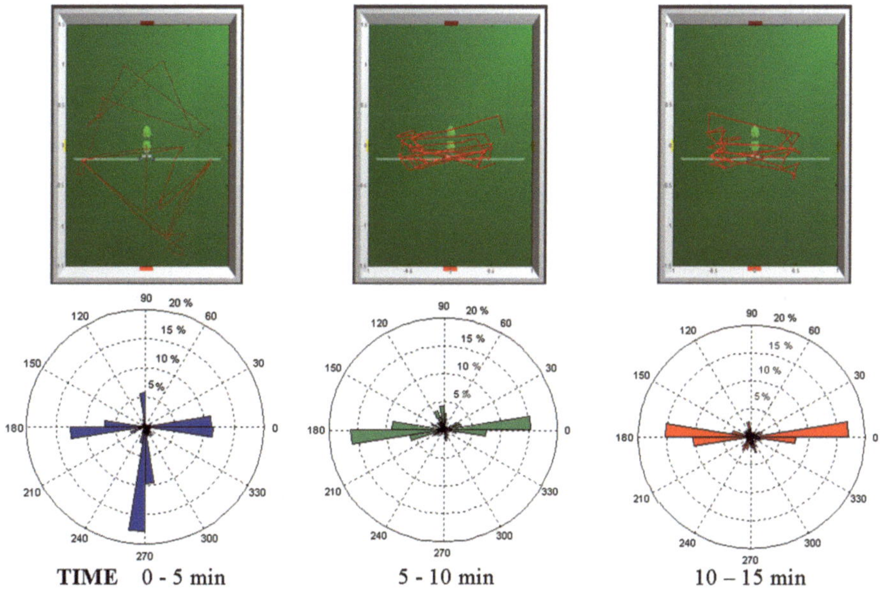


**Fig. 11.30** The behaviour of the robot in terms of followed trajectories and gaze distribution, is acquired in windows of 5 min. The red squared landmark in the down and upper part of the arena are dangerous and the robot learns to avoid objects with these features

In the following some tests are reported, for details see [22].

Working with the same set-up used for visual learning, the robot, depending on the received punishment signals, modifies its behaviour as shown in Fig. 11.30 where the robot activity is summarized in windows of 5 min for a total of 15 min of

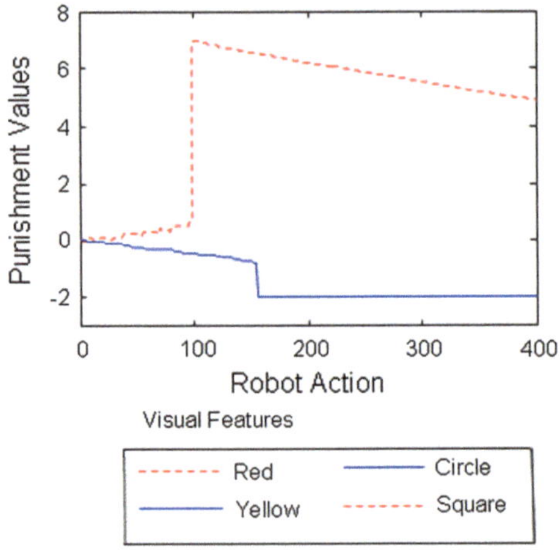


**Fig. 11.31** Evolution of the weights associated to the features, a high value corresponds to dangerous features. The behaviour presents an hysteretic shape, the red color feature and the square shape reach high level of dangerousness

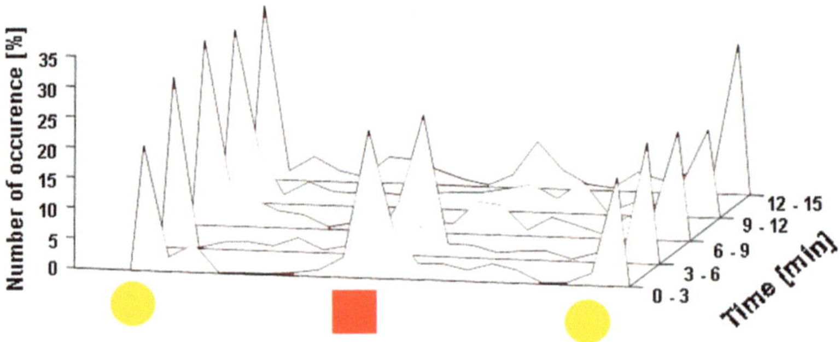


**Fig. 11.32** Evolution of the gaze distribution in time. After a few punishment events, the robot avoids to devote its attention to the red squared object

simulations. The robot trajectories, together with the gaze direction for each time window are reported. The robot, during an initial exploration phase, learns to avoid the red squares that are associated to a punishment.

The FB network, where meanings are associated to visual features, evaluates, depending on the received punishment, a punishment value for each feature. In the case of the arena reported in Fig. 11.30 where two pairs of objects are considered (i.e. red square and yellow circle) the trend of the feature punishment values are reported in Fig. 11.31. The time evolution of the gaze direction distribution is shown in Fig. 11.32 where it is evident that in a few minutes the simulated robot learns to avoid the red square. A very similar behaviour is reached also when the information about the object shape is missing (e.g. all the objects have the same rectangular shape) as shown in Fig. 11.33 where the gaze direction in time is similar to the previous case. In Fig. 11.34 the evolution of the punishment values is also reported, it is evident that the dangerous feature is now represented only by the yellow color while the square shape is no longer considered dangerous in this scenario.

Another interesting case is obtained when only the shapes are different, for instance putting in the arena red circles and squares. Only the square objects produce a punishment to the robot. The gaze direction is reported in Fig. 11.35 and the behaviour is similar to the other illustrated cases, whereas the trends of the punishment values associated to the visual features reported in Fig. 11.36 show, for the red feature, a series of oscillations not visible in the previous case (for the square shape) that depend on the detection sequence during the simulation and on the general preference for the color feature with respect to the other cues that was implemented in resemblance with the biological case.

Finally the role of the MB3 block, that works as a gateway modulating the information generated by the FB, is evident in the case of a dilemma. The robot after learning in the arena shown in Fig. 11.30, is facing a new scene where the visual features associated to the objects are contradictory if compared with the first arena, in fact a yellow square and a red circle are now present: the color was switched with the shape. In Fig. 11.37 the behaviour of the robot is shown through the gaze direction obtained in three different simulations that differ for the value of the color saturation. When the saturation is high ($CS = 1$) the robot has a preference for the color (i.e. red), decreasing the saturation ($CS = 0.8$) the robot has no preference and tries to escape from all the objects, finally for low level of color saturation ($CS = 0.5$) a preference for the shape arises. A similar experiment was performed with the Rover and the results during the different stages are reported in Fig. 11.38. When the level of saturation is high there is an high preference for the color, avoiding that one that has been associated to punishments during the learning stage, whereas for low level of color saturation, the robot prefers the shape with a relevant changes in its behaviour.

Going deeper in details analyzing the critical case of $CS = 0.8$, it is possible to evaluate the temporal evolution of the robot behavior as shown in Fig. 11.39. During the first 5–10 minutes the robot has no preference for the objects while, thanks to a forgetting factor, during time the robot starts to approach objects with square shapes and later on also red objects if the punishment signal is no longer present in the environment.

A very interesting result obtained in [21] consists in replicating the Dilemma experiment in a mutant fly with defects in the MB block. The results obtained while changing the level of color saturation are given in Fig. 11.40, in our case the block MB(2) is not active it is interesting to analyze that while for low level of saturation the behaviour is very similar, when the color saturation is high the robot is not able to decide which feature is dangerous with respect to the other and tries to avoid both without solving the dilemma.

To summarize the results and compare the behaviour of wild type and mutant brain-based robots, a preference index has been considered. The index is similar to that one used in the biological experiments in [21], and is calculated considering the time spent observing each object compared with the entire simulation time:

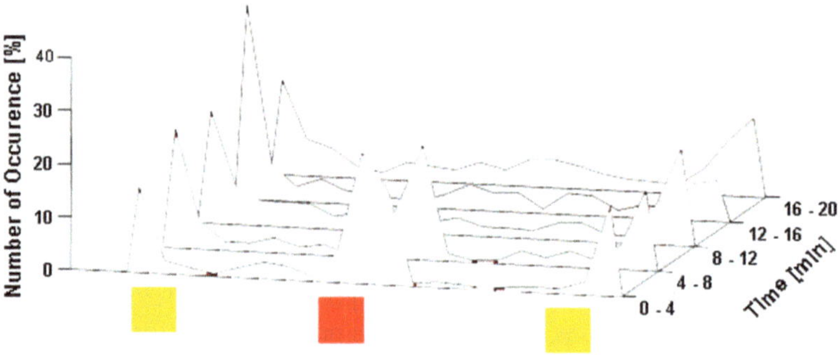


**Fig. 11.33** Evolution of the gaze distribution in time when the objects have features in common (shape) and differ from the color

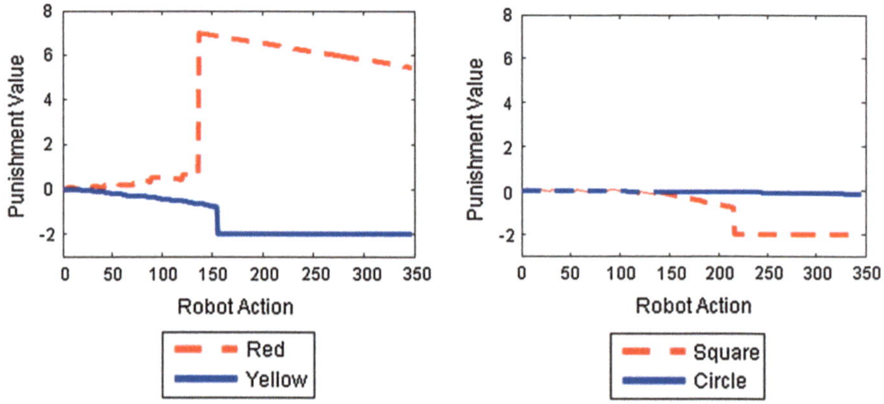


**Fig. 11.34** Evolution of the punishment value when the objects have the same shape but different colors

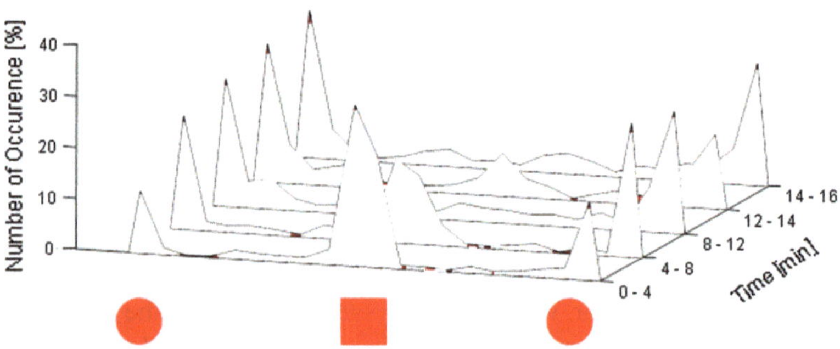


**Fig. 11.35** Evolution of the gaze distribution in time when the objects have the same colors but different shapes

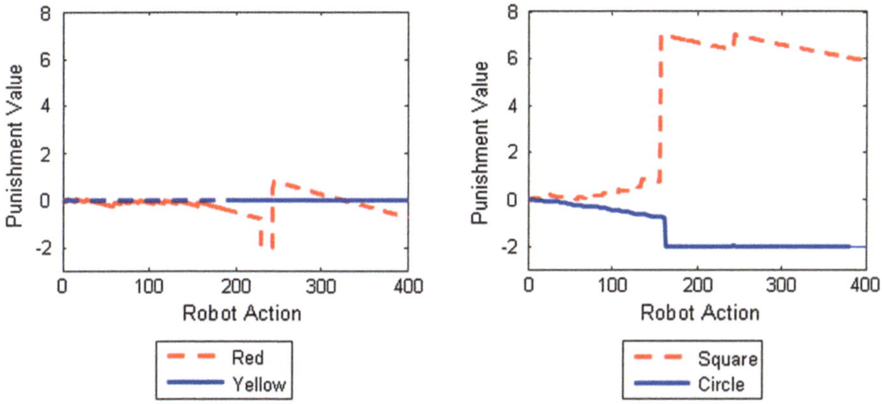


**Fig. 11.36** Evolution of the punishment value when the objects have the same colors but different shapes

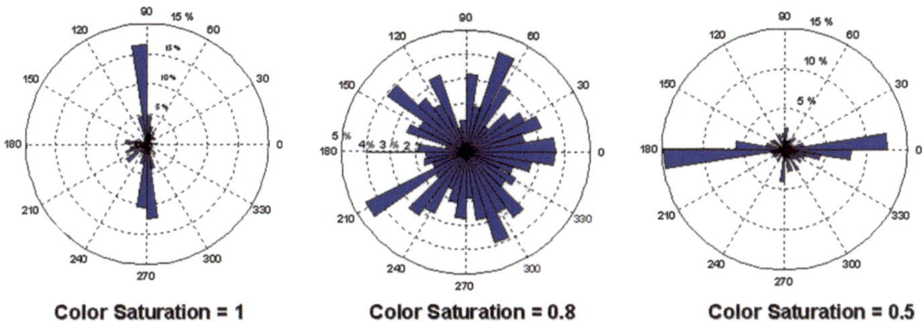


**Fig. 11.37** Dilemma test in presence of the MB block. After a learning phase performed in the arena in Fig. 11.30, a testing is performed changing the color feature between the two objects. When the color saturation is high, the robot has a remarked preference for the color, reducing the color saturation (CS = 0.8) no preference appears whereas further reducing the color saturation (CS = 0.5) a preference for the shape appears

$$PI = \frac{t_{YS} - t_{RC}}{t_{tot}} \tag{11.6}$$

where $t_{YS}$ is the time spent observing the yellow squared object, $t_{RC}$ is the time spent observing the red circular object and $t_{tot}$ is the duration of the experiment.

Figure11.41 shows the trend of the Preference index both in the case of a real fly (Fig. 11.41a) and in the robot experiments (Fig. 11.41b). The index is obtained performing a set of 10 simulations of 5 minutes for the different levels of color saturation. The results show that the robot controlled with an insect-like brain structure is able to make a clear decision among the two objects depending either on the color (for $CS > 0.8$) or on the shape ($CS < 0.8$) while the mutant structure, where the effect of the MB block on the decision process is missing, is not able to decide in

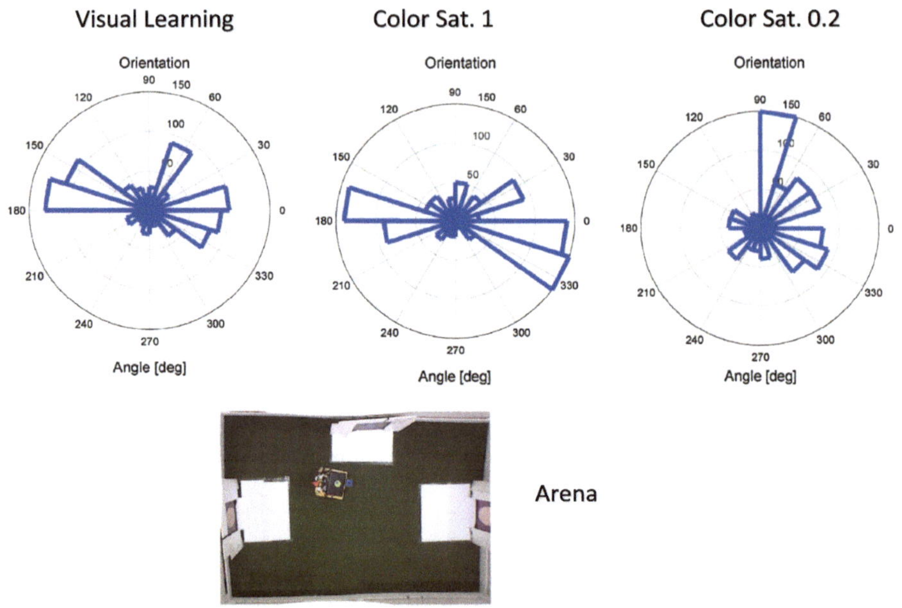


**Fig. 11.38** Results obtained in the robot experiment. During the learning stage the robot is punished when trying to reach the landmark placed on the top of the arena, maintaining a high level of color saturation the decision does not change whereas when the color saturation is decreased to CS = 0.3, the preference from the color is shifted toward the shape

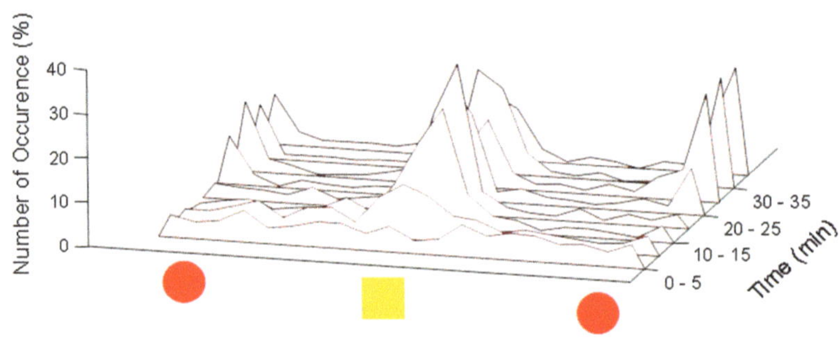


**Fig. 11.39** Dilemma test in presence of the MB block. When the color saturation is about CS = 0.8, the robot shows no preference but due to a forgetting factor a preference for the shape appears in time and persists if the punishment signal is removed

case of high level of color saturation. Finally also in the mutant experiment when the degradation of the color is too much, the preference for the shape takes the lead.

A light difference between our set-up and the biological one where a fly-simulator is used, consists in the type of movements allowed to the agent: our robot can navigate in the environment and not only turn like for the real fly. Moreover the vision field of the simulated camera is smaller with respect to the real case (i.e. about 80°). For

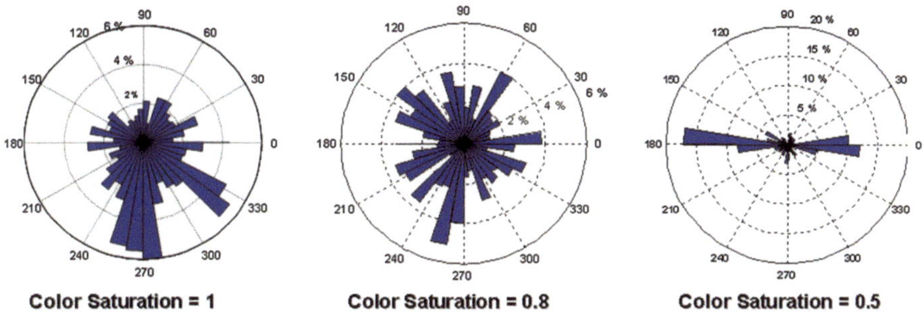


**Fig. 11.40** Dilemma test after ablating the MB block. The presence of an high color saturation is not enough to force the robot to make a choice

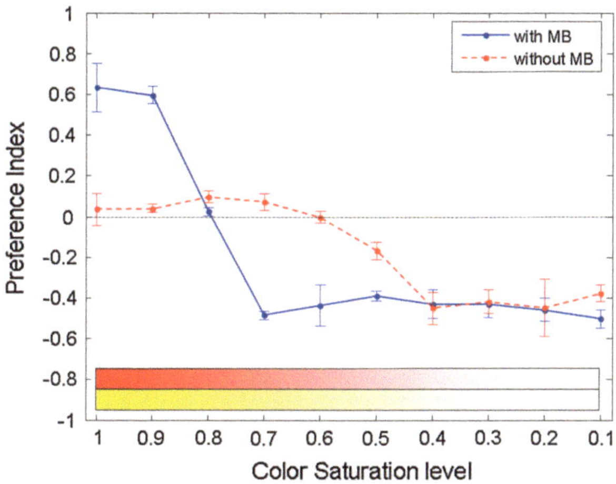


**Fig. 11.41** Representation of the robot behaviour through the preference index in presence of a dilemma (see [21] for details). Significative differences are visible when the MB block is eliminated from the control loop

these reasons we can extract from the simulations more information; for instance we can analyze the time spent in performing the three possible behaviours: attention devoted to one or the other object and exploration. Figure 11.42 shows, in the case of a wild type structure the percentage of time dedicated to each behaviour that can be compared with the mutant results reported in Fig. 11.43. It is interesting to notice the great amount of time used for exploration (over 50 %) when the mutant fly is not able to take a clear decision in presence of a dilemma ($CS > 0.5$).

For the robot experiments the Reward/Punishment signal has been assigned using both the low level color sensors that detect a colored spot on the ground and a sound sensor that can recognize frequency patterns in the incoming sound signals. In both cases we are using simple solutions that mimic the real insect behaviour, in fact the low level color sensor resembles the gustatory (reward) or nociceptor (punishment) sensor whereas the sound pattern recognition is a capability present in several insect

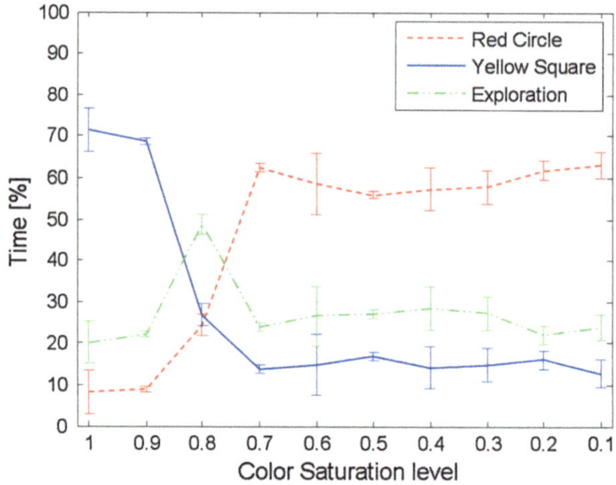


**Fig. 11.42** Time spent by the robot in performing different behaviours in the dilemma case (exploration, attention devoted to the red circle and attention devoted to the yellow square) when the MB3 block is active

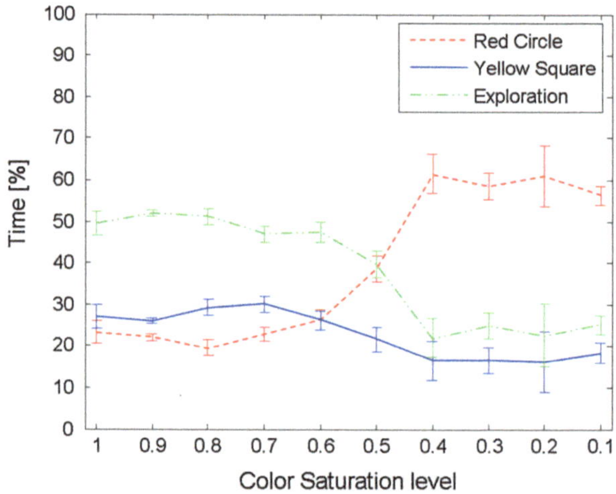


**Fig. 11.43** Time spent by the robot in performing different behaviours in the dilemma case (exploration, attention devoted to the red circle and attention devoted to the yellow square) when the MB3 block is ablated

species like in crickets that recognize predators (e.g. bats) and insect from the same species listening to the calling song for mating.

### *11.7.1 Attention and Expectation*

Insects are really surprising beings, able to clearly elicit apparently high level cognitive functions like attention. This capability was well assessed experimentally in

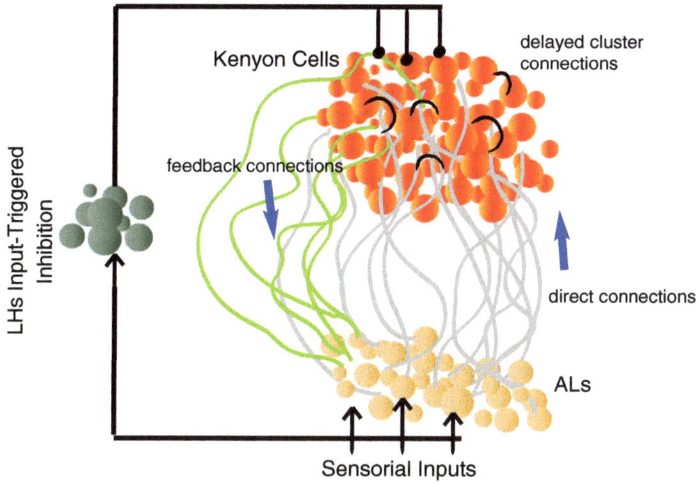


**Fig. 11.44** The Antennal Lobes (ALs) model is directly connected to the sensorial inputs (Antennas). We can assume that the activity of each AL neuron codifies a particular value for a specific feature. Neurons in the ALs are organized in groups. Synapses from the Antennal Lobes model to the Kenyon Cells (KCs) layer are randomly chosen, with a given probability of connectivity. The KCs layer is a toroidal lattice, with local excitatory and global inhibitory synapses. The Lateral Horns (LHs) periodically inhibit the KCs neurons. Feedback plastic connections link clusters in the MBs to the ALs neurons. Local connections within the ALs and within the KCs are not reported. The delayed plastic synapses in the KCs layer allow a temporal link between subsequent object presentations

[23]. During the efforts spent in delivering an MB model able to elicit both traditional odor conditioning features and attention processes, a behavior emerged, which is related to our concept of expectation. Indeed this capability is not yet proved to reside in insects, but nevertheless the computational results can open the way to a new wave of insect experiments in this direction. Biological experiments identify the MBs as the neuropile mainly responsible for the implementation of these functionalities. Taking inspiration from the insect brain and from the different behavioral aspects and learning capabilities of the MBs, a neural architecture, mainly based on spiking neurons, was designed.

The neural model here proposed (Fig. 11.44) combines behavioral evidences on attention capabilities in insects, including flies, with the state of the art on the neural structures responsible for these processes. It includes also a biologically relevant learning rule which boosts the system capabilities: expectation is the emergent aspect of this new architecture.

The core of the architecture consists of three mutually connected networks. The input layer is derived from modeling the capabilities of the antennal lobe but can be generalized and extended, from a functional point of view, also to other sensory modalities like vision that is generally processed by the Central Complex. This input lattice is able to decompose and represent olfactory/visual stimuli through features.

We can assume to have a layer able to codify the interesting features of objects. As in the insect Antennal Lobes, each glomerulus is able to detect a peculiar component of an odor, in our model each neuron (or group of neurons) in this layer encodes a particular aspect related to a detected object. Each group codifies a kind of feature (examples of visual features could be color, shape, etc.), and neurons in the same group codify different values of that feature (i.e. different colors or shapes). Moreover, neurons in the same group are linked together through inhibitory synapses. This topology implies that, when the ALs layer is stimulated, after a short transient time, only one neuron in each group remains excited, according to a Winner Takes All (WTA) topology. Neurons in different groups are linked together through plastic synapses, reinforced when neurons are firing together, according to the Hebbian paradigm. Therefore, when an object is presented to the network, the ALs model decomposes that object in its relevant features, and the corresponding neuron of each group is excited through a constant input current. This feature implies a processing at the input layer which, through local inhibition and WTA, leads to a simple but effective clustering of the input entities (odors or visual objects) according to their constituent features. Neurons in different groups are linked together through plastic synapses, reinforced when neurons are firing together, according to the Hebbian rule. This process reinforces organization among the neurons representing the different features of a presented object, leading them to reach a phase locked firing state. These mechanisms remind some way the Winnerless competition principle [24, 25], modeled after ALs studies. The learnable connections between groups of neurons allow the model to temporarily retain the presented objects through all its features and to reconstruct the features in case of incomplete or noisy detection. The ALs layer projects excitatory synapses to the MBs model. Each neuron of the ALs is connected with each neuron of the MBs model with a given probability of connection analogously to the biological counterpart.

The MBs model is based on a competitive topology that we call Self Organizing Spiking Layer (SOSL), the core of the whole neural model equivalent to the Kenyon Cells (KCs) in the MBs. The SOSL is a spherical lattice in which each neuron is connected through fast excitatory synapses with all the neurons of its neighborhood, and through fast inhibitory synapses with all the other neurons of the lattice. The main capability of the SOSL is a spontaneous clustering of neural activity, driven by a WTA-like structure. In this way, information coming from the ALs are compressed together into a cluster. In addition to this, a slow and delayed diffusion of the neural activity within the SOSL is present. In fact, another set of plastic delayed synapses links each neuron of the SOSL to the other neurons of the same lattice, in order to have the possibility to temporally link different clusters. These synapses are subject to the STDP learning algorithm, that allows to discover and retain temporal causality among clusters. Simulation results show that these connections have the interesting capability to generate, within the SOSL layer, expectation and short term prediction capabilities [26, 27]. Plastic feedback connections are also present from the MBs model to the ALs model. These connections were never present in any bioinspired model up to now, since only very recently were discovered [28]. Their role is precious to boost the model performance and two main functions have been identified: they are useful to create an expectation-based depolarization of the neurons in the ALs, they are also essential to reconstruct the expected object. The actual model hypothesizes massive feedback connections from SOSL to ALs neurons, even if a probability distribution could maintain the same performance in case of large scale implementation. When a cluster is elicited in the MBs model, the synaptic connections between the neurons of the MB cluster and those neurons which are firing in the AL (due to the synchronous presence of the corresponding input) are raised, according to the Hebbian paradigm. The Lateral Horns model contribution, after the neurobiological evidence of connections to the MBs in the fly, and according to the mainly inhibitory role in the locust, has been modeled as a time-driven resetting system, that generates an inhibitory wave to the MBs. In this way, the SOSL integrates the information coming from the ALs in a given time window and, after the emergence of a cluster, the neural activity of the network is inhibited by the LHs wave. One crucial issue is the synchronization between the delay of the synaptic clusters-linking connections in the SOSL and the LHs induced inhibitory action onset time. In our model the LH reset wave is propagated to the SOSL neurons somas, leaving the possibility to the last emerged spikes to proceed along the axon and contribute to the delayed information processing among the clusters.

An interesting simulation consists in the presentation of a first object, followed by another object with an incomplete feature set presented at the second step. In particular, information about the feature related to the second group of neurons in ALs is missing. In this case, thanks to the previous learning cycles, the KCs layer is able to reconstruct the lacking feature: feedback connections from the MBs excite the ALs neurons that codify that object and synaptic connections among neurons in different ALs groups allow the reconstruction of the complete object. No competition does take place in the ALs layer, but top-down connections allow the recalling of the lost information. An example of this reconstruction mechanism is shown in Fig. 11.45.

The expectation capability herewith presented can be further exploited for modeling a simple sequence learning paradigm. In fact, if during the learning phase a short number (say three) of objects is presented, three winning clusters are subsequently elicited and the central neuron for each cluster reinforces its connection to the central neuron for the winning cluster associated to the subsequent object. So, during the testing phase the presentation of the first object is able to elicit the sequence of winning clusters in the KCs layer, which reconstruct at the AL layer, the sequence of corresponding objects. The sequence of formed clusters following this procedure is depicted in Fig. 11.46. At the end of each phase, the network is reset by the LHs.

Exploiting the presence of noise in the system [27], an interesting property of the network can emerge. In fact the contribution of noise can be useful to consolidate the acquired knowledge during a resting phase, when no sensorial inputs are given to the network. A simulation divided in two phases was prepared: the first one is a learning phase, in which the network creates the association between clusters and objects. In particular, four sequences of two objects have been presented to the network. At the end of the training phase no physical input is presented at the ALs layer, but it is assumed that the KCs layer is subject to noise which was supposed

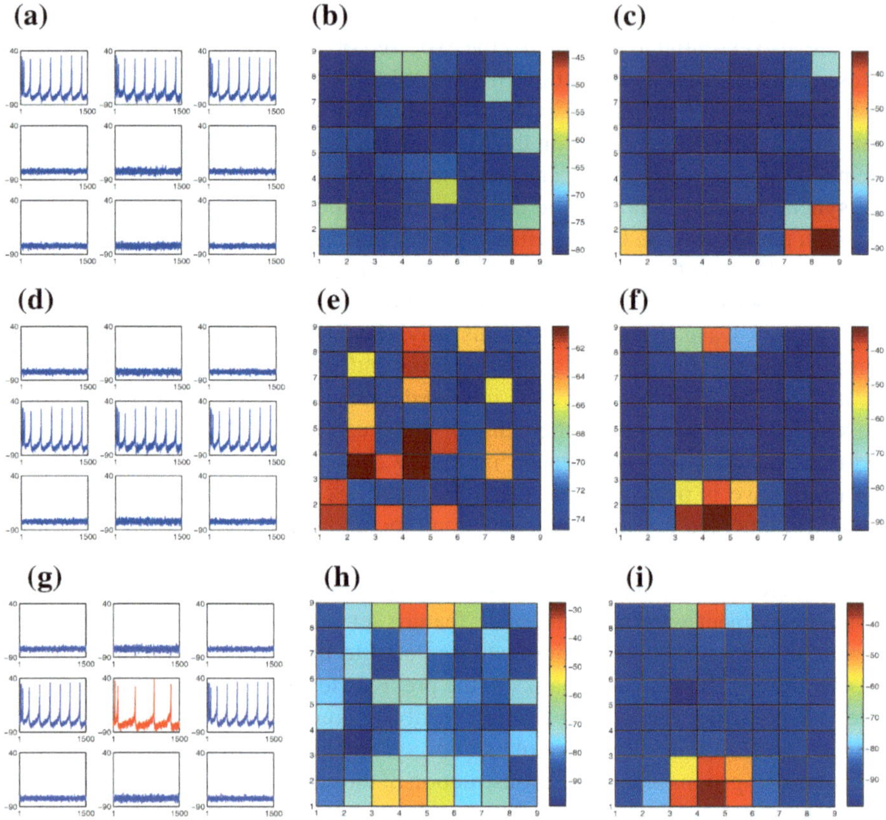

**Fig. 11.45** The network is able to reconstruct the missing feature using expectation mechanisms. When the first object is presented the ALs layer is stimulated (**a**) and a cluster emerges in the KCs layer. The mean of the KCs membrane potential at the beginning (**b**) and the end of the simulation (**c**) (for a total of 1500 simulation steps) is reported. During this process the plastic synapses in the ALs change according to the Hebbian rule as well as the feedback connections from KCs to ALs. When a second object is presented again, the ALs process the new features information (**d**) and the KCs layer converges to a new cluster (**e**, **f**). During this process the delayed connections in the KCs layer are modulated accordingly and the two consecutive clusters (**c**) and (**f**) will be linked together. Finally, after the learning phase, if the first object is presented, it is possible to reconstruct the missing feature of the second object using the acquired knowledge (**g**) and the correct cluster emerges (**h**, **i**)

to be gaussian with zero mean and variance $\sigma_{noise} = 10$. This noisy disturbance will onset a dynamical transient in the KCs neurons until a cluster will emerge over the others. If this cluster had formerly been trained to represent a given object, this will be recalled at the ALs layer, like an "evoked object". A new learning cycle will then arise, in which not only this object will be consolidated, but also all the other objects eventually expected after this one in an already learned sequence. New imagined solutions could also be experienced during this "sleeping phase" and,

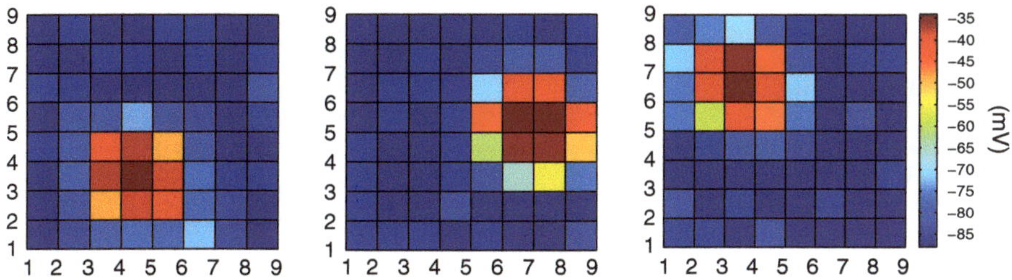


**Fig. 11.46** During the learning phase, three objects are presented to the network. In the test phase each cluster representing a given object excites the cluster that represents the following object in the sequence. Objects can be reconstructed thanks to the feedback connections from the KCs layer to the ALs

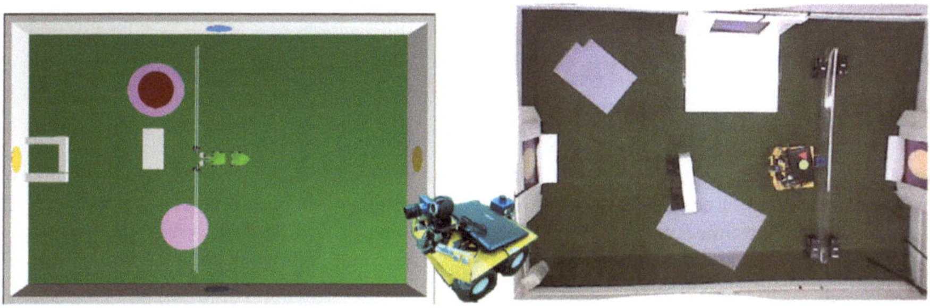

**Fig. 11.47** Simulated and real arena used for the experiment where the integration among the different blocks is emphasized

in more complex networks including also more environmental details, the system could discover new strategies to efficiently solve difficult situations starting from what already learned during the "awake phase".

## 11.8 Integration Experiment

The different blocks of the insect brain architecture already discussed in details in the previous sections, can work together in a general complex environment to determine the in an adaptive way the robot behaviour. To have an idea about the capability of the architecture, a general arena has been designed and both dynamical simulations and real experiments have been performed. The virtual and real environments are reported in Fig. 11.47; the system can interact with visual landmarks, obstacles and odor plumes (i.e. circles on the ground). Moreover rewarding and punishment signals can be provided using specific sound patterns. Moreover during the experiment specific blocks can be altered mimicking the mutant fly behaviour.

The summary of the simulated robot behaviour is shown in 11.48 where the density plot for the dwelling time, the evolution of the robot behaviour in time

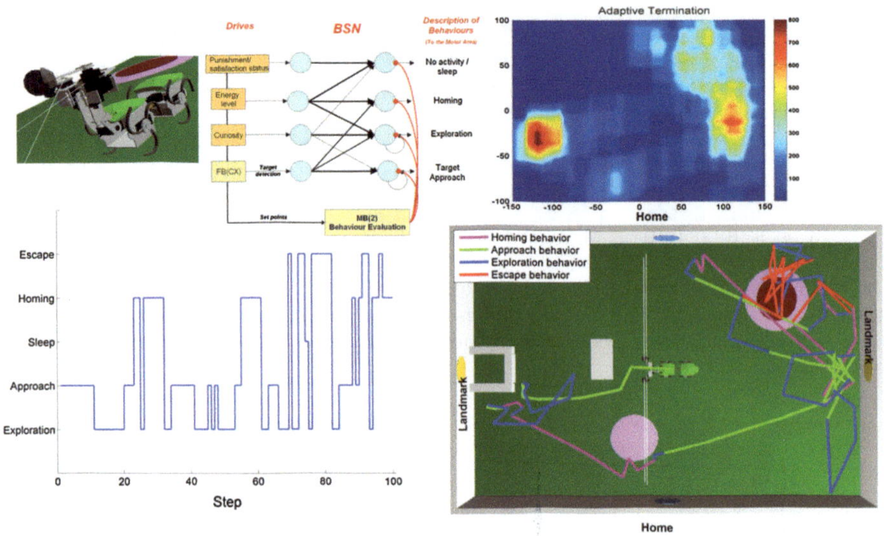


**Fig. 11.48** Behaviour of the simulated Tribot robot in a complex arena with visual landmarks, obstacles and odor plumes (i.e. circles on the ground). The simulated robot controlled with the whole insect brain architecture navigates in the environment for 100 robot steps. The density plot for the dwelling time and the time evolution of the active robot behaviour are shown

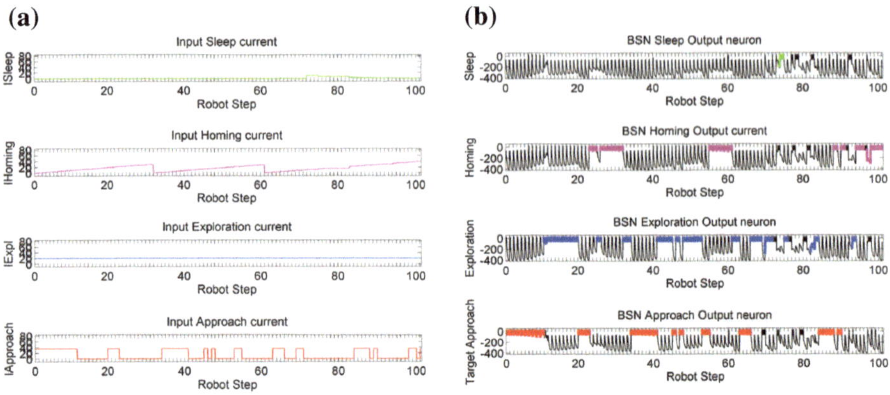


**Fig. 11.49** Time evolution of the **a** input and **b** output signals elaborated by the BSN block

and the complete trajectory are reported. The switching between robot behaviours are determined by the BSN, on the basis of the context and the previous acquired experience. The input/output signals for the BSN are shown in Fig. 11.49 and the important role concerning behavior evaluation performed by the MB(2) block is depicted in Fig. 11.50.

Finally the same experiment was performed by using the roving platform. Figure 11.51 shows the experimental set-up together with the emerged behaviours

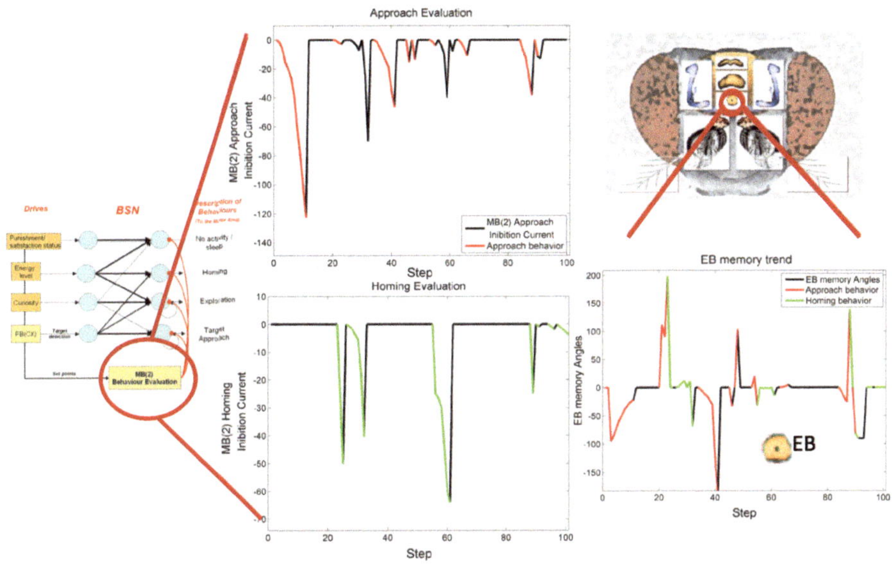


Fig. 11.50 Inhibitory effect generated by the MB(2) block on the BSN on the basis of the evaluation of the ongoing behavior

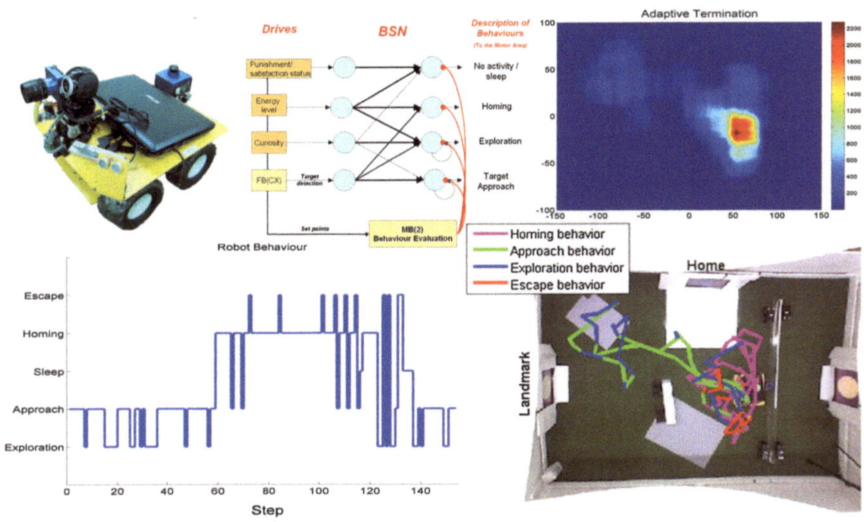


**Fig. 11.51** Behaviour of the Rover in the real arena with visual landmarks, obstacles and odor plumes (i.e. circles on the ground). The roving robot controlled with the whole insect brain architecture navigates in the environment and the followed trajectory is reported. The density plot for the dwelling time and the time evolution of the active robot behaviour are also shown

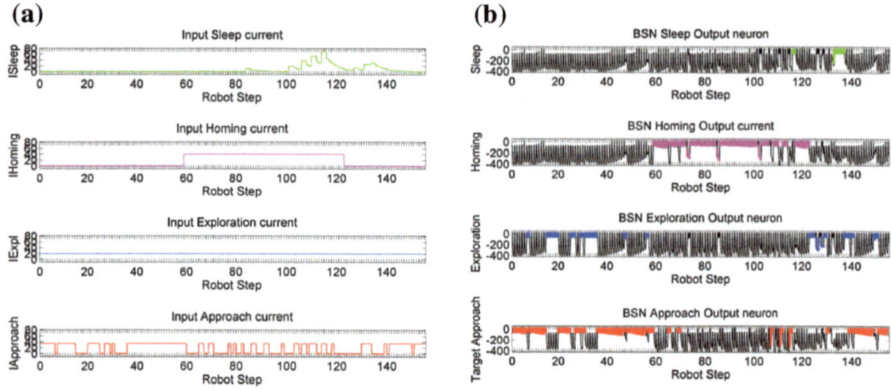


**Fig. 11.52** Time evolution of the (**a**) input and (**b**) output signals elaborated by the BSN block

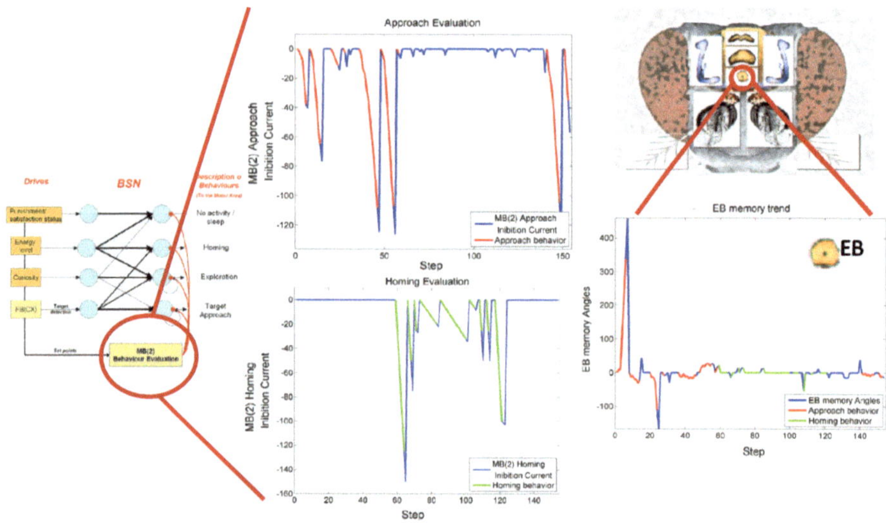


**Fig. 11.53** Inhibitory effect generated by the MB(2) block on the BSN on the basis of the evaluation of the ongoing beahviour

and the trajectory followed by the robot during the experiment. Other results in terms of signals elaborated through the BSN are shown in Figs. 11.52 and 11.53. Moreover Fig. 11.54 shows the time evolution of the spiking network used to model the olfactory learning on the robot. For sake of simplicity low level color sensors, able to detect the colored patches on the ground have been used instead of odor sensors. Specific sound patterns have been used as punishment signals for the robot.

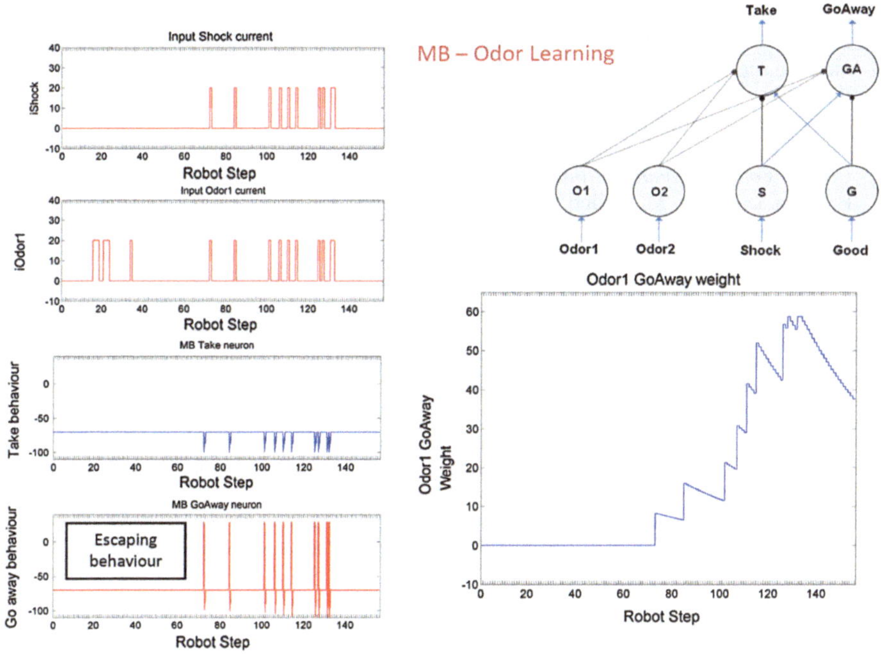


**Fig. 11.54** Time evolution of the spiking network used to model the "olfactory learning" on the robot. Low level color sensors, able to detect the colored patches on the ground have been used instead of odor sensors for sake of simplicity. Specific sound patterns have been used as punishment signals for the robot

## 11.9 Conclusions

The presented results show the final experiments performed to evaluate the functionalities of the proposed insect brain model, compared with the data acquired with both wild type and mutant flies. The information acquired from the experiments have been used to further improve the capabilities of the insect brain model validating the role and interconnections between the basic blocks constituting the insect brain architecture. Moreover new experiments can be designed to unravel the dynamics still hidden verifying hypotheses on the reflexive and cognitive processing mechanisms in insects.

## References

1. J.D. Armstrong, J.I. van Hemert, Towards a virtual fly brain. Phil. Trans. R. Soc. A **367**, 2387–2397 (2009)
2. P. Arena, C. Berg, L. Patané, R. Strauss, P.S. Termini, An insect brain computational model inspired by drosophila melanogaster: architecture description, in *WCCI 2010 IEEE World*

*Congress on Computational Intelligence* (Barcelona, Spain, 2010), pp. 831–837
3. P. Arena, L. Patané, P.S. Termini, An insect brain computational model inspired by drosophila melanogaster: simulation results, in *WCCI 2010 IEEE World Congress on Computational Intelligence* (Barcelona, Spain, 2010), pp. 838–845
4. P. Arena, S. De Fiore, L. Patané, M. Pollino, C. Ventura, Insect inspired unsupervised learning for tactic and phobic behavior enhancement in a hybrid robot, in *WCCI 2010 IEEE World Congress on Computational Intelligence* (Barcelona, Spain, 2010), pp. 2417–2424
5. P. Arena, L. Patané, M. Pollino, and C. Ventura, Tribot: a hybrid robot for cognitive algorithm implementation, in *18th IEEE Workshop on Nonlinear Dynamics of Electronic Systems (NDES)*, (2010), pp. 1–6
6. P. Arena, L. Fortuna, M. Frasca, L. Patané, M Pollino, An autonomous mini-hexapod robot controlled through a cnn-based cpg vlsi chip, in *Proceedings of the 10th IEEE International Workshop on Cellular Neural Networks and their Applications*, Istanbul, Turkey, 28 (2006)
7. P. Arena, M. Cosentino, L. Patané, A. Vitanza, Sparkrs4cs: a software/hardware framework for cognitive architectures, in *5th SPIE's International Symposium on Microtechnologies* (Czech Republic, Prague, 2011), pp. 1–12
8. P. Arena, L. Patané, P.S. Termini, A. Vitanza, R. Strauss, Software/hardware issues in modelling insect brain architecture, in *4th International Conference on Intelligent Robotics and Applications (ICIRA)*, Aachen, Germany, (2011)
9. M. Mronz, R. Strauss, Visual motion integration controls attractiveness of objects in walking flies and a mobile robot, in *International Conference on Intelligent Robots and Systems* (Nice, France, 2008), pp. 3559–3564
10. L. Alba, P. Arena, S. De Fiore, L. Patané, R. Strauss, G. Vagliasindi, Implementation of a drosophila-inspired model on the eye-ris platform, in *Proceedings of the IEEE International Conference of CNNA*, Berkley (2010)
11. P. Arena, S. De Fiore, L. Patané, L. Alba, R. Strauss, Drosophila-inspired visual orientation model on the eye-ris platform: experiments on a roving robot, in *Proceedings of the SPIE Microthecnologies Conference*, Prague
12. K.G. Gotz, R. Biesinger, Centrophobism in drosophila melanogaster. physiological approach to search and search control. J. Comp. Physiol. A **156**, 329–337 (1987)
13. S. De Belle, M.B. Sokolowski, Heredity of rover/sitter: alternative foraging strategies of drosophila melanogaster larvae. Heredity **59**, 73–83 (1987)
14. H. Schwegler, T. Wittmann, Path integration—a network model. Biol. Cybern. **73**, 569–575 (1995)
15. D.A. Redish, D.S. Touretzky, H.S. Wan, The sinusoidal array: atheory of representation for spatial vectors, in *Proceedings of Computational Neurosciences Conference* Kluwer Academic Press, (1993)
16. D. Rumelhart, J. McClelland, *Parallel distributed processing* (MIT Press, Cambridge, MA, 1986)
17. L. Chittka, P. Skorupski, N.E. Raine, Speed-accuracy tradeoffs in animal decision making. Trends Ecol. Evol. **24**, 400–407 (2009)
18. P. Arena, S. De Fiore L. Patané, M. Pollino, C. Ventura, Stdp-based behavior learning on tribot robot, in *Proceedings of IEEE/RSJ International Conference SPIE* (2009)
19. P. Arena, L. Fortuna, M. Frasca, L. Patané, Learning anticipation via spiking networks: application to navigation control. IEEE Trans. Neural Networks **20**(2), 202–216 (2009)
20. P. Arena, L. Patané, Simple sensors provide inputs for cognitive robots. Instrum. Meas. Mag. **12**(3), 13–20 (2009)
21. S. Tang, A. Guo, Choice behavior of drosophila facing contradictory visual cues. Science **294**, 1543–1547 (2001)
22. P. Arena, L. Patané, P.S. Termini, Decision making processes in the fruit fly: a computational model, in *Frontiers in Artificial Intelligence and Applications—Proceedings of the 21st Italian Workshop on Neural Nets*, vol. 234 (Seville, Spain, 2011), pp. 284–291
23. B. van Swinderen, Attention in drosophila. Int. Rev. Neurobiol. **99**, 51–85 (2011)

24. M. Rabinovich, A. Volkovskii, P. Lecanda, R. Huerta, H.D.I Abarbanel, G. Laurent, Dynamical encoding by networks of competing neuron groups: winnerless competition. Phys. Rev. Lett. **37**, 068102 (2001)
25. P. Arena, L. Fortuna, D. Lombardo, L. Patané, M. Velarde, The winnerless competition paradigm in cellular nonlinear networks: models and applications. Int. J. Circ. Theor. Appl. **37**, 505–528 (2009)
26. P. Arena, L. Patané, V. Stornanti, P.S. Termini, B. Zaepf, and R. Strauss, Modelling the insect mushroom bodies: application to a delayed match to sample task. Neural Networks (special issue on Autonomous Learning), doi: 10.1016/j.neunet.2012.11.013, 2012
27. P. Arena, L. Patané, P.S. Termini, Learning expectation in insects: a recurrent spiking neural model for spatio-temporal representation. Neural Networks **32**, 35–45 (2012)
28. P. Arena, L. Patané, P.S. Termini, Modeling attentional loop in the insect mushroom bodies, in *International Joint Conference on Neural Networks (IJCNN)*, (2012), pp. 7–12

MIX
Papier aus verantwortungsvollen Quellen
Paper from responsible sources
FSC® C105338
FSC
www.fsc.org

If you have any concerns about our products,
you can contact us on
**ProductSafety@springernature.com**

In case Publisher is established outside the EU,
the EU authorized representative is:
**Springer Nature Customer Service Center GmbH**
**Europaplatz 3, 69115 Heidelberg, Germany**

Printed by Libri Plureos GmbH
in Hamburg, Germany